AF805869

Latin American Mathematics Series

Latin American Mathematics Series – UFSCar subseries

Series Editors

César R. de Oliveira, Center of Exact Sciences and Technology, Federal University of São Carlos, São Carlos, Brazil

Ruy Tojeiro, ICMC, Universidade de São Paulo, São Carlos, Brazil

Advisory Editors

Shiferaw Berhanu, Department of Mathematics, Temple University, Philadelphia, USA

Ugo Bruzzo, SISSA - Scuola Internazionale Superiore di Studi Avanzati, Trieste, Italy

Irene Fonseca, Department of Math Sci, Carnegie Mellon University, Pittsburgh, USA

Published under the Latin American Mathematics Series, which was created to showcase the new, vibrant mathematical output that is emerging from this region, this subseries aims to gather high-quality monographs, graduate textbooks, and contributing volumes based on mathematical research conducted at/with the Federal University of São Carlos, a technological pole located in the State of São Paulo, Brazil. Submissions are evaluated by an international editorial board and undergo a rigorous peer review before acceptance.

Ruy Tojeiro
Editor

Advances in Submanifold Theory and Related Topics

In Honor of Marcos Dajczer on his 75th Birthday

Springer

Editor
Ruy Tojeiro
ICMC
Universidade de São Paulo
São Carlos, São Paulo, Brazil

ISSN 2524-6755 ISSN 2524-6763 (electronic)
Latin American Mathematics Series
ISSN 2524-6755 ISSN 2524-6763 (electronic)
Latin American Mathematics Series – UFSCar subseries
ISBN 978-3-032-00917-3 ISBN 978-3-032-00918-0 (eBook)
https://doi.org/10.1007/978-3-032-00918-0

Mathematics Subject Classification: 53C42, 53C21, 53C25, 53A10

© The Editor(s) (if applicable) and The Author(s), under exclusive license to Springer Nature Switzerland AG 2025

This work is subject to copyright. All rights are solely and exclusively licensed by the Publisher, whether the whole or part of the material is concerned, specifically the rights of translation, reprinting, reuse of illustrations, recitation, broadcasting, reproduction on microfilms or in any other physical way, and transmission or information storage and retrieval, electronic adaptation, computer software, or by similar or dissimilar methodology now known or hereafter developed.
The use of general descriptive names, registered names, trademarks, service marks, etc. in this publication does not imply, even in the absence of a specific statement, that such names are exempt from the relevant protective laws and regulations and therefore free for general use.
The publisher, the authors and the editors are safe to assume that the advice and information in this book are believed to be true and accurate at the date of publication. Neither the publisher nor the authors or the editors give a warranty, expressed or implied, with respect to the material contained herein or for any errors or omissions that may have been made. The publisher remains neutral with regard to jurisdictional claims in published maps and institutional affiliations.

This Springer imprint is published by the registered company Springer Nature Switzerland AG
The registered company address is: Gewerbestrasse 11, 6330 Cham, Switzerland

If disposing of this product, please recycle the paper.

Preface

This is a special volume in honor of Professor Marcos Dajczer on his 75th birthday. It collects articles by specialists in Submanifold theory, Geometric analysis, and Riemannian geometry, reflecting some of Marcos's interests. The articles are aimed at both experienced readers and graduate students alike.

It is a great pleasure to edit this volume, be one of its contributors, and write its preface. Besides being my supervisor and main collaborator over the last three decades, Marcos has become a real friend, and I miss our long discussions every summer over all these years in his office at IMPA—Associação Nacional Instituto de Matemática Pura e Aplicada, at Rio de Janeiro, Brazil. Marcos is currently living with his family in Murcia (Spain), after a lifetime as a professor at IMPA.

Marcos's contribution to the development of Submanifold theory in Brazil is invaluable, not only for his outstanding work in the field but also for his influence on the work of new generations of geometers.

Marcos graduated in Mathematics from the Universidad de Buenos Aires in 1975 and received his Ph.D. in Mathematics from IMPA in 1980, under the supervision of Prof. M. do Carmo. From 1980 to 1983 he worked as an Assistant Professor at IMPA, a period in which he obtained a post-doctoral fellowship at the State University of New York in Stony Brook (1982–83). This was the start of a fruitful collaboration with Prof. D. Gromoll, leading him back to Stony Brook several times later as a visiting professor. In 1986, Marcos became an Associate Professor at IMPA, and a Professor in 1988.

As I write this preface, MathSciNet numbers 143 articles under Marcos's authorship, with as many as 31 co-authors. The first of such articles was written in 1979 as an announcement in the Anais da Academia Brasileira de Ciências of the results of his PhD thesis. These would appear later in a more complete form in the article "Reduction of the codimension of regular isometric immersions," published in the *Mathematische Zeitschrift* in 1982. However, these numbers will likely have changed significantly until this book is published, for Marcos is still an active researcher, with no less than 15 papers published in the last three years.

Providing a rough description of Marcos's accomplishments is a real challenge. I do not intend to be complete in doing so, and I apologize to him and many of his co-authors for the inevitable relevant omissions.

Marcos's work ranges on a variety of distinct topics. However, I dare say that his main interest is in the isometric and conformal deformation problem for submanifolds, including its infinitesimal version, a topic that owes him a lot for its current revival of interest. About 35 of his papers deal with this subject, so I can only single out a few of them. Yet in his post-doc stage at Stony Brook, together with D. Gromoll he introduced a useful parametrization of hypersurfaces with a positive index of relative nullity, which they called the *Gauss parametrization*, and used it to study rigidity aspects of hypersurfaces. The results appeared in the article "Gauss parametrizations and rigidity aspects of submanifolds," published in the *Journal of Differential Geometry* in 1985. Two years later, his seminal paper "Conformal rigidit," in collaboration with M. do Carmo, was published in the *American Journal of Mathematics*. In that paper, the Main lemma on flat bilinear forms was established and applied to derive rigidity theorems for isometric and conformal immersions of Riemannian manifolds into Euclidean space in low codimension. Flat bilinear forms had been previously introduced by J. D. Moore as an outgrowth of Cartan's theory of exteriorly orthogonal quadratic forms. That the Main lemma could not be improved concerning the dimensions involved was only settled in the paper "A counterexample to a conjecture on flat bilinear forms," published in the *Proceedings of the American Mathematical Society* in 2004 and written in collaboration with L. Florit, his former student and afterward colleague at IMPA.

Marcos's work on the isometric deformation problem for submanifolds also included our very first joint paper "On compositions of isometric immersion," published in 1992 in the *Journal of Differential Geometry*. Besides describing the structure of isometric immersions of the n-dimensional sphere into Euclidean space $\mathbb{R}^{n+p}$ up to codimension $p = n - 2$, that paper paved the way to several significant generalizations. These culminated in the crucial notion of genuine deformations of a submanifold, developed in the fundamental article Genuine deformations of submanifolds, a joint work with L. Florit that was published in the *Communications in Analysis and Geometry* in 2004.

Among the very few results of global nature on this topic, one should mention Marcos's joint papers with D. Gromoll, "Rigidity of complete Euclidean hypersurfaces," published in the *Journal of Differential Geometry* in 1990, and "Isometric deformations of compact Euclidean submanifolds in codimension" 2, which appeared in the *Duke Mathematical Journal* in 1995.

A nice compilation of his contributions to the study of *infinitesimal* isometric and conformal deformations of submanifolds has recently appeared in a book in collaboration with M. I. Jimenez, his last PhD student at IMPA. The book was published by IMPA in 2021 as an entire issue of *Ensaios Matemáticos*, and contains several recent articles he wrote with M. I. Jimenez himself and with T. Vlachos from Ioannina University (Greece), one of his main collaborators in the last decade. It also includes an earlier paper with L. Rodríguez from 1990, which appeared in the *Annales de L'Institut Fourier*.

In his post-doc stage at Stony Brook in 1982–83, Marcos was also introduced to other of his favorite research topics. Namely, the study of isometric immersions of Kaehler manifolds into Euclidean space, the so-called *real Kaehler submanifolds*, and later in other ambient spaces. About 15 of his articles are on this subject, including some very recent ones. We mention his first article on the subject, "Real Kaehler submanifolds and uniqueness of the Gauss map," written in collaboration with Detlef Gromoll in 1985 and published in the *Journal of Differential Geometry*, and his last one, "Real Kaehler submanifolds in codimension up to four," with S. Chion, published in 2023 in the *Revista Matemática Iberoamericana*. In between, among many other important contributions, he combined this topic with the previous one in a paper with another of his main collaborators, L. Rodríguez, his colleague at IMPA, with whom he wrote, among others, "Rigidity of real Kaehler submanifolds," published in 1986 in the *Duke Mathematical Journal*.

Another distinct line of investigation in Marcos's work is related to the *Ribaucour transformation*. His first article on the subject appeared as improvements of some results in part of my doctoral thesis *Isometric immersions between spaces of constant curvature*, from 1991. In the article "An extension of the classical Ribaucour transformation," published in the *Proceedings of the London Mathematical Society* in 2002, the classical Ribaucour transformation for surfaces in $\mathbb{R}^3$ was extended for submanifolds of arbitrary dimension and codimension. The primary aim was to construct explicit examples of local isometric immersions of the n-dimensional sphere $\mathbb{S}^n$ into $\mathbb{R}^{2n-1}$ not having the structure of a composition of an umbilical inclusion of $\mathbb{S}^n$ into $\mathbb{R}^{n+1}$ with a local isometric immersion of $\mathbb{R}^{n+1}$ into $\mathbb{R}^{2n-1}$. However, the theory had a much wider range of applications. Noteworthy was the description of all submanifolds with a flat normal bundle in terms of a commuting family of Hessian matrices on an open subset of Euclidean space $\mathbb{R}^m$. This was one of the main results of our article "The vectorial Ribaucour transformation for submanifolds and applications," in collaboration with L. Florit, published in the *Transactions of the American Mathematical Society* in 2007.

A class of submanifolds that also attracted Marcos's attention was that of the conformally flat ones. Among his several papers on this topic, I mention his article "Compact conformally flat hypersurfaces," with M. do Carmo and F. Mercuri, published in the *Transactions of the American Mathematical Society* in 1985, his highly cited "Rotation hypersurfaces in spaces of constant curvature," with M. do Carmo, published in the *Transactions of the American Mathematical Society* in 1983, his description of "Conformally flat Riemannian manifolds as hypersurfaces of the light cone," with A.C. Asperti, which appeared in the *Canadian Mathematical Bulletin* in 1989, and his joint paper with L. Florit, "On conformally flat submanifolds," published in the *Communications in Analysis and Geometry* in 1996, where they have studied conformally flat Euclidean submanifolds with codimension two.

Minimal submanifolds were also among Marcos's topics of interest. Indeed, at least 18 of his papers dealt with problems related to this class of submanifolds. His main interest was in the higher dimensional case. In one of his first papers on the subject, he and M. do Carmo extended, in the article "Necessary and sufficient conditions for existence of minimal hypersurfaces in spaces of constant curvature," published in the *Boletim da Sociedade Brasileira de Matemática* in 1981, the classical necessary and sufficient conditions established by Ricci-Curbastro for a Riemannian metric on a surface to be the first fundamental form of a minimal immersion of that surface into the Euclidean space. It is worth noticing that one of the articles in this volume, authored by L. Ambrozio, proposes new versions of Ricci-Curbastro results in the context of the theory of constant mean curvature surfaces in three-dimensional space forms that meet umbilical surfaces orthogonally along their boundary components, and discusses generalizations of those results inspired by do Carmo and Dajczer theorem.

Among his many articles related to minimal submanifolds, it is noteworthy to quote his paper "The Weierstrass representation for complete minimal real Kaehler submanifolds of codimension two," published in the *Inventiones Mathematicae* in 1995, in collaboration with D. Gromoll. I also highlight his work with L. Florit on "Austere submanifolds," published in the *Illinois Journal of Mathematics* in 2001, and his series of papers on "Complete minimal submanifolds with nullity" in Euclidean space, the sphere and hyperbolic space, in collaboration with T. Kasioumis, A. Savas-Halilaj and T. Vlachos, published in the *Mathematische Zeitschrift* in 2017, in the *Commentarii Mathematici Helvetici* in 2018, and in the *Journal of Geometric Analysis* in 2019. Two-dimensional minimal surfaces have appeared nicely in the description of all superconformal surfaces, as well as in the classification of submanifolds of codimension two that attain the equality in the so-called DDVV inequality, which have given rise to articles with myself in the *Mathematische Zeitschrift* and in the *Mathematical Proceedings of the Cambridge Philosophical Society* in 2009.

As anticipated at the beginning of this preface, Marcos's work ranges on a wide variety of themes. In the impossibility of quoting them all, I conclude by mentioning his many papers on graphs of different kinds (like Killing graphs, for instance) with relevant geometric properties and in several distinct ambient spaces. Most of these works have been developed in collaboration with J. H. Lira, L. J. Alías, and J. Ripoll, the last two of which also contribute to this volume. Below are some of those papers:

"Uniqueness of constant mean curvature surfaces properly immersed in a slab," published in collaboration with L. J. Alías in the *Commentarii Mathematici Helvetici* in 2006

"Killing graphs with prescribed mean curvature and Riemannian submersions," published in collaboration with J. H. de Lira in the *Annales de l'Institut Henri Poincaré - Analyse non linéaire* in 2009

"Constant mean curvature hypersurfaces with single-valued projections on planar domains," published in collaboration with J. Ripoll in the *Journal of Differential Equations* in 2011

"Entire bounded constant mean curvature Killing graphs," published in collaboration with J. H. de Lira in the *Journal des Mathématiques Pures et Appliquées* in 2015

"An interior gradient estimate for the mean curvature equation of Killing graphs," published in collaboration with J. H. de Lira and J. Ripoll in the *Journal d'Analyse Mathématique* in 2016

Among Marcos's awards and titles, it is worth mentioning his membership in the **Academia Brasileira de Ciências** (2002) and the **Third World Academy of Sciences** (2005), and his fellowship by the **J. S. Guggenheim Foundation** (1996).

I finish the above examples of Marcos's contributions by quoting his book *Submanifold Theory Beyond an Introduction*, published in collaboration with myself in 2019 by Springer-Verlag in the Universitext series, which focuses on general properties of isometric and conformal immersions of Riemannian manifolds into space forms, and has the isometric and conformal deformation problem for submanifolds of arbitrary dimension and codimension as one of its main themes. Several relevant classes of submanifolds are also discussed, including constant curvature submanifolds, submanifolds of nonpositive extrinsic curvature, conformally flat submanifolds, and real Kaehler submanifolds. The book fills a gap in the literature, being the first textbook to address a substantial proportion of its contents.

São Carlos, Brazil
November 2023

Ruy Tojeiro

Acknowledgements I am very grateful to all of those who have contributed to this volume (in alphabetical order by their first surnames): Ari Aiolfi, Luis José Alías, Lucas Ambrozio, Caroline Assmann, Esteban Calviño-Louzao, Sergio Chion, Giulio Colombo, Andrzej Derdzinski, Eduardo García-Río, Ixchel Gutiérrez-Rodríguez, Katsuei Kenmotsu, Paolo Piccione, Marco Rigoli, Jaime Ripoll, Graham Smith, and Ramón Vázquez-Lorenzo. They have taken seriously the task of making a high-level book to honor Marcos Dajczer on his 75th birthday, with significant contributions to their respective research areas. I am sure every article in this book would be worth publishing in a high-standard journal. Thanks also goes to Fernando Manfio and Alan Sousa França, for their help in editing this book, and to all of the anonymous referees, for carefully analyzing the articles, considering their relevance, correctness, and suitability. Finally, I thank Robinson dos Santos, from Springer-Verlag, Cesar Rogério de Oliveira, and the remaining editors of the UFSCar subseries of the Latin American Mathematics series of Springer, for their interest and support in the process of publishing this volume within that subseries.

Contents

The Dirichlet Problem for the Minimal Surface Equation on Unbounded Helicoidal Domains of $\mathbb{R}^m$

Ari Aiolfi, Caroline Assmann, and Jaime Ripoll

Dedicated to Professor Marcos Dajczer on his 75th birthday.

Abstract We consider a helicoidal group G in $\mathbb{R}^{n+1}$ and unbounded G-invariant $C^{2,\alpha}$-domains $\Omega \subset \mathbb{R}^{n+1}$ whose helicoidal projections are exterior domains in $\mathbb{R}^n$, $n \geq 2$. We show that for all $s \in \mathbb{R}$, there exists a G-invariant solution $u_s \in C^{2,\alpha}\left(\overline{\Omega}\right)$ of the Dirichlet problem for the minimal surface equation with zero boundary data which satisfies $\sup_{\partial\Omega} |\text{grad}\, u_s| = |s|$. Additionally, we provide further information on the behavior of these solutions at infinity.

1 Introduction

The Dirichlet problem for the minimal surface equation (mse) in $\mathbb{R}^m$, $m \geq 2$, namely,

$$\begin{cases} \mathcal{M}(u) := \text{div}\left(\frac{\text{grad}\, u}{\sqrt{1+|\text{grad}\, u|^2}}\right) = 0 \text{ in } \Omega, \ u \in C^2(\Omega) \cap C^0\left(\overline{\Omega}\right) \\ \qquad\qquad\qquad\qquad u|_{\partial\Omega} = \varphi \end{cases} \tag{1}$$

where $\Omega \subset \mathbb{R}^m$ is a C^2-domain and $\varphi \in C^0(\partial\Omega)$ is given a priori, has been intensively explored in the last decades. One of the most general answers to the Dirichlet problem (1) for bounded domains was given by H. Jenkins and J. Serrin in [10]. They showed that (1) is solvable for arbitrary $\varphi \in C^0(\partial\Omega)$ if and only if Ω is mean convex. Moreover, they noted that if $\varphi \in C^2(\partial\Omega)$, a bound on the oscillation of φ in terms of the second-order norm of φ should be enough to ensure the solvability of (1) on arbitrary bounded domains (Theorem 2 of [10]).

A. Aiolfi · C. Assmann (✉) · J. Ripoll
Departamento de Matemática, Universidade Federal de Santa Maria, Santa Maria, RS, Brazil
e-mail: ari.aiolfi@ufsm.br; jaime.ripoll@ufrgs.br

© The Author(s), under exclusive license to Springer Nature Switzerland AG 2025
R. Tojeiro (ed.), *Advances in Submanifold Theory and Related Topics*,
Latin American Mathematics Series – UFSCar subseries,
https://doi.org/10.1007/978-3-032-00918-0_1

The study of the Dirichlet problem for the mse on unbounded domains began with J. C. C. Nitsche in the so-called *exterior domains* , that is, when $\mathbb{R}^m \backslash \Omega$ is relatively compact (Section 4 of [16]). Since then several authors continue the investigation of the Dirichlet problem for the mse in exterior domains [1, 3, 13–15, 17, 18, 20].

The Dirichlet problem (1) for more general unbounded domains reduces, to authors' knowledge, to few works: when $m = 2$, Rosenberg and Sa Earp [7] proved that (1) has a solution if $\Omega \subset \mathbb{R}^2$ is convex subset distinct of a half plane for any continuous boundary data φ. In the half plane case, in the case that φ is bounded, Collin and Krust [3] proved that if Ω is a convex set distinct of a half plane, then the solution is unique, and, if Ω is a half plane, then there is a unique solution with linear growth. For an arbitrary dimension m, Z. Jin, in [11], proved that (1) has a solution if Ω is a mean convex domain contained in some special like parabola-shaped region or in the complement of a cone in $\mathbb{R}^m$. More recently, N. Edelen and Z. Wang proved that if $\Omega \subsetneq \mathbb{R}^n$ is an open convex domain (e.g., a half-space) and $\varphi \in C^0(\partial\Omega)$ is a linear function, then any solution of (1) must also be linear.

In our work, we obtain an extension of the exterior Dirichlet problem for the minimal surface equation in $\mathbb{R}^m$, $m \geq 3$, in the following sense: we say that a domain is $k-$bounded, $0 \leq k \leq m$, if it is bounded in k orthogonal directions of the space $\mathbb{R}^m$ (as a direction, we mean an equivalence class of parallel lines of $\mathbb{R}^m$). As so, a domain of $\mathbb{R}^m$ is relatively compact if and only if it is $m-$bounded. In our main results, we study the Dirichlet problem for the mse on certain domains Ω of $\mathbb{R}^m$, $m \geq 3$, which complement $\mathbb{R}^m \backslash \Omega$ is $(m-1)-$bounded, and for zero boundary data. We recall that Theorem 3.5 of [3] proves the existence of solutions of the Dirichlet problem for the mse on a strip, a special $1-$bounded domain of $\mathbb{R}^2$, for arbitrary continuous bounded boundary data.

To state precisely our theorems, we need to recall a result of the third author and F. Tomi (Theorem 2 of [20]) which asserts that if Ω is G-invariant $C^{2,\alpha}$-domain for $m \geq 3$, where G is a subgroup of $ISO(\mathbb{R}^m)$ that acts freely and properly on $\mathbb{R}^m$, such that $P(\Omega)$ is a bounded and mean convex domain, then (1) has an unique G-invariant solution for any $G-$invariant boundary data $\varphi \in C^{2,\alpha}(\partial\Omega)$, where $P : \mathbb{R}^m \longrightarrow \mathbb{R}^m/G$ is the projection through the orbits of G and $\mathbb{R}^m/G$ is endowed with a metric such that P becomes a Riemannian submersion.

Related to the above result, we would like to mention that the use of a Lie group of symmetries to study minimal surfaces was first considered by Wu-yi Hsiang and Blaine Lawson in [8]. Although proving distinct facts, we can say that Proposition 3 of [20] has the same spirit of Theorem 2 of [8]. Also related to these results, there is the idea of using Lie groups of symmetries to the study of minimal graphs (constant mean curvature graphs and more general PDEs too), as Killing graphs in warped products. This technique was first considered by Marcos Dajczer and the third author of this work in [4] and, since then, many works have been done extending and generalizing the results of [4], as [5, 6], and [9].

Let $\lambda \in \mathbb{R}$, $a \in \mathbb{R}\backslash\{0\}$ and $i, j, k \in \{1, \ldots, n+1\}$ be given with any two i, j and k distinct. Consider the helicoidal like group $G \equiv G^{i,j,k}_{\lambda,a}$ in $\mathbb{R}^{n+1}$, $n \geq 2$, determined by the one-parameter subgroup of isometries $G = \{\varphi_t\}_{t\in\mathbb{R}}$, where $\varphi_t : \mathbb{R}^{n+1} \longrightarrow \mathbb{R}^{n+1}$ is given by

$$\varphi_t(p) = \alpha(p)\, e_i + \beta(p)\, e_j + \gamma(p)\, e_k + \sum_{l\neq i,j,k} x_l e_l, \tag{2}$$

where $p = (x_1, \ldots, x_{n+1})$, $\{e_i\}_{i=1}^{n+1}$ is usual orthonormal basis of $\mathbb{R}^{n+1}$,

$$\alpha(p) = x_i \cos\lambda t + x_j \sin\lambda t,\ \beta(p) = x_j \cos\lambda t - x_i \sin\lambda t$$

and $\gamma(p) = x_k + at$.

Let $\pi : \mathbb{R}^{n+1} \longrightarrow \{x_k = 0\} \equiv \mathbb{R}^n$ be the *helicoidal* projection determined by G, that is,

$$\pi(p) = \widehat{\alpha}(p)\, e_i + \widehat{\beta}(p)\, e_j + \sum_{l\neq i,j,k} x_l e_l, \tag{3}$$

where

$$\widehat{\alpha}(p) = x_i \cos\frac{\lambda x_k}{a} - x_j \sin\frac{\lambda x_k}{a},\ \widehat{\beta}(p) = x_j \cos\frac{\lambda x_k}{a} + x_i \sin\frac{\lambda x_k}{a}.$$

Set

$$M := (\mathbb{R}^n, \langle , \rangle_G), \tag{4}$$

where $\langle\ ,\ \rangle_G$ is the metric such that π becomes a Riemannian submersion (clearly G acts freely and properly in $\mathbb{R}^{n+1}$ and $\{x_k = 0\} \equiv \mathbb{R}^n$ is a slice relative to the orbits $Gp = \{\varphi_t(p), t \in \mathbb{R}\}$). One may see that the map $\psi : \mathbb{R}^{n+1}/G \to \mathbb{R}^n$ given by $\psi = \pi \circ P^{-1}$ is well defined and is an isometry with the metrics mentioned above. From the isometric submersion theory, it follows that M is complete.

Let $\Omega \subset \mathbb{R}^{n+1}$ be a G-invariant domain of class $C^{2,\alpha}$ and set $\Lambda = \pi(\Omega)$. Let $d_E(p) = d_E(p, \partial\Omega)$, $p \in \Omega$, be the (Euclidean) distance in $\mathbb{R}^{n+1}$ to $\partial\Omega$ and let $d(q) = d_M(q, \partial\Lambda), q \in \Lambda$, be the distance in M to $\partial\Lambda$. Given $u \in C^2(\Omega) \cap C^0(\overline{\Omega})$ and $\varphi \in C^0(\partial\Omega)$, G-invariant functions, that is, $u = v \circ \pi$ for some $v \in C^2(\Lambda) \cap C^0(\overline{\Lambda})$ and $\varphi = \psi \circ \pi$ for some $\psi \in C^0(\partial\Lambda)$, it follows from Proposition 3 of [20] that u is solution of (1) (relative to $m = n + 1$) if, and only if

$$\begin{cases} \mathfrak{M}(v) := \operatorname{div}_M\left(\frac{\nabla v}{\sqrt{1+|\nabla v|^2}}\right) - \frac{1}{\sqrt{1+|\nabla v|^2}}\langle \nabla v, J\rangle_M = 0 \text{ on } \Lambda \\ v|_{\partial\Lambda} = \psi \end{cases} \tag{5}$$

where ∇ and div_M are the gradient and divergence in M, respectively, and

$$J(\pi(p)) = d\pi_p\left(\overrightarrow{H}_{Gp}(p)\right),\ p \in \mathbb{R}^{n+1}, \tag{6}$$

where $\overrightarrow{H}_{Gp}$ is the mean curvature vector of the 1-dimensional submanifold Gp of $\mathbb{R}^{n+1}$. Moreover, $\left|\overline{\nabla}u\right| = |\nabla v| \circ \pi$, where $\overline{\nabla}$ denotes the gradient in $\mathbb{R}^{n+1}$.

When Λ is a bounded and mean convex domain, it is proved in [20], as mentioned above, that there is a unique G-invariant solution $u \in C^{2,\alpha}\left(\overline{\Omega}\right)$ of (1) for all G-invariant $\varphi \in C^{2,\alpha}(\partial\Omega)$.[1] In this note, we work with the case where $\Lambda = \pi(\Omega)$ is an exterior domain in $\mathbb{R}^n$ and the boundary data is zero. We observe that Λ is an exterior domain in M if and only if Ω is $n-$bounded.

Regarding the condition of zero boundary data, we recall an old but quite suggestive result to our work of Osserman that proves that in $\mathbb{R}^2$ with the Euclidean metric, there is a boundary data on a disk D for which there is no solution to the exterior Dirichlet problem in $\mathbb{R}^2\backslash D$. This strongly suggests, since $K_M > 0$, that the zero boundary data condition cannot be rejected either in our case. But we don't have a counterexample.

In order to state our main results, we remember that, relative to an exterior domain $\mathbb{R}^n\backslash\overline{\mathfrak{B}}_\rho(p_0)$ in $\mathbb{R}^n$, $n \geq 2$, where $\mathfrak{B}_\rho(p_0)$ is an open ball of $\mathbb{R}^n$ centered at $p_0 \in \mathbb{R}^n$ and of radius $\rho > 0$, the function

$$v_\rho(p) := \rho \int_1^{\frac{\tau}{\rho}} \frac{dt}{\sqrt{t^{2(n-1)} - 1}}, \; \tau = |p - p_0|, \; p \in \mathbb{R}^n - \mathfrak{B}_\rho(p_0), \tag{7}$$

Fig. 1 A G-invariant domain in $\mathbb{R}^3$

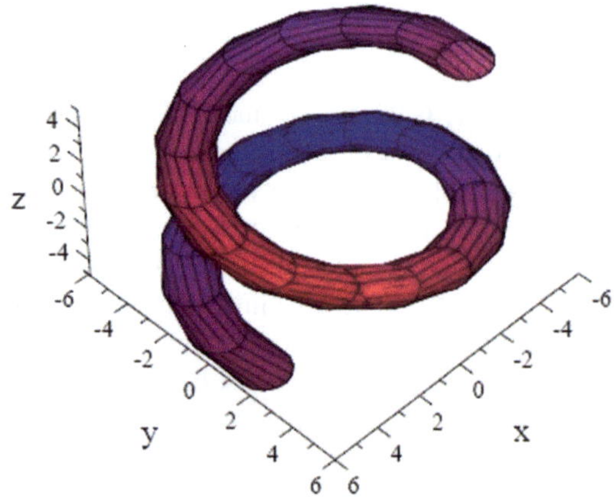

[1] The white region in Fig. 1 is a G-invariant domain $\Omega \subset \mathbb{R}^3$ (with $\lambda = a = 1$), whose boundary is the surface $\Psi : [0, 2\pi] \times \mathbb{R} \longrightarrow \mathbb{R}^3$ given by

$$\Psi(t, z) = (\cos t \cos z + (5 + \sin t) \sin z, (5 + \sin t) \cos z - \cos t \sin z, z).$$

Note that $\mathbb{R}^3\backslash\overline{\Omega}$ is an example of a two-bounded domain in $\mathbb{R}^3$.

is the solution relative to the Dirichlet problem (1) which satisfies

$$\lim_{p \to \partial \mathfrak{B}_\rho(p_0)} \left| \nabla v_\rho (p) \right| = \infty, \lim_{|p| \to \infty} \left| \nabla v_\rho (p) \right| = 0$$

(a half-catenoid). If $n \geq 3$, v_ρ is bounded and its height at infinity, which we denote by $h(n, \rho)$, is

$$h(n, \rho) = \rho h(n, 1) = \rho \int_1^\infty \frac{dt}{\sqrt{t^{2(n-1)} - 1}}. \tag{8}$$

In all of the results from now on, $G \equiv G_{\lambda,a}^{i,j,k}$ is as defined in (2), with λ, a, i, j, k fixed.

We prove:

Theorem 1 *Let $\Omega \subset \mathbb{R}^{n+1}$, $n \geq 2$, be a G-invariant and $C^{2,\alpha}$-domain such that $\pi(\Omega)$ is an exterior domain of $\mathbb{R}^n = \{x_k = 0\}$, where G and π are as defined in (2) and (3), respectively. Let $\varrho > 0$ be the radius of smallest geodesic ball of $\mathbb{R}^n$ which contain $\partial\pi(\Omega)$, centered at origin of $\mathbb{R}^n$ if $\lambda \neq 0$. Then, given $s \in \mathbb{R}$, there is a G-invariant solution $u_s \in C^{2,\alpha}\left(\overline{\Omega}\right)$ of the Dirichlet problem (1) with $u_s|_{\partial\Omega} = 0$, such that:*

i) $\sup_{\overline{\Omega}} \left|\nabla u_s\right| = \sup_{\partial\Omega} \left|\nabla u_s\right| = |s|$.

ii) u_s is unbounded if $n = 2$ and $s \neq 0$ and, if $n \geq 3$, either

$$\sup |u_s| \leq h(n, \varrho)$$

or there is a complete, non-compact, properly embedded n-dimensional submanifold $N \subset \Omega$, such that

$$\lim_{d_E(p) \to +\infty} u_s|_N (p) = h(n, \varrho),$$

where $h(n, \varrho)$ is given by (8).

iii) u_s satisfies

$$\lim_{d_E(p) \to \infty} \left|\nabla u_s (p)\right| = 0$$

if $\lambda = 0$ or $s = 0$ or, if $\lambda \neq 0$, $3 \leq n \leq 6$ and u_s is bounded.

Additional information relative to set of G-invariant solutions of the Dirichlet problem (1) also are obtained under the assumption that $M \backslash \pi(\Omega)$ satisfies the interior sphere condition of radius $r > 0$, that is, for each $q \in \partial\pi(\Omega)$, there is a geodesic sphere S_q of M of radius r contained in $M \backslash \pi(\Omega)$ such that S_q is tangent to $\partial\pi(\Omega)$ at q and r is maximal with this property.

Given $a \in \mathbb{R}\backslash\{0\}$, $\lambda \in \mathbb{R}$, $n \geq 3$, and $r > 0$, set

$$C = C(r, n, \lambda, a) := \frac{2|a|(n-1) + |\lambda| r}{2|a| r}. \tag{9}$$

Let $\varsigma > C$ be the solution of the equation:

$$\cosh\left(\frac{\mu}{\sqrt{\mu^2 - C^2}}\right) = \frac{\mu}{C}, \mu > C, \tag{10}$$

and set

$$\mathcal{L} = \mathcal{L}(r, n, \lambda, a) := \begin{cases} \frac{1}{\sqrt{\varsigma^2 - C^2}} \text{ if } \lambda \neq 0 \\ h(n, r) \text{ if } \lambda = 0 \end{cases}, \tag{11}$$

where $h(n, r)$ is given by (8).

Theorem 2 *Let $\Omega \subset \mathbb{R}^{n+1}$, $n \geq 3$, be a G-invariant and $C^{2,\alpha}$-domain such that $\pi(\Omega)$ is an exterior domain of $\mathbb{R}^n = \{x_k = 0\}$, where G and π are as defined in (2) and (3), respectively. Assume that $M\backslash\pi(\Omega)$ satisfies the interior sphere condition of radius $r > 0$, where M is the n-dimensional Riemannian manifold given by (4). Let $\mathcal{L} = \mathcal{L}(r, n, \lambda, a)$ be as defined in (11), where λ and a are given in the definition of G. Then, given $c \in [0, \mathcal{L}]$, there is a G-invariant solution $u_c \in C^2(\Omega) \cap C^0(\overline{\Omega})$ of (1) with $u_c|_{\partial\Omega} = 0$, such that*

$$\lim_{d_E(p)\to\infty} u_c(p) = c.$$

In particular, if $c \in [0, \mathcal{L})$, then $u_c \in C^{2,\alpha}(\overline{\Omega})$.

We note that our approach is not applicable to general boundary data. Moreover, we were not able to prove that the solutions u_s obtained in Theorem 1 have a limit at infinity and this could be the subject of a future research.

2 Preliminaries

We first observe that, relative to the PDE given in (5), the maximum and comparison principles work (see Section 3 of [20]).

In this section, we give further information on M, and we provide the basic results to construct barriers relative to the Dirichlet problem (5) when the boundary data is zero. We shall use the meaning of the indexes i and j as defined in (2).

Lemma 1 *Let G be as defined in (2). Given $p = (x_1, \ldots, x_{n+1}) \in \mathbb{R}^{n+1}$, the orbit Gp has constant curvature:*

$$H_{Gp} = \frac{\lambda^2 \sqrt{x_i^2 + x_j^2}}{\lambda^2 (x_i^2 + x_j^2) + a^2}. \tag{12}$$

In particular

$$\sup_{p \in \mathbb{R}^{n+1}} H_{Gp} = \frac{|\lambda|}{2\,|a|}, \tag{13}$$

and the supremum is attended, if $\lambda \neq 0$, at those orbits through the points p in $\mathbb{R}^{n+1}$ such that $\left|(x_i, x_j)\right| = |a| \,/\, |\lambda|$.

Proof An arch length parametrization of Gp is given by

$$\gamma_p(s) = F(p)\, e_i + G(p)\, e_j + H(p)\, e_k + \sum_{l \neq i,j,k} x_l e_l.$$

where

$$F(p) = x_i \cos(A(p)\, s) + x_j \sin(A(p)\, s),\ G(p) = x_j \cos(A(p)\, s) - x_i \sin(A(p)\, s)$$

and

$$H(p) = x_k + \frac{a A(p)\, s}{\lambda},$$

with

$$A(p) = \frac{\lambda}{\sqrt{\lambda^2 (x_i^2 + x_j^2) + a^2}}. \tag{14}$$

The mean curvature vector of Gp is then

$$\overrightarrow{H}_{Gp}(\gamma_p(s)) = \gamma_p''(s) = -A^2(p) \left[F(p)\, e_i + G(p)\, e_j \right], \tag{15}$$

and, consequently, the mean curvature of Gp in $\mathbb{R}^{n+1}$ is given by (12). Since the mean curvature of the orbit Gp only depends on the Euclidean distance of $x_i e_i + x_j e_j$ to the origin, setting $\sigma(p) = \left|(x_i, x_j)\right|$, we have that $\xi : [0, \infty) \longrightarrow \mathbb{R}$ given by

$$\xi(\sigma) = \frac{\lambda^2 \sigma}{\lambda^2 \sigma^2 + a^2}, \tag{16}$$

has a maximal absolute point at $\sigma_0 = |a| / |\lambda|$ if $\lambda \neq 0$ and the result follows. □

Since $\pi : \mathbb{R}^{n+1} \longrightarrow M$ is a Riemannian submersion, given two orthonormal vector fields $X, Y \in \chi(M)$ and their respective horizontal lifts $\overline{X}, \overline{Y}$, we know that

$$K_M(X, Y) = K_{\mathbb{R}^{n+1}}\left(\overline{X}, \overline{Y}\right) + \frac{3}{4}\left|\left[\overline{X}, \overline{Y}\right]^v\right|^2_{\mathbb{R}^{n+1}}$$

where K and $\left[\overline{X}, \overline{Y}\right]^v$ mean, respectively, the sectional curvature and the vertical component of $\left[\overline{X}, \overline{Y}\right]$, that is, the component which is tangent to the orbits Gp, $p \in \mathbb{R}^{n+1}$. As $K_{\mathbb{R}^{n+1}}\left(\overline{X}, \overline{Y}\right) = 0$, it follows that $K_M(X, Y) \geq 0$ and, therefore, $Ric_M \geq 0$ (straightforward, but quite extensive calculations give us that, in fact, $K_M > 0$ with $K_M \to 0$ at infinity).

Lemma 2 *Let Λ be an exterior domain in M. Denote by ν the horizontal lift of ∇d, where $d = d_M(., \partial\Lambda)$. Then*

$$\langle \nabla d, J \rangle_M \circ \pi = \left\langle \nu, \vec{H}_G \right\rangle \tag{17}$$

where $\langle , \rangle$ is the Euclidean metric and J is given by (6). In particular

$$-H_{Gp}(p) \leq \langle \nabla d, J \rangle_M (\pi(p)),$$

for all $p \in \pi^{-1}(\Lambda)$, where H_{Gp} is given by (12).

Proof As $\vec{H}_G$ and ν are the horizontal lift of J and ∇d, respectively, we have $J(\pi(p)) = d\pi_p\left(\vec{H}_{Gp}(p)\right)$ and $\nabla d(\pi(p)) = d\pi_p(\nu(p))$. Since π is a Riemannian submersion

$$d\pi_p|_{[T_p\mathbb{R}^{n+1}]^h} : \left[T_p\mathbb{R}^{n+1}\right]^h \longrightarrow T_{\pi(p)}M$$

is an isometry, where $\left[T_p\mathbb{R}^{n+1}\right]^h$ means the horizontal vector space relative to Gp at p and, from this, we have (17). In particular, $\left\langle \nu, \vec{H}_{Gp}\right\rangle$ is constant along Gp. Note that $|\nu| = 1$ since $|\nabla d| = 1$. Thus

$$-H_{Gp}(p) = -\left|\vec{H}_{Gp}(p)\right| \leq \left\langle \nu, \vec{H}_{Gp}\right\rangle(p)$$

and the result follows. □

Proposition 1 *Let G, π, and $M = (\mathbb{R}^n, \langle , \rangle_G)$ as defined in (2), (3), and (4), respectively, and assume $n \geq 3$ and $\lambda \neq 0$. Let o be an arbitrary but fixed point of M and let $\Lambda = M\backslash\overline{B_r(o)}$, where $B_r(o)$ is the open geodesic ball of M of radius r centered at o. Let $b \in \mathbb{R}$ satisfying $b > C$, where C is given by (9). Consider $\psi : [0, \infty) \longrightarrow \mathbb{R}$ given by*

$$\psi(t) = \frac{1}{b}\cosh^{-1}(1 + bt) \tag{18}$$

and $\Lambda_0 := \{q \in \Lambda; d(q) \leq t_0\}$, where

$$t_0 = \frac{b - C}{bC}. \tag{19}$$

Then, $w = \psi \circ d : \Lambda_0 \to \mathbb{R}$ is such that $w \in C^2(\Lambda_0) \cap C^0\left(\overline{\Lambda_0}\right)$, $w|_{\partial\Lambda} = 0$, $w > 0$ on Λ_0

$$\lim_{d(q)\to 0} |\nabla w(q)| = +\infty$$

and $\mathfrak{M}(w) \leq 0$ on Λ_0, where $\mathfrak{M}$ is the operator defined in (5).

Proof Let $\varphi \in C^2((0,\infty)) \cap C^0([0,\infty))$ to be determined *a posteriori* and consider the function $w : \Lambda \subset M \longrightarrow \mathbb{R}$ given by $w(q) = (\varphi \circ d)(q)$. Straightforward calculations give us that in Λ

$$div_M\left(\frac{\nabla w}{\sqrt{1 + |\nabla w|^2}}\right) = g(d)\,\Delta d + g'(d)$$

and

$$\frac{1}{\sqrt{1 + |\nabla w|^2}}\langle \nabla w, J\rangle_M = g(d)\langle \nabla d, J\rangle_M,$$

where

$$g(d) := \frac{\varphi'(d)}{\sqrt{1 + [\varphi'(d)]^2}}, \tag{20}$$

Δ is the Laplacian in M and "$'$" means $\frac{\partial}{\partial d}$. Thus, $\mathfrak{M}(w) \le 0$ in Λ if and only if

$$g(d)\,\Delta d + g'(d) - g(d)\,\langle \nabla d, J\rangle_M \le 0 \text{ in } \Lambda. \tag{21}$$

From the Laplacian comparison theorem, since $Ric_M \ge 0$ and $\dim M = n$, we have

$$\Delta d(q) \le \frac{n-1}{d(q)+r} \le \frac{n-1}{r},\ q \in \Lambda. \tag{22}$$

Now, we assume that our φ satisfies $\varphi(0) = 0$ and $\varphi'(d) > 0$ for $d > 0$ (consequently, $\varphi(d) > 0$ for $d > 0$). From (20), it follows that $g(d) > 0$ for $d > 0$ and, from (22), we conclude that if

$$\frac{(n-1)\,g(d)}{r} + g'(d) - g(d)\,\langle \nabla d, J\rangle_M \le 0 \text{ in } \Lambda \tag{23}$$

then we have (21). From Lemma 2 and (13), we see that

$$-\frac{|\lambda|}{2\,|a|} \le -H_{Gp}(p) \le \langle \nabla d, J\rangle_M \circ \pi(p)$$

and, then, if

$$\frac{g'(d)}{g(d)} \le -C \text{ in } \Lambda, \tag{24}$$

where C is given by (9), then we have (23). From (20), we see that (24) is equivalent to

$$\frac{\varphi''(d)}{\varphi'(d)} \le -C\left(1 + \left[\psi'(d)\right]^2\right) \text{ in } \Lambda. \tag{25}$$

We will assume from now on that our φ satisfies

$$\lim_{d \to 0} \varphi'(d) = +\infty.$$

A function φ which satisfies all the requirements demanded until now is given by

$$\varphi(t) = \alpha \cosh^{-1}(1 + bt)$$

$t \geq 0$, with α, b positive constants to be determinate, where, $t = d(q)$, $q \in \Lambda$. Assuming such one φ, we see that (25) is equivalent to

$$\frac{-b(1+bt)}{\left((1+bt)^2 - 1 + \alpha^2 b^2\right)} \leq -C. \tag{26}$$

We assume α, b such that $\alpha b = 1$. Thus, we have (26) if

$$t \leq \frac{b-C}{bC}, C < b.$$

Thus, assuming $b > C$, setting

$$t_0 := \frac{b-C}{bC},$$

considering the neighborhood of $\partial\Lambda$ in Λ given by

$$\Lambda_0 := \{q \in \Lambda; d(q) \leq t_0\},$$

and the function

$$\psi(t) = \frac{1}{b}\cosh^{-1}(1+bt), t \geq 0,$$

we have that

$$w(q) = \psi \circ d(q), q \in \Lambda_0, \tag{27}$$

satisfies

$$\mathfrak{M}(w) \leq 0$$

in Λ_0 (note that $\exp_{\partial\Lambda} : \partial\Lambda \times [0, t_0] \longrightarrow \Lambda_0$, $\exp_{\partial\Lambda}(p, t) = \exp_p t\eta(p)$, where η is the unit vector field normal to $\partial\Lambda$ that points to Λ, is a diffeomorphism, since Λ is the exterior of a geodesic ball in M). The other conclusions follow directly from the definition of ψ. □

Corollary 1 *Assume the same hypotheses of Proposition 1. Let $\varsigma > C$ be the solution of the equation (10), where $C = C(r, n, \lambda, a)$ is given by (9). The function w given in Proposition 1 with the biggest height at $\partial\Lambda_0 \backslash \partial\Lambda$ is obtained taking $b = \varsigma$. In particular, for such w*

$$\sup_{\overline{\Lambda}_0} w = \sup_{\partial\Lambda_0 - \partial\Lambda} w = \mathcal{L},$$

where $\mathcal{L}$ is given by (11), and setting $W : \overline{\Lambda} \subset M \longrightarrow \mathbb{R}$ given by

$$W(q) = \begin{cases} w(q), \text{ if } q \in \Lambda_0 \\ \mathcal{L} \text{ if } q \in \Lambda - \Lambda_0 \end{cases}, \tag{28}$$

we have $W \in C^0\left(\overline{\Lambda}\right)$ and radial with relation to the point o, $\mathfrak{M}(W) = 0$ on $\Lambda \backslash \partial\Lambda_0$, $W|_{\partial\Lambda} = 0$, with

$$\lim_{d(q)\to 0} |\nabla W(q)| = +\infty.$$

Proof Given $\mu > C$, take b as in Proposition 1 given by $b = \mu$. The correspondent t_0 is

$$t_0 = \frac{\mu - C}{\mu C}$$

and, at t_0, from (27), we have $\psi(t_0) = \mu^{-1}\cosh^{-1}\left(\mu C^{-1}\right)$. The function

$$f(\mu) = \frac{1}{\mu}\cosh^{-1}\left(\frac{\mu}{C}\right), \mu > C,$$

clearly satisfies $f(\mu) \to 0$ when $\mu \to C$ and $f(\mu) \to 0$ when $\mu \to +\infty$ and the absolute maximal point of f in (C, ∞) is reached at the point ς which is solution of (10) and $f(\varsigma) = \left(\varsigma^2 - C^2\right)^{-1/2}$. The other conclusions follow from the definition of W and from the fact that Λ is the exterior of a geodesic ball of M center at o. □

Remark 1 We observe that if $\lambda = 0$, then G is the group of the translations in e_k-direction. In this case, we have $M \equiv \mathbb{R}^n$, and the domain Λ, as in the hypothesis of Proposition 1, is the exterior of the geodesic ball of $\mathbb{R}^n$ of radius r. Then, v_r given by (7) is a solution of (5) if the boundary data is zero. In particular, if $n \geq 3$, its height at infinity is $h(n, r)$, where $h(n, r)$ is given by (8).

Proposition 2 *Let G, π, and $M = (\mathbb{R}^n, \langle,\rangle_G)$, $n \geq 2$, as defined in (2), (3), and (4), respectively. Let $\Lambda_\rho := M \backslash \overline{\mathfrak{B}_\rho(0)}$, where $\mathfrak{B}_\rho(0)$ is the open geodesic ball of $\mathbb{R}^n$ of radius ρ centered at origin of $\mathbb{R}^n$. Then, $v_\rho \in C^2\left(\Lambda_\rho\right) \cap C^0\left(\Lambda_\rho\right)$ given by (7) is a nonnegative solution of the Dirichlet problem (5) relative to Λ_ρ with $v_\rho|_{\partial\Lambda_\rho} = 0$, which is unbounded if $n = 2$ and satisfies $\left|\nabla v_\rho\right| \circ \pi = \left|\overline{\nabla} v_\rho\right|$*

$$\lim_{d(q)\to 0} \left|\nabla v_\rho(q)\right| = \infty, \lim_{d(q)\to\infty} \left|\nabla v_\rho(q)\right| = 0 \tag{29}$$

and

$$\lim_{d(q)\to\infty} v_\rho(q) = h(n, \rho) \text{ if } n \geq 3 \tag{30}$$

where $d = d_M\left(., \partial\Lambda_\rho\right)$, $h(n, \rho)$ is given by (8).

Proof Let $v_\rho : \mathbb{R}^n \backslash \mathfrak{B}_\rho(0) \to \mathbb{R}$ be the function given by (7), $\mathbb{R}^n \equiv \{x_k = 0\}$. As the graph of v_ρ is a minimal graph in $\mathbb{R}^{n+1}$, it follows that the graph of

$$u_\rho(x_1, \ldots, x_{n+1}) := v_\rho(x), \ x = \sum_{i \neq k} x_i e_i \in \mathbb{R}^n \backslash \mathfrak{B}_\rho(0),$$

is a minimal graph in $\mathbb{R}^{n+2}$. In particular, $\left|\overline{\nabla} u_\rho(x_1, \ldots, x_{n+1})\right| = \left|\overline{\nabla} v_\rho(x)\right|$. Note that, setting $\Omega_\rho \subset \mathbb{R}^{n+1}$ the domain of u_ρ, we have Ω_ρ a G-invariant domain such that its helicoidal projection $\pi(\Omega)$ on $\mathbb{R}^n$ coincides with the image of the its orthogonal projection on $\mathbb{R}^n$ in this case. From (3), we see that $|x| = |\pi(x_1, \ldots, x_{n+1})|$. As v_ρ is radial, it follows that

$$u_\rho(x_1, \ldots, x_{n+1}) = \left(v_\rho \circ \pi\right)(x_1, \ldots, x_{n+1})$$

and, then, u_ρ is a G-invariant function with $u_\rho|_{\partial\Omega_\rho} = 0$. It follows from Proposition 3 of [20] that $v_\rho \in C^2\left(\Lambda_\rho\right) \cap C^0\left(\Lambda_\rho\right)$ is a solution of the Dirichlet problem (5) relative to Λ_ρ with $v_\rho|_{\partial\Lambda_\rho} = 0$ and, moreover

$$\left|\nabla v_\rho\right| \circ \pi = \left|\overline{\nabla} u_\rho\right|.$$

The other conclusions follow immediately from the definition of u_ρ and from the properties satisfied by v_ρ, taking into account that $d(q) \to 0$ ($d(q) \to \infty$) if and only if $d_E\left(q, \partial\Lambda_\rho\right) \to 0$ ($d_E\left(q, \partial\Lambda_\rho\right) \to \infty$). □

3 Proof of the Main Results

The proof of Theorem 1 and Theorem 2 follows directly from the results of this section, by using Proposition 3 of [20].

Proposition 3 *Let G, π, and $M = (\mathbb{R}^n, \langle,\rangle_G)$ as defined in (2), (3), and (4), respectively. Let U be an exterior $C^{2,\alpha}$-domain in M, $U = \pi(\Omega)$, where $\Omega \subset \mathbb{R}^{n+1}$ is a G-invariant $C^{2,\alpha}$-domain and let $d = d_M(., \partial U)$. Let $\varrho > 0$ be the radius of the smallest geodesic ball of $\mathbb{R}^n$ which contain $M \backslash U$, centered at origin of $\mathbb{R}^n$ if $\lambda \neq 0$. Given $s \geq 0$, there is a nonnegative solution $\vartheta_s \in C^{2,\alpha}\left(\overline{U}\right)$ of the Dirichlet problem (5) with $\vartheta_s|_{\partial U} = 0$, such that*

$$\sup_{\overline{U}} |\nabla \vartheta_s| = \sup_{\partial U} |\nabla \vartheta_s| = s,$$

ϑ_s is unbounded if $n = 2$ and $s \neq 0$ and, if $n \geq 3$, either

$$\sup_{\overline{U}} \vartheta_s \leq h\,(n, \varrho) \tag{31}$$

or there is a complete, non-compact, properly embedded $(n-1)$-dimensional submanifold Σ of M, $\Sigma \subset U$, such that

$$\lim_{d(q)\to+\infty} \vartheta_s\,|_\Sigma\,(q) = h\,(n, \varrho)\,, \tag{32}$$

where $h\,(n, \varrho)$ is given by (8). Moreover

$$\lim_{d(q)\to\infty} |\nabla \vartheta_s\,(q)| = 0$$

if $\lambda = 0$ or $s = 0$ or if $\lambda \neq 0$, $3 \leq n \leq 6$, and ϑ_s is bounded.

Proof As mentioned in Remark 1, if $\lambda = 0$, then $M \equiv \mathbb{R}^n$, and the result is already contemplated in Theorem 1 of [1] for $n \geq 3$ and in Theorem 1 of [19] if $n = 2$. The case $s = 0$ is trivial. Assume then $s > 0$ and $\lambda \neq 0$.
Let $\rho > 0$ be such that $\partial U \subset \mathfrak{B}_\rho\,(0)$, where $\mathfrak{B}_\rho := \mathfrak{B}_\rho\,(0)$ is the open geodesic ball of $\mathbb{R}^n$, centered at origin of $\mathbb{R}^n$ and of radius ρ. Let $\Lambda_\rho = M \backslash \overline{\mathfrak{B}_\rho\,(0)}$. From Proposition 2 , there is $v_\rho \in C^\infty\left(\Lambda_\rho\right)$ solution of (5) relative to Λ_ρ, with $v_\rho|_{\partial\Lambda_\rho} = 0$, satisfying what is stated in (29). Since

$$\lim_{d_M(q,\partial\Lambda_\rho)\to\infty} \left|\nabla v_\rho\,(q)\right| = 0, \tag{33}$$

we can choose $k > \rho$ such that

$$\left|\nabla v_\rho\right|_{\partial\mathfrak{B}_k(0)} \leq \frac{s}{2}, \tag{34}$$

where $\mathfrak{B}_k\,(0)$ is the open geodesic ball of $\mathbb{R}^n$ centered at origin and of radius k. Let $U_k = \mathfrak{B}_k\,(0) \cap U$ and define

$$T_k := \left\{ \begin{array}{c} t \geq 0\,;\, \exists\; w_t \in C^{2,\alpha}\left(\overline{U}_k\right);\, \mathfrak{M}\,(w_t) = 0 \\ \sup_{\overline{U}_k} |\nabla w_t| \leq s,\;\; w_t|_{\partial U} = 0,\, w_t|_{\partial\mathfrak{B}_k(0)} = t \end{array} \right\}. \tag{35}$$

Note that the constant function $w_0 \equiv 0$ on $\overline{U}_k$ satisfies all the condition in (35), and then, $T_k \neq \emptyset$. Moreover, $\sup T_k < \infty$ since

$$\sup_{\overline{U}_k} |\nabla w_t| \leq s$$

for all $t \in T_k$. Now, since the maximum principle and comparison principle are applicable relative to the operator $\mathfrak{M}$, we can use the same approach used in the proof of Theorem 1 of [1] to show that $t_k := \sup T_k \in T_k$

$$\sup_{\overline{U}_k} \|\nabla w_{t_k}\| = s, \quad \sup_{\partial \mathfrak{B}_k} |\nabla w_{t_k}| \le s/2$$

and, since $\partial U_k = \partial U \cup \partial \mathfrak{B}_k(0)$, to conclude that

$$\sup_{\overline{U}_k} |\nabla w_{t_k}| = \sup_{\partial U} |\nabla w_{t_k}| = s. \tag{36}$$

The proof of these facts follows essentially the same steps of the aforementioned theorem (see p. 3067 and 3068 of [1]), and then, we will not do it here. Now, taking $k \to \infty$ and from diagonal method, we obtain a subsequence of (w_{t_k}) which converges uniformly in the C^2 norm in compact subsets of $\overline{U}$ to a function $\vartheta_s \in C^{2,\alpha}(\overline{U})$ satisfying $\mathfrak{M}(\vartheta_s) = 0$ in U, $\vartheta_s|_{\partial U} = 0$, which is nonnegative and such that $\sup_{\overline{U}} |\nabla \vartheta_s| = \sup_{\partial U} |\nabla \vartheta_s| = s$. In particular, from regularity elliptic PDE theory [12], we have $\vartheta_s \in C^\infty(U)$.

We will show now that if $n = 2$, then ϑ_s is unbounded and, if $n \ge 3$, we have either (31) or (32).

Let $\varrho > 0$ be the radius of the smallest open geodesic ball of $\mathbb{R}^n$ which contain $M \backslash U$, centered at origin of $\mathbb{R}^n$ and denote such ball by $\mathfrak{B}_\varrho(0)$. We have $\partial U \subset \overline{\mathfrak{B}_\varrho(0)}$, and we can conclude that $\partial U \cap \partial \mathfrak{B}_\varrho \neq \emptyset$. Let $\Lambda_\varrho = M \backslash \overline{\mathfrak{B}_\varrho(0)}$. From Proposition 2, there is $v_\varrho \in C^\infty(\Lambda_\varrho)$, $v_\rho|_{\partial \Lambda_\varrho} = 0$, solution of (5), satisfying what is stated in (29) and (30) relative to Λ_ϱ if $n \ge 3$. Otherwise, if $n = 2$, v_ϱ is unbounded and satisfies the equalities in (29).

Let $q_0 \in \partial U \cap \partial \mathfrak{B}_\varrho$. Since

$$\lim_{d_M(q, \partial \Lambda_\varrho) \to 0} |\nabla v_\varrho(q)| = +\infty, \sup_{\overline{U}} |\nabla \vartheta_s| = \sup_{\partial U} |\nabla \vartheta_s| = s < +\infty$$

and $v_\varrho(q_0) = \vartheta_s(q_0) = 0$, it follows that there is an open set V_{q0} in $U \cap \Lambda_\varrho$, with $q_0 \in \partial V_{q0}$, such that $\vartheta_s < v_\varrho$ in V_{q0}. We claim that V_{q0} is unbounded. Suppose that V_{q0} is bounded. Since $\vartheta_s|_{\partial V_{q0}} = v_\varrho|_{\partial V_{q0}}$, it follows that $\vartheta_s|_{\overline{V}_{q0}}$ and $v_\varrho|_{\overline{V}_{q0}}$ are distinct solutions to the Dirichlet problem:

$$\begin{cases} \mathfrak{M}(f) = 0 \text{ in } V_{q0},\ f \in C^2(V_{q0}) \cap C^0(\overline{V}_{q0}) \\ f|_{\partial V_{q0}} = \vartheta_s|_{\partial V_{q0}} \end{cases}, \tag{37}$$

a contradiction, since a solution of (37) is unique if the domain is bounded. It follows that V_{q0} is unbounded. Note that we can have two possibility for ∂V_{q0}: either ∂V_{q0} is bounded (in this case $V_{q0} = \Lambda_\varrho$) or ∂V_{q0} is unbounded (in this case, setting $\Sigma = \partial V_{q0}$, we have $\Sigma \subset \overline{U} \cap \overline{\Lambda}_\varrho$ a complete $(n-1)$-dimensional manifold of M). Assume first that $n \ge 3$. In this case v_ϱ is bounded. If ∂V_{q0} is bounded, as in this case $V_{q0} = \Lambda_\varrho$, this means that $\vartheta_s < v_\varrho$ on Λ_ϱ and so we have (31). If ∂V_{q0} is unbounded, as $\vartheta_s|_{\partial V_{q0}} = v_\varrho|_{\partial V_{q0}}$, we conclude that

$$\lim_{d(q)\to+\infty} \vartheta_s|_{\partial V_{q_0}}(q) = h(n, \varrho).$$

Assume now that $n = 2$. Note first that on $\Lambda_\varrho \subset \mathbb{R}^2$, it is well known that, for all $s > 0$, there is a half-catenoid $v_{\varrho,s} \in C^2\left(\overline{\Lambda_\varrho}\right)$ which is unbounded (logarithmic growth), which satisfies $v_{\varrho,s}|_{\partial\Lambda_\varrho} = 0$ and $\left|\overline{\nabla} v_{\varrho,s}\right| = s$ on $\partial\Lambda_\varrho$ $(= \partial\mathfrak{B}_\varrho)$. The same arguments used in Proposition 2 give us that $v_{\varrho,s}$ is solution for the Dirichlet problem (5) for zero boundary data relative to Λ_ϱ and satisfies $\left|\nabla v_{\varrho,s}\right| = s$ on $\partial\Lambda_\varrho$, since $\mathfrak{B}_\varrho$ is centered at origin of $\mathbb{R}^2$. As $\sup_{\partial U}|\nabla\vartheta_s| = s$ and $\partial U \cap \partial\mathfrak{B}_\varrho \neq \emptyset$, there is $0 < s' < s$ such that $v_{\varrho,s'} \in C^2\left(\overline{\Lambda_\varrho}\right)$ as described above satisfies $v_{\varrho,s'} < \vartheta_s$ in some open set $V \subset U \cap \Lambda_\varrho$. The same arguments used before to prove that V_{q_0} is unbounded give us that V is unbounded, and we see that if ∂V is bounded, then $V = \Lambda_\varrho$. Thus, if ∂V is bounded, we have $v_{\varrho,s'} < \vartheta_s$ on Λ_ϱ, and then, ϑ_s is unbounded. If ∂V is unbounded, as $v_{\varrho,s'}|_{\partial V} = \vartheta_s|_{\partial V}$, we have

$$\lim_{d(q)\to+\infty} \vartheta_s|_{\partial V}(q) = +\infty$$

since

$$\lim_{d(q)\to+\infty} v_{\varrho,s'}|_{\partial V}(q) = +\infty.$$

and, therefore, ϑ_s is unbounded.

Now, we will prove the last affirmation of the proposition.

As $\vartheta_s \in C^{2,\alpha}\left(\overline{U}\right)$ is a solution of the Dirichlet problem (5) with $\vartheta_s|_{\partial U} = 0$, it follows from Proposition 3 of [20] that $u = \vartheta_s \circ \pi \in C^{2,\alpha}\left(\overline{\Omega}\right)$ satisfies $\mathcal{M}(u) = 0$ in $\overline{\Omega}$ with $u|_{\partial\Omega} = 0$ and $\left|\overline{\nabla} u(p)\right| = |\nabla(\vartheta_s \circ \pi)(p)|$, $p \in \Omega$, where $\mathcal{M}$ is the operator defined in (1). Note that we have necessarily Ω unbounded with $d_E(p) \to \infty$ if only if $d(\pi(p)) \to \infty$, where d_E is the Euclidean distance in $\mathbb{R}^{n+1}$ to $\partial\Omega$. Suppose that

$$\lim_{d_E(p)\to\infty} \left|\overline{\nabla} u(p)\right| \neq 0.$$

Then, there is $\varepsilon > 0$ and a sequence (p_n) in Ω, with $d_E(p_n) \to \infty$ when $n \to \infty$ such that $\left|\overline{\nabla} u(p_n)\right| \geq \varepsilon$ for all n large enough, $n \geq n_0$. For each $n \in \mathbb{N}$, define

$$\Omega_n = \left\{p \in \mathbb{R}^{n+1};\ p + p_n \in \Omega\right\}$$

and consider the sequence of functions $u_n : \Omega_n \subset \mathbb{R}^{n+1} \longrightarrow \mathbb{R}$ given by $u_n(p) = u(p + p_n)$. Note that $0 \in \Omega_n$ for all n since $(p_n) \subset \Omega$. Also

$$\mathbb{R}^{n+1} = \bigcup_{n\in\mathbb{N}} \Omega_n.$$

Indeed, given $w \in \mathbb{R}^{n+1}$, if the sequence $(w + p_n)_n$ was contained in $\mathbb{R}^{n+1} - \Omega$, since $\pi\left(\mathbb{R}^{n+1} \backslash \Omega\right)$ is compact, we would have $d_E(w + p_n) \leq R$ for all n, for some $R > 0$, a contradiction, since $d_E(p_n) \to \infty$. It follows that, as $u_n(0) = u(p_n)$, for all $n \geq n_0$, we have $\left|\overline{\nabla} u_n(0)\right| \geq \varepsilon$. Note that (u_n) is uniformly bounded since, by hypothesis, $n \geq 3$ and ϑ_s is bounded. Then, (u_n) has a subsequence $\left(u_{n_k}\right)$ which converges uniformly on compact subsets of $\mathbb{R}^{n+1}$ to a bounded function $\widetilde{u}$ defined on the whole $\mathbb{R}^{n+1}$ which satisfies $\mathcal{M}(\widetilde{u}) = 0$. Assume that dimension of M satisfies $3 \leq n \leq 6$. Then, $\Omega \subset \mathbb{R}^m$, $4 \leq m \leq 7$. From Bernstein theorem extended to $\mathbb{R}^m$, $2 \leq m \leq 7$, by Simons [21]—which is false for $m \geq 8$ [2]—it follows that $\widetilde{u}$ has to be constant. Therefore, we cannot have $\left|\overline{\nabla} u_{n_k}(0)\right| \geq \varepsilon$ for all $n_k \geq n_0$, a contradiction. Hence

$$\lim_{d_E(p) \to \infty} \left|\overline{\nabla} u(p)\right| = 0.$$

From Proposition 3 of [20], we have $\left|\overline{\nabla} u\right| = |\nabla \vartheta_s| \circ \pi$ and, therefore

$$\lim_{d(q) \to \infty} |\nabla \vartheta_s(q)| = 0.$$

□

Remark 2 If $\lambda = 0$ and $n = 2$, it was proved in [19] that setting

$$\vartheta_\infty(p) := \lim_{s \to \infty} \vartheta_s(p), \, p \in U, \tag{38}$$

$\vartheta_\infty \in C^2(U) \cap C^0\left(\overline{U}\right)$ and is an unbounded solution of the Dirichlet problem (5) with $\vartheta_\infty|_{\partial U} = 0$, which satisfies

$$\lim_{d_E(p) \to \infty} \left|\overline{\nabla} u(p)\right| = 0.$$

If $\lambda = 0$ and $n \geq 3$, it was proved in [1] that ϑ_∞, as defined in (38), is in $C^2(U)$, is a bounded solution of the Dirichlet problem (5) and its graph is contained in a $C^{1,1}$-manifold $\Upsilon \subset \overline{U} \times \mathbb{R}$ such that $\partial \Upsilon = \partial \Omega$.

Proposition 4 *Let G, π, and $M = (\mathbb{R}^n, \langle , \rangle_G)$, $n \geq 3$, as defined in (2), (3), and (4), respectively. Let U be an exterior $C^{2,\alpha}$-domain in M, $U = \pi(\Omega)$, where $\Omega \subset \mathbb{R}^{n+1}$ is a G-invariant $C^{2,\alpha}$-domain and let $d = d_M(., \partial U)$. Assume that $M \backslash U$ satisfies the interior sphere condition of radius $r > 0$. Let $\mathcal{L} = \mathcal{L}(r, n, \lambda a)$ be as given in (11), where λ and a are given in the definition of G. Then, given $c \in [0, \mathcal{L}]$, there is $w_c \in C^2(U) \cap C^0\left(\overline{U}\right)$ solution of the Dirichlet problem (5) relative to U, with $w_c|_{\partial U} = 0$, which satisfies*

$$\lim_{d(q) \to \infty} w_c(q) = c.$$

In particular, if $c \in [0, \mathcal{L})$, then $w_c \in C^{2,\alpha}\left(\overline{U}\right)$.

Proof If $\lambda = 0$, the result is already contemplated in Theorem 1 of [1]. Assume $\lambda \neq 0$. Given $c \in [0, \mathcal{L})$, define

$$F = \left\{ \begin{array}{c} f \in C^0\left(\overline{U}\right);\ f \text{ is subsolution relative to } \mathfrak{M} \\ f|_{\partial U} = 0 \text{ and } \lim\limits_{d(q)\to\infty} \sup f(q) \leq c \end{array} \right\}. \tag{39}$$

Note that $F \neq \emptyset$ since $f_0 \equiv 0 \in F$. From comparison principle, we have $f \leq c$ for all $f \in F$. From Perron method applied relative to operator $\mathfrak{M}$ [12, Section 2.8], [20, Section 3], we conclude that

$$w_c(q) := \sup\{f(q);\ f \in F\}, q \in \overline{U},$$

is in $C^\infty(U)$ and satisfies $\mathfrak{M}(w_c) = 0$ in U. We will show now that

$$\lim_{d(q)\to\infty} w_c(q) = c. \tag{40}$$

Consider $\alpha > 0$ such that $\mathfrak{B}_\alpha(0)$, the geodesic ball of $\mathbb{R}^n$ center at origin of $\mathbb{R}^n$, of radius $\alpha > 0$, contain $M\backslash U$ and be such that $v_\alpha(\infty) > c$, where v_α is as defined in (7) (note that $v_\alpha(\infty) = \alpha h(n, 1)$ and $h(n, 1) > 0$). Define now $f \in C^0\left(\overline{U}\right)$ by

$$f(q) = \begin{cases} 0 \text{ if } q \in \overline{U} \cap \mathfrak{B}_\alpha(0) \\ \max\{0, v_\alpha(q) - (v_\alpha(\infty) - c)\}, \text{ if } q \in M\backslash\mathfrak{B}_\alpha(0)\,. \end{cases}$$

From Proposition 2, we have f a nonnegative (generalized) subsolution to the Dirichlet problem (5) (for zero boundary data) relative to U, which satisfies

$$\lim_{d(q)\to\infty} f(q) = c. \tag{41}$$

It follows that $f \in F$ and that $f \leq w_c \leq c$. Then, we have (40).

Given $q_0 \in \partial U$, by hypothesis, there is a geodesic open ball of M, say B_r, of radius $r > 0$, contained in $M\backslash U$ and such that ∂B_r is tangent to ∂U ($= \partial(M\backslash U)$) at q_0. From Corollary 1, there is $W \in C^0(M\backslash B_r)$ a (generalized) supersolution relative to the operator $\mathfrak{M}$ on $M\backslash B_r$ such that $c \leq W(\infty) = \mathcal{L}$, with $W|_{\partial B_r} = 0$, which is C^1 in a neighborhood of ∂B_r in $M\backslash B_r$ and such that

$$\lim_{d_M(q,\partial B_r)\to 0} |\nabla W(q)| = +\infty.$$

From the comparison principle, since $\overline{U} \subset M\backslash B_r$, it follows that, on $\overline{U}$, we have $0 \leq w_c \leq W$. As q_0 is arbitrary, we conclude that $w_c \in C^0\left(\overline{U}\right)$ with $w_c|_{\partial U} = 0$. Assume that $0 \leq c < \mathcal{L}$. Let $\delta = (c + \mathcal{L})/2$. Let $L(\sigma) := \mathcal{L}(\sigma, n, \lambda, a)$, $\sigma \in (0, r]$, where $\mathcal{L}$ is given by (11). Since $L \in C^0(0, r]$, either there is $\sigma_0 \in (0, r)$

such that $L(\sigma_0) = \delta$, or $\delta < L(\sigma)$ for all $\sigma \in (0,r)$. Take $r' = \sigma_0$ in the first case and r' any point in $(0,r)$ in the second case. Let $B_{r'}$ be the open geodesic ball of M, $B_{r'} \subset B_r$, with the same center of B_r. Consider the correspondent $t_0 > 0$ and the function $w \in C^2(\Lambda_0) \cap C^0\left(\overline{\Lambda}_0\right)$, $w|_{\partial B_{r'}} = 0$, where $\Lambda_0 = \{q \in M \backslash B_{r'}; d_M(q, \partial B_{r'}) \leq t_0\}$ is such that $\mathfrak{M}(w) \leq 0$, as given in Corollary 1. If the second case occurs, the height of this w is greater than δ at its correspondent distance t_0 to $\partial B_{r'}$ and, as w is radial with respect to the center of $B_{r'}$, there is $t_0' < t_0$ such that, at the distance t_0' of $\partial B_{r'}$, the height of w is δ. In any case, there is $0 < t_0' \leq t_0$ such that, setting

$$W_{r'}(q) = \begin{cases} w(q) \text{ if } q \in \Lambda_0' \\ \delta \text{ if } q \in M \backslash \Lambda_0' \end{cases}$$

where

$$\Lambda_0' := \left\{q \in \Lambda; d(q) \leq t_0'\right\}.$$

satisfies the same properties of the function W as given in Corollary 1. Note that it is possible to translate the graph of $W_{r'}$ in the $\partial/\partial t$-direction (e_k-direction) in a way that its height at infinity is in $[c, \delta)$ and such that Γ the intersection of the hypersurface resulting of this displacement with $\{t \geq 0\}$ is such that

$$\partial\Gamma = \Gamma \cap B_r = \partial B_{r''},$$

with $B_{r''} \subset B_r$ a geodesic open ball of M with the same center of B_r and radius r'', being $r' < r'' < r$. Now, move Γ keeping $\partial\Gamma$ on M and the center of $\partial\Gamma$ on the geodesic of M linking the center of B_r to $q_0 \in \partial U$, until $\partial\Gamma$ touches ∂U at q_0 and call $\widetilde{\Gamma}$ this final hypersurface. Observe that such displacement is an isometry in $M \times \mathbb{R}$. Denote by $\widetilde{B}_{r''}$ the geodesic ball contained in B_r of radius r'' such that $\partial \widetilde{B}_{r''} = \partial\widetilde{\Gamma}$. We have then that $W_{r''} : M \backslash \widetilde{B}_{r''} \longrightarrow \mathbb{R}$ is a (generalized) supersolution relative to $\mathfrak{M}$. Moreover, since our translation in $\partial/\partial t$ direction is small enough, $W_{r''}$ satisfies $\mathfrak{M}(W_{r''}) = 0$ in $\overline{\Lambda}_0''$, where Λ_0'' is a neighborhood in U such that $q_0 \in \partial\Lambda_0''$. In particular, $W_{r''} \in C^\infty\left(\overline{\Lambda}_0''\right)$. From the comparison principle, since $0 \leq w_c|_{\partial U} \leq W_{r''}|_{\partial U}$ and $c \leq W_{r''}(\infty)$, we conclude that $0 \leq w_c \leq W_{r''}$ on $\overline{U}$. As $w_c(q_0) = W_{r''}(q_0)$ and $W_{r''} \in C^\infty\left(\overline{\Lambda}_0''\right)$, it follows that $w_c \in C^1\left(\overline{U}\right)$, and from elliptic PDE regularity theory [12], it follows that $w_c \in C^{2,\alpha}\left(\overline{U}\right) \cap C^\infty(U)$. □

References

1. A. Aiolfi, D. Bustos, J. Ripoll: On the existence of foliations by solutions to the exterior Dirichlet problem for the minimal surface equation, Proceedings of the AMS vol. 150, no.7, p. 3063-3073 (2022).

2. E. Bombieri, E. De Giorgi, E. Giusti: Minimal cones and the Bernstein problem. Invent. Math. 7, p. 243–268 (1969).
3. P. Collin, R. Krust: Le problème de Dirichlet pour l'équation des surfaces minimales sur des domaines non bornés Bull. Soc. Math. France, 119, no. 4, p. 443-462 (1991).
4. M. Dajczer, J. Ripoll: An extension of a theorem of Serrin to graphs in warped products, J. Geom. Anal., 15 193–205 (2005).
5. M. Dajczer, P. Hinojosa, J.H de Lira: Killing graphs with prescribed mean curvature, Calc. Var. Partial Differ. Eq. 33, p. 231–248 (2008).
6. M. Dajczer, J.H de Lira: Killing graphs with prescribed mean curvature and Riemannian submersions, Ann. Inst. H. Poincaré (Anal. Non Linéaire), vol. 26, no 3, p. 763-775 (2009).
7. R. Sa Earp, H, Rosenberg: The Dirichlet problem for the minimal surface equation on unbounded planar domains, J. Math. Pures Appl., 68, p. 163-183 (1989).
8. Wu-yi Hsiang, B. Lawson, Jr: Minimal submanifolds of low cohomogeneity, Journal of Differential Geometry, p. 1-38 (1971).
9. JB Casteras, E. Heinonen, I. Holopainen, J. Lira: Asymptotic Dirichlet problems in warped products, Math. Z. 295, 211–248 (2020).
10. H. Jenkins, J. Serrin: The Dirichlet problem for the minimal surface equation in higher dimensions. J. Reine Angew. Math. 229, p. 170–187 (1968).
11. Z. Jin: Growth rate and existence of solutions to Dirichlet problems for prescribed mean curvature equation on unbounded domains, Electronic J. of Diff. Equations, 24, p. 1–15 (2008)
12. D. Gilbarg, N. Trudinger: Elliptic Partial Differential Equations of Second Order, Springer, Berlin, 1998.
13. E. Kuwert: On solutions of the exterior Dirichlet problem for the minimal surface equation, Ann. Inst. H. Poincaré (Anal. Non Liéaire), vol. 10, no. 4, p. 445-451 (1993).
14. R. Krust: Remarques sur le problème extérieur de Plateau, Duke Math. J., Vol. 59, pp. 161-173 (1989).
15. N. Kutev and F. Tomi: Existence and nonexistence for the exterior Dirichlet problem for the minimal surface equation in the plane, Differential Integral Equations, 11, no. 6, p. 917–928 (1998).
16. J. C. C. Nitsche: Vorlesungen über Minimalflächen, Grundlehren der mathematischen Wissenschaften 199, Springer-Verlag, Berlin (1975).
17. R. Osserman: A Survey of Minimal Surfaces, Van Nostrand Reinhold Math. Studies 25, New York, 1969.
18. J. Ripoll: Some characterization, uniqueness and existence results for euclidean graphs of constant mean curvature with planar boundary, Pacific Journal of Mathematics, Vol. 198, N. 1, 175-196 (2001).
19. J. Ripoll, F. Tomi: On solutions to the exterior Dirichlet problem for the minimal surface equation with catenoidal ends, Adv. Calc. Var. 7, no. 2, p. 205–226 (2014).
20. J. Ripoll, F. Tomi: Group invariant solutions of certain partial differential equations, Pacific J. of Math 315 (1), p. 235-254 (2021).
21. Simons, J.: Minimal varieties in Riemannian manifolds. Ann. of Math., 88, p. 62–105 (1968).

Maximum Principles for Weakly 1-Coercive Operators with Applications to Capillary and Prescribed Mean Curvature Graphs

Luis J. Alías, Giulio Colombo, and Marco Rigoli

Dedicated to Professor Marcos Dajczer on his 75th birthday.

Abstract In this paper, we establish maximum principles for weakly 1-coercive operators L on complete, non-compact Riemannian manifolds M. In particular, we search for conditions under which one can guarantee that solutions u of differential equations of the form $L(u) \geq f(u)$ satisfy $f(u) \leq 0$ on M. The case of weakly p-coercive operators with $p > 1$, including the p-Laplacian and in particular the Laplace-Beltrami operator for $p = 2$, has been considered in our recent paper [2]. As a consequence of the main results, we infer comparison principles for that kind of operators. Furthermore, we apply them to geometric situations dealing with the mean curvature operator, which is a typical weakly 1-coercive operator. We first consider the case of C^1 operators L acting on functions u of class C^2 and, in the last section of the paper, we show how our results can be extended to the case of less regular operators L acting on functions u which are just continuous and locally $W^{1,1}$ regular.

L. J. Alías (✉)
Departamento de Matemáticas, Universidad de Murcia, Espinardo, Murcia, Spain
e-mail: ljalias@um.es

G. Colombo
Dipartimento di Matematica e Applicazioni "R. Caccioppoli", Università degli Studi di Napoli "Federico II", Naples, Italy
e-mail: giulio.colombo@unina.it

M. Rigoli
Dipartimento di Matematica "F. Enriques", Università degli Studi di Milano, Milano, Italy
e-mail: marco.rigoli@unimi.it

© The Author(s), under exclusive license to Springer Nature Switzerland AG 2025
R. Tojeiro (ed.), *Advances in Submanifold Theory and Related Topics*,
Latin American Mathematics Series – UFSCar subseries,
https://doi.org/10.1007/978-3-032-00918-0_2

1 Introduction

In what follows, $(M, \langle\ ,\ \rangle)$ will always denote a connected Riemannian manifold of dimension $m \geq 2$. In its simplest form, the validity of the weak maximum principle (WMP) for the Laplace-Beltrami operator Δ on M states that if u is (to simplify the matter) a C^2-function on M such that

$$u^* = \sup_M u < +\infty \tag{1}$$

then, there exists a sequence $\{x_k\} \subset M$ with the following properties:

$$u(x_k) > u^* - \frac{1}{k} \quad \text{and} \quad \Delta u(x_k) < \frac{1}{k} \tag{2}$$

for each $k \in \mathbb{N}$. This is equivalent to the stochastic completeness of M, see [22], and the latter can be analytically expressed in the following terms: for each $\lambda > 0$, the only solution $u \in C^2(M)$ of

$$\Delta u \geq \lambda u \quad \text{on } M \tag{3}$$

satisfying

$$0 \leq u \leq u^* < +\infty \tag{4}$$

is the trivial solution $u \equiv 0$ (see [17]). In the above formulation, allowing a weak interpretation of (3), we can avoid the request $u \geq 0$; indeed, for u a solution of (3)

$$u_+ = \max\{u, 0\}$$

is still a (weak) solution of (3) on M. Note that, since the WMP for Δ on $(M, \langle\ ,\ \rangle)$ is equivalent to stochastic completeness, geodesic completeness is not necessary for its validity; thus, the WMP for Δ holds on $\mathbb{R}^2 \setminus \{0\}$ with the Euclidean (uncomplete) metric.

Replacing the right-hand side of (3) with $f(u)$ for some continuous $f : \mathbb{R} \to \mathbb{R}$, in case f satisfies

$$\liminf_{t \to +\infty} \frac{f(t)}{t^\sigma} > 0 \tag{5}$$

for some $\sigma > 1$, it can be seen that assumptions which guarantee the validity of the WMP for Δ also guarantee that a solution u of

$$\Delta u \geq f(u) \tag{6}$$

first satisfies $u^* < +\infty$ and then application of the WMP implies $f(u^*) \leq 0$; see Theorems 4.1 and 4.2 of [5] for very general operators others than Δ and further applications. Please note that in condition (4.29) in [5], due to a typo, a "log" is missing just before the integral over B_r. We emphasize here that assumption (5) plays an essential role in establishing $u^* < +\infty$.

Once we assume geodesic completeness of the metric $\langle\,,\,\rangle$, there are at least three interesting types of assumptions that guarantee the validity of the WMP for the Laplace-Beltrami operator (and, in fact, for many other operators). The first one is based on Khasminskii's test but, in what follows, we will not deal with it; we refer the interested reader to [6] and [5]. The second condition reads

$$\liminf_{R\to+\infty} \frac{1}{R^2} \log \operatorname{vol} B_R < +\infty \tag{7}$$

and the third one involves the positive part $u_+ = \max\{u, 0\}$ of u via the request

$$\liminf_{R\to+\infty} \frac{1}{R^2} \log \int_{B_R} u_+^q < +\infty \tag{8}$$

for some $q > 0$.

The aim of this paper is to present conditions similar to (7) and (8) to guarantee that solutions u of differential inequalities of the form

$$Lu \geq f(u)$$

satisfy $f(u) \leq 0$ on M, without requiring the validity of (5), when L is a weakly 1-coercive operator as defined in Sect. 2 below (for instance, the mean curvature operator falls in this class). The case of weakly p-coercive operators with $p > 1$, including the p-Laplacian and in particular the Laplace-Beltrami operator for $p = 2$, has been considered in the recent paper [2].

In what follows, we mainly present two kind of results; see, for instance, Theorem 1 and Theorem 15. In the first case, the proof has a geometric flavor in the vein of the paper [1] and requires more regularity; in the second case, we refine our approach to make it less demanding from the point of view of regularity. As a consequence of the main results, we infer comparison principles. Furthermore, we apply them to geometric situations dealing with the mean curvature operator, which is a typical weakly 1-coercive operator. Comparison principles and related uniqueness results for the mean curvature operator have been considered by many authors especially in the Euclidean setting. Just to mention some references, we refer the reader to Concus and Finn [9, 10], Finn [12, 13], Finn and Hwang [14], and Hwang [18] for results on the Euclidean space $\mathbb{R}^m$, while for results on general Riemannian manifolds, we recall Pigola et al. [21].

2 Weakly 1-Coercive Operators

Let $(M, \langle\, , \rangle)$ be a complete, non-compact, Riemannian manifold of dimension $m \geq 2$. Given a non-empty open set $\Omega \subseteq M$, we denote by $TM|_\Omega = \cup_{x\in\Omega} T_x M$ the restriction of the tangent bundle of M to Ω and by $W^{1,1}_{\text{loc}}(\Omega)$ the space of measurable functions $u : \Omega \to \mathbb{R}$ such that for any relatively compact open subset $\Omega_0 \subseteq \Omega$ the restriction $u|_{\Omega_0}$ belongs to the Sobolev space $W^{1,1}(\Omega_0)$, characterized as the completion of $C^1(\Omega_0)$ with respect to the Sobolev norm:

$$\|u\|_{W^{1,1}(\Omega_0)} = \int_{\Omega_0} (|u| + |\nabla u|) \qquad \forall\, u \in C^1(\Omega_0)\,.$$

In this paper, we consider quasilinear differential operators L in divergence form weakly defined on functions $u \in W^{1,1}_{\text{loc}}(\Omega)$ by

$$Lu = \operatorname{div}(A(x, u, \nabla u)) \tag{9}$$

or, more generally, by

$$Lu = w^{-1}\operatorname{div}(w\, A(x, u, \nabla u))\,. \tag{10}$$

Here, $A : \mathbb{R} \times TM|_\Omega \to TM|_\Omega$ is a *Carathéodory-type (generalized) bundle map*, that is, a function such that

$$A(x, s, \xi) \in T_x M \qquad \forall\, x \in \Omega,\ s \in \mathbb{R},\ \xi \in T_x M$$

and whose representation $\tilde{A} : \psi(U) \times \mathbb{R} \times \mathbb{R}^m \to \mathbb{R}^m$ in any chart $\psi : U \to \mathbb{R}^m$ (with $U \subseteq \Omega$) satisfies the Carathéodory conditions, that is,

- $\tilde{A}(y, \cdot, \cdot) : \mathbb{R} \times \mathbb{R}^m \to \mathbb{R}^m$ is continuous for a.e. $y \in \psi(U)$.
- $\tilde{A}(\cdot, s, v) : \psi(U) \to \mathbb{R}^m$ is measurable for every $(s, v) \in \mathbb{R} \times \mathbb{R}^m$.

Throughout the paper, we assume A to be *weakly-1-coercive*, that is, we require that there exists a positive constant $k > 0$ such that

$$\langle A(x, s, \xi), \xi\rangle \geq 0 \qquad \forall\, x \in \Omega,\ s \in \mathbb{R},\ \xi \in T_x M \tag{11}$$

$$A(x, s, 0) = 0 \qquad \forall\, x \in \Omega,\ s \in \mathbb{R} \tag{12}$$

$$|A(x, s, \xi)| \leq k \qquad \forall\, x \in \Omega,\ s \in \mathbb{R},\ \xi \in T_x M\,. \tag{13}$$

A Carathéodory-type generalized bundle map A satisfying only (11) and (12) is usually said to be *weakly elliptic*. We adopt the said terminology for bundle maps A satisfying all three conditions above because we regard (13) as the limit case for $p \to 1^+$ of the so-called weakly-p-coercivity (with $1 < p < +\infty$) condition for weakly elliptic bundle maps:

$$|A(x, s, \xi)| \leq k\langle A(x, s, \xi), \xi\rangle^{\frac{p-1}{p}}.$$

If A is weakly-1-coercive, then the differential operator L associated to A weakly defined by (9) is also said to be weakly-1-coercive, and the constant $k > 0$ appearing in (13) is said to be a *coercivity constant* for L. To be precise, L is defined as the map

$$L : W^{1,1}_{\mathrm{loc}}(\Omega) \to W^{1,1}_c(\Omega)^*$$

sending each $u \in W^{1,1}_{\mathrm{loc}}(\Omega)$ to the continuous linear functional on $W^{1,1}_c(\Omega)$ given by

$$W^{1,1}_c(\Omega) \ni \varphi \mapsto -\int_\Omega \langle A(x, u, \nabla u), \nabla\varphi\rangle .$$

This is well defined because the structural conditions on A ensure that for every $u \in W^{1,1}_{\mathrm{loc}}(\Omega)$ the function $\Omega \ni x \mapsto A(x, u(x), \nabla u(x))$ is a bounded measurable vector field on Ω. For $u \in W^{1,1}_{\mathrm{loc}}(\Omega)$ and $f \in L^\infty_{\mathrm{loc}}(\Omega)$, we write

$$Lu = f \qquad \text{(weakly) in } \Omega$$

if the functional Lu can be represented as integration against f, that is, if

$$-\int_\Omega \langle A(x, u, \nabla u), \nabla\varphi\rangle = \int_\Omega f\varphi \qquad \forall\varphi \in W^{1,1}_c(\Omega). \tag{14}$$

More generally, in case $f : \Omega \to \mathbb{R}$ is a measurable function satisfying $f_+ \in L^\infty_{\mathrm{loc}}(\Omega)$, then we write

$$Lu \geq f \qquad \text{(weakly) in } \Omega$$

for a given function $u \in W^{1,1}_{\mathrm{loc}}(\Omega)$ if

$$-\int_\Omega \langle A(x, u, \nabla u), \nabla\varphi\rangle \geq \int_\Omega f\varphi \qquad \forall\varphi \in W^{1,1}_c(\Omega),\ \varphi \geq 0 .$$

The meaning of $Lu \leq f$, with $f : \Omega \to \mathbb{R}$ measurable and such that $f_- \in L^\infty_{\mathrm{loc}}(\Omega)$, is defined analogously.

Of course, in case $A : \mathbb{R}\times TM|_\Omega \to TM|_\Omega$ is C^1 regular and $u \in C^2(\Omega)$, then the weak definition of Lu given above agrees with the strong one, that is, $A(x, u, \nabla u)$ is a C^1 vector field and its divergence $f = \mathrm{div}(A(x, u, \nabla u))$ is pointwise well defined and continuous in Ω and satisfies (14).

If together with a weakly-1-coercive Carathéodory bundle map A we also consider a function $w \in L^\infty_{\mathrm{loc}}(\Omega)$ satisfying $w > 0$ a.e. in Ω, then the *weighted* operator L weakly defined by (10), that is, implicitely determined by

$$wLu = \mathrm{div}(w\,A(x, u, \nabla u)) \qquad \forall\, u \in W^{1,1}_{\mathrm{loc}}(\Omega)\,,$$

will still be called a weakly-1-coercive operator (with coercivity constant $k > 0$). Similarly to what we did above, if $u \in W^{1,1}_{\mathrm{loc}}(\Omega)$ and $f \in L^\infty_{\mathrm{loc}}(\Omega)$, then we write

$$Lu = f \qquad \text{(weakly) in } \Omega$$

for L as above if

$$-\int_\Omega w\langle A(x, u, \nabla u), \nabla\varphi\rangle = \int_\Omega wf\varphi \qquad \forall\, \varphi \in W^{1,1}_c(\Omega)\,, \tag{15}$$

and in case $f : \Omega \to \mathbb{R}$ is measurable and such that $f_+ \in L^\infty_{\mathrm{loc}}(\Omega)$ then we also write

$$Lu \geq f \qquad \text{(weakly) in } \Omega$$

if

$$-\int_\Omega w\langle A(x, u, \nabla u), \nabla\varphi\rangle \geq \int_\Omega wf\varphi \qquad \forall\, \varphi \in W^{1,1}_c(\Omega),\ \varphi \geq 0\,,$$

and a similar definition can be given for the reversed inequality $Lu \leq f$ in case $f_- \in L^\infty_{\mathrm{loc}}(\Omega)$. Again, in case $A : \mathbb{R}\times TM|_\Omega \to TM|_\Omega$ is C^1 regular and $u \in C^2(\Omega)$, if we also have $w \in C^1(\Omega)$, then the weak definition of Lu coincides with the strong one, meaning that the "weighted" divergence

$$f = w^{-1}\mathrm{div}(w\,A(x, u, \nabla u)) \equiv \mathrm{div}(A(x, u, \nabla u)) + \frac{\langle \nabla w, A(x, u, \nabla u)\rangle}{w}$$

is a pointwise well-defined and continuous function in Ω satisfying (15).

3 Maximum Principles for Weakly-1-Coercive Operators

Let $(M, \langle\ ,\ \rangle)$ be a complete Riemannian manifold of dimension $m \geq 2$. Throughout this section, L will always be a (possibly weighted) weakly-1-coercive operator associated to a C^1 bundle map A and acting on functions u of class C^2. In particular, $A(x, u, \nabla u)$ will always be a C^1 vector field, and all differential equalities and inequalities can be interpreted in a strong, pointwise sense.

In order to state our first result, we have to introduce two definitions. Let $o \in M$ be a fixed point. We denote with r the Riemannian distance function from o in M, that is, we write $r(x) = \mathrm{dist}(o, x)$ for $x \in M$, and we denote with $B_R = B_R(o)$ the geodesic ball of radius $R > 0$ centered at o. Let $\mu \in [0, 1]$ be a given a parameter. First, we define

$$c_\mu(M, o) = \begin{cases} \liminf\limits_{R\to+\infty} \dfrac{(1-\mu)\log \mathrm{vol}(B_R)}{R^{1-\mu}} & \text{if } 0 \le \mu < 1 \\ \liminf\limits_{R\to+\infty} \dfrac{\log \mathrm{vol}(B_R)}{\log R} & \text{if } \mu = 1\,. \end{cases} \tag{16}$$

As a consequence of the triangle inequality, the value of $c_\mu(M, o)$ is actually independent on the choice of the base point $o \in M$, so we will simply write $c_\mu(M)$ for $c_\mu(M, o)$. Moreover, we always have

$$c_\mu(M) \ge 0$$

because, $\log \mathrm{vol}(B_R)$ being a nondecreasing function of $R > 0$, we have

$$c_\mu(M) \ge \begin{cases} \liminf\limits_{R\to+\infty} \dfrac{(1-\mu)\log \mathrm{vol}(B_1)}{R^{1-\mu}} = 0 & \text{if } 0 \le \mu < 1 \\ \liminf\limits_{R\to+\infty} \dfrac{\log \mathrm{vol}(B_1)}{\log R} = 0 & \text{if } \mu = 1\,. \end{cases}$$

We will comment further below on exact values and lower bounds on $c_\mu(M)$ for some particular classes of manifolds. Secondly, we will say that a positive continuous function $b : M \to (0, +\infty)$ *satisfies condition* (B) for a couple of given parameters $\mu \in [0, 1]$ and $\beta > 0$ when the following circumstance is verified:

$$\begin{cases} \liminf\limits_{r(x)\to+\infty} r(x)^\mu b(x) \ge \beta & \text{in case } 0 < \mu \le 1 \\ b(x) \ge \beta \quad \forall x \in M & \text{in case } \mu = 0\,. \end{cases} \tag{B}$$

Again, for a given continuous positive function $b(x)$ on M, the validity of condition (B) for a given pair of parameters $\mu \in [0, 1]$ and $\beta > 0$ does not depend on the choice of the base point $o \in M$ with respect to which the distance $r(x)$ is measured. In case $\mu \in (0, 1]$, condition (B) amounts to saying that for each $\beta^* < \beta$ there exists $R_0 = R_0(\beta^*) > 0$ such that

$$b(x) \ge \beta^* r(x)^{-\mu} \qquad \text{for } r(x) > R_0\,,$$

that is, $b(x)$ does not decay faster than $\beta r(x)^{-\mu}$ as x goes to infinity in M along any diverging sequence. In particular, in case $\mu \in (0, 1]$ condition (B) is satisfied, for a given $\beta > 0$, by any continuous $b : M \to (0, +\infty)$ such that

$$b(x) \geq \beta r(x)^{-\mu} \qquad \text{for } r(x) \gg 1\,.$$

Our first result in this setting is the following:

Theorem 1 *Let $(M, \langle\ ,\ \rangle)$ be a complete manifold, $A : \mathbb{R} \times TM \to TM$ a C^1 weakly-1-coercive generalized bundle map with coercivity constant $k > 0$ and let L be the corresponding weakly-1-coercive operator given by* (9).

Let $f \in C(\mathbb{R})$ be nondecreasing and $b \in C(M)$ be positive and satisfy condition (B) *for parameters $\mu \in [0, 1]$ and $\beta > 0$. If $u \in C^2(M)$ satisfies*

$$Lu \geq b(x) f(u) \qquad \text{on } M, \tag{17}$$

then

$$f(u) \leq \frac{k}{\beta} c_\mu(M) \qquad \text{on } M. \tag{18}$$

In particular, if $c_\mu(M) = 0$, then

$$f(u) \leq 0 \qquad \text{on } M.$$

Remark 1 Regarding the values of $c_\mu(M)$, a few remarks are in order, in particular to highlight some relevant cases in which $c_\mu(M) = 0$. If $M = \mathbb{R}^m$ is the Euclidean space, then $\mathrm{vol}(B_R) = \omega_m R^m$, where ω_m denotes the volume of the Euclidean unit ball of the same dimension. As a consequence

$$c_\mu(\mathbb{R}^m) = \begin{cases} 0 & \text{if } 0 \leq \mu < 1 \\ m & \text{if } \mu = 1\,. \end{cases} \tag{19}$$

More generally, if $(M, \langle\ ,\ \rangle)$ has at most polynomial volume growth, that is,

$$\limsup_{R \to +\infty} \frac{\mathrm{vol}(B_R)}{R^N} < +\infty$$

for some real number $N > 0$, then

$$c_\mu(M) \begin{cases} = 0 & \text{if } 0 \leq \mu < 1 \\ \leq N & \text{if } \mu = 1\,. \end{cases}$$

Even more, if M has finite volume, then $\mathrm{vol}(B_R) \le \mathrm{vol}(M) < +\infty$ for every $R > 0$ and $c_\mu(M) = 0$ for every $\mu \in [0, 1]$. If $M = \mathbb{H}^m$ is the hyperbolic space of constant sectional curvature -1, then it is known that $\mathrm{vol}(B_R) \sim c_m e^{(m-1)R}$ as $R \to +\infty$, where $c_m = \omega_{m-1}2^{1-m}$, and then

$$c_\mu(\mathbb{H}^m) = \begin{cases} m-1 & \text{if } \mu = 0 \\ +\infty & \text{if } 0 < \mu \le 1 \,. \end{cases} \tag{20}$$

More generally, by Bishop-Gromov volume comparison theorem, if $(M, \langle\, , \rangle)$ is complete and

$$\mathrm{Ric}(\nabla r, \nabla r) \ge -(m-1)\frac{B^2}{(1+r^2)^{\alpha/2}} \qquad \text{with } 0 < \alpha \le 2\,, B \ge 0 \tag{21}$$

then we can estimate $c_\mu(M)$ from above in cases where $0 \le \mu \le \frac{\alpha}{2}$. Using Theorem 3.11 in [6] (and also taking into account the subsequent discussion there), we have

- If $0 < \alpha < 2$, then

$$c_\mu(M) \begin{cases} = 0 & \text{if } 0 \le \mu < \dfrac{\alpha}{2} \\ \le (m-1)B & \text{if } \mu = \dfrac{\alpha}{2}\,. \end{cases}$$

- If $\alpha = 2$, then

$$c_\mu(M) \begin{cases} = 0 & \text{if } 0 \le \mu < 1 \\ \le (m-1)\dfrac{1+\sqrt{1+4B^2}}{2} + 1 & \text{if } \mu = 1\,. \end{cases}$$

Also, if (21) holds with $\alpha = 0$ and $B \ge 0$, we still can estimate $c_0(M) \le (m-1)B$ for $\mu = 0$. Note that these estimates are sharp in view of (19) and (20) for $\mu = 0$.

Proof of Theorem 1 Set $c^* = \frac{k}{\beta}c_\mu(M)$ and suppose, by contradiction, that (18) is false. Choose $\varepsilon > 0$ small enough so that

$$\Omega_\varepsilon = \{x \in M : f(u(x)) > c^* + \varepsilon\} \ne \emptyset\,.$$

Since f is continuous and monotone nondecreasing, there exists $s_\varepsilon \in \mathbb{R} \cup \{-\infty\}$ such that

$$\Omega_\varepsilon = \{x \in M : u(x) > s_\varepsilon\}$$

and Ω_ε is open by continuity of u. Consider the vector field $X : M \to TM$ given by

$$X = A(x, u(x), \nabla u(x)) .$$

The assumptions on A and u ensure that X is of class C^1 and that $|X| \le k$ on M. Therefore, since M is complete, we know that X has a flow $\Psi(t, x)$ well defined for every $t \in \mathbb{R}$ and $x \in M$. We observe now that $\Psi : [0, +\infty) \times \Omega_\varepsilon \to \Omega_\varepsilon$. Actually, for any $x \in \Omega_\varepsilon$, denote $\gamma(t) = \Psi(t, x)$. Then

$$\frac{\mathrm{d}}{\mathrm{d}t} u(\gamma(t)) = \langle \nabla u, \dot{\gamma}(t) \rangle_{\gamma(t)} = \langle \nabla u, X_{\gamma(t)} \rangle_{\gamma(t)} = \langle \nabla u, A(\gamma(t), u, \nabla u) \rangle_{\gamma(t)} \ge 0 ,$$

which implies $u(\gamma(t)) \ge u(\gamma(0)) = u(x) > s_\varepsilon$ for every $t \in [0, +\infty)$. Thus, $\gamma(t)$ lies in Ω_ε for every $t \in [0, +\infty)$.

Now, fix $p \in \Omega_\varepsilon$ and $\delta > 0$ such that $B_\delta(p) \subseteq \Omega_\varepsilon$ and $\partial B_\delta(p)$ is smooth. For every $t \in [0, +\infty)$, let

$$U_t = \Psi(t, B_\delta(p)) = \Psi_t(B_\delta(p))$$

and

$$g(t) = \mathrm{vol}(U_t) = \int_{\Psi_t(B_\delta(p))} \mathrm{d}v = \int_{B_\delta(p)} \Psi_t^*(\mathrm{d}v).$$

Since $\overline{B_\delta(p)}$ is compact, we can differentiate under the integral and, using that $\mathcal{L}_X(\mathrm{d}v) = \mathrm{div} X \mathrm{d}v$, where $\mathcal{L}_X$ denotes the Lie derivative with respect to X, we get

$$\begin{aligned} g'(t) &= \left.\frac{\mathrm{d}}{\mathrm{d}\tau}\right|_{\tau=0} \int_{\Psi_{\tau+t}(B_\delta(p))} \mathrm{d}v = \left.\frac{\mathrm{d}}{\mathrm{d}\tau}\right|_{\tau=0} \int_{\Psi_t(B_\delta(p))} \Psi_\tau^*(\mathrm{d}v) \\ &= \int_{\Psi_t(B_\delta(p))} \left.\frac{\mathrm{d}}{\mathrm{d}\tau}\right|_{\tau=0} \Psi_\tau^*(\mathrm{d}v) = \int_{U_t} \mathrm{div} X \mathrm{d}v. \end{aligned}$$

Since $\Psi(t, \Omega_\varepsilon) \subseteq \Omega_\varepsilon$, we have $U_t \subseteq \Omega_\varepsilon$ and, by (17), we have

$$g'(t) = \int_{U_t} \mathrm{div} X \mathrm{d}v = \int_{U_t} Lu \mathrm{d}v \ge \int_{U_t} b(x) f(u) \mathrm{d}v > (c^* + \varepsilon) \int_{U_t} b(x) \mathrm{d}v . \tag{22}$$

By continuity, we can choose $\beta^* \in \mathbb{R}$ satisfying

$$\beta > \beta^* > \frac{c^*}{c^* + \varepsilon} \beta .$$

Note that, by the definition of c^*, the second inequality amounts to

$$\frac{(c^* + \varepsilon)\beta^*}{k} > c_\mu(M)\,. \tag{23}$$

Because of the assumptions on b and since $\beta^* < \beta$, we can choose $T > 0$ such that

$$b(x) \geq \beta^* r(x)^{-\mu} \qquad \forall\, x \notin \overline{B_T}\,. \tag{24}$$

Then, there exists $R_0 > 0$ such that

$$\inf_{B_R} b(x) \geq \beta^* R^{-\mu} \qquad \forall\, R \geq R_0\,. \tag{25}$$

Indeed, it suffices to choose $R_0 > T$ such that $\beta^* R_0^{-\mu} \leq \min_{\overline{B_T}} b > 0$. In that case and for every $R > R_0$, if $x \in B_R \setminus \overline{B_T}$, then we know that $b(x) \geq \beta^* r(x)^{-\mu} \geq \beta^* R^{-\mu}$. On the other hand, if $x \in \overline{B_T}$, then $b(x) \geq \min_{\overline{B_T}} b \geq \beta^* R_0^{-\mu} \geq \beta^* R^{-\mu}$. Summing up, for every $R > R_0$ and for every $x \in B_R$, we have $b(x) \geq \beta^* R^{-\mu}$, that is, (25).

As a consequence of $|X| \leq k$ and of the triangle inequality

$$d(p, x) < \delta + kt \qquad \forall x \in U_t. \tag{26}$$

To see it, let $x = \Psi(t, y) \in U_t$ with $y \in B_\delta(p)$. Then

$$d(p, x) = d(p, \Psi(t, y)) \leq d(p, y) + d(y, \Psi(t, y)) < \delta + d(y, \Psi(t, y)),$$

with

$$d(y, \Psi(t, y)) \leq \int_0^t \left|\frac{\partial \Psi}{\partial s}(s, y)\right| \mathrm{d}s = \int_0^t |X_{\Psi(s,y)}| \mathrm{d}s \leq kt,$$

which implies (26). Therefore

$$U_t \subseteq B_{\delta+kt}(p) \subseteq B_{r(p)+\delta+kt} \qquad \forall\, t \geq 0\,. \tag{27}$$

From (22), we have

$$g'(t) \geq (c^* + \varepsilon) \inf_{U_t} b(x)\, g(t) \qquad \forall\, t \geq 0\,.$$

Setting $R_1 = \max\{R_0, r(p) + \delta\}$, from (25) and (27), we get

$$\inf_{U_t} b(x) \geq \inf_{B_{r(p)+\delta+kt}} b(x) \geq (R_1 + kt)^{-\mu} \beta^*,$$

and then

$$g'(t) \geq (c^* + \varepsilon)\beta^*(R_1 + kt)^{-\mu} g(t) \qquad \forall t \geq 0,$$

equivalently

$$\frac{g'(t)}{g(t)} \geq (c^* + \varepsilon)\beta^*(R_1 + kt)^{-\mu} \qquad \forall t \geq 0.$$

Integrating this inequality, we obtain

$$\log(g(t)) \geq \log(g(0)) + (c^* + \varepsilon)\beta^* \int_0^t (R_1 + ks)^{-\mu}\, \mathrm{d}s. \tag{28}$$

Since $U_t \subseteq B_{\delta+kt}(p) \subseteq B_{R_1+kt}$, we have $g(t) \leq \mathrm{vol}(B_{R_1+kt})$, thus letting $R = R_1 + kt$ and changing the integration variable in (28), we get

$$\begin{aligned}
\log \mathrm{vol}(B_R) &\geq \log(g(t)) \\
&\geq \log(g(0)) + \frac{(c^* + \varepsilon)\beta^*}{k} \int_{R_1}^{R} s^{-\mu}\, \mathrm{d}s \\
&= \log(g(0)) + \frac{(c^* + \varepsilon)\beta^*}{k} \begin{cases} \left(\dfrac{R^{1-\mu}}{1-\mu} - \dfrac{R_1^{1-\mu}}{1-\mu} \right) & \text{if } 0 \leq \mu < 1 \\ (\log R - \log R_1) & \text{if } \mu = 1 \end{cases}
\end{aligned}$$

for every $R \geq R_1$. In other words, for all $R \geq R_1$, we, respectively, have

$$\frac{(1-\mu)\log \mathrm{vol}(B_R)}{R^{1-\mu}} \geq \frac{(c^* + \varepsilon)\beta^*}{k} \left(1 - \frac{R_1^{1-\mu}}{R^{1-\mu}} \right) + \frac{(1-\mu)\log(g(0))}{R^{1-\mu}}$$

in case $0 \leq \mu < 1$, or

$$\frac{\log \mathrm{vol}(B_R)}{\log R} \geq \frac{(c^* + \varepsilon)\beta^*}{k} \left(1 - \frac{\log R_1}{\log R} \right) + \frac{\log(g(0))}{\log R}$$

in case $\mu = 1$. In both cases, letting $R \to +\infty$ and applying the definition (16) of $c_\mu(M)$, we conclude from here that

$$c_\mu(M) \geq \frac{(c^* + \varepsilon)\beta^*}{k},$$

contradicting (23). □

We can "localize" Theorem 1 as follows: if $\Omega \subsetneq M$ is a non-empty open set and $\mu \in [0, 1]$ is a given parameter, then fixing $o \in \Omega$ we define

$$c_\mu(\Omega) = \begin{cases} \liminf\limits_{R\to+\infty} \dfrac{(1-\mu)\log \mathrm{vol}(B_R \cap \Omega)}{R^{1-\mu}} & \text{if } 0 \le \mu < 1 \\[2ex] \liminf\limits_{R\to+\infty} \dfrac{\log \mathrm{vol}(B_R \cap \Omega)}{\log R} & \text{if } \mu = 1 \end{cases} \tag{29}$$

where $B_R = B_R(o)$. As in case $\Omega = M$, the resulting value of $c_\mu(\Omega)$ is independent of the choice of the base point $o \in \Omega$. Moreover, the set function c_μ is nonnegative and monotonic nondecreasing with respect to inclusion, that is, for every $\mu \in [0, 1]$ and for any pair of non-empty open sets $\Omega_1, \Omega_2 \subseteq M$ with $\Omega_1 \subseteq \Omega_2$, we have

$$0 \le c_\mu(\Omega_1) \le c_\mu(\Omega_2) \le c_\mu(M)\,.$$

Theorem 2 *Let $(M, \langle\, ,\, \rangle)$ be a complete manifold, $\Omega \subsetneq M$ a non-empty open set, $A : \mathbb{R} \times TM|_\Omega \to TM|_\Omega$ a C^1 weakly-1-coercive generalized bundle map with coercivity constant $k > 0$ and let L be the corresponding operator given by* (9)*. Let $f \in C(\mathbb{R})$ be nondecreasing and $b \in C(M)$ be positive and satisfy condition* (B) *for given parameters $\mu \in [0, 1]$ and $\beta > 0$. If $u \in C(\overline{\Omega}) \cap C^2(\Omega)$ satisfies*

$$Lu \ge b(x) f(u) \qquad on\ \Omega\,, \tag{30}$$

then

$$f(u) \le \max\left\{\frac{k}{\beta} c_\mu(\Omega),\ \sup_{\partial\Omega} f(u)\right\} \quad on\ \Omega\,. \tag{31}$$

In particular, if $c_\mu(\Omega) = 0$, then

$$f(u) \le \max\left\{0,\ \sup_{\partial\Omega} f(u)\right\} \quad on\ \Omega. \tag{32}$$

Proof of Theorem 2 If $\sup_{\partial\Omega} f(u) = +\infty$, then the conclusion is trivially satisfied; otherwise, we proceed in a way similar to that of Theorem 1 and set $c^* = \max\left\{\frac{k}{\beta} c_\mu(\Omega), c\right\}$, with $c = \sup_{\partial\Omega} f(u) \in \mathbb{R}$. Note that $c^* \ge 0$, since $c_\mu(\Omega) \ge 0$ as remarked above. Suppose, by contradiction, that (31) is false and choose $\varepsilon > 0$ small enough so that

$$\Omega_\varepsilon = \{x \in \Omega : f(u(x)) > c^* + \varepsilon\} \ne \emptyset$$

which is open by continuity of f and u. Let us see that $\overline{\Omega_\varepsilon} \subseteq \Omega$. Indeed, let $\tilde{x} \in \overline{\Omega_\varepsilon}$ then, for some $\{x_j\}_{j\in\mathbb{N}} \subset \Omega_\varepsilon$, $\tilde{x} = \lim x_j$ and then $f(u(\tilde{x})) = \lim f(u(x_j)) \ge$

$c^* + \varepsilon > c$, which means that $\tilde{x} \notin \partial\Omega$ and hence $\tilde{x} \in \Omega$. Therefore, there exists $s_\varepsilon \in \mathbb{R}$ such that

$$\Omega_\varepsilon = \{x \in \Omega : u(x) > s_\varepsilon\}.$$

Consider the vector field $X : \Omega \to TM$ defined on $\Omega \subsetneq M$ by

$$X = A(x, u(x), \nabla u(x)) .$$

The assumptions on A and u ensure that X is of class $C^1(\Omega)$ and that $|X| \le k$ on Ω. Although X is not defined on all of M, we observe that X still has a well-defined flow $\Psi : [0, +\infty) \times \Omega_\varepsilon \to \Omega_\varepsilon$. Indeed, we know that for any $x \in \Omega_\varepsilon$ there exists a maximal time $T_x \in (0, +\infty]$ such that the initial value problem

$$\begin{cases} \gamma(0) = x \in \Omega_\varepsilon \\ \dot{\gamma}(t) = X_{\gamma(t)} \end{cases} \tag{33}$$

has a (unique) solution $\gamma : [0, T_x) \to \Omega$. We will show that $T_x = +\infty$. Toward this aim, we observe that since

$$\frac{\mathrm{d}}{\mathrm{d}t} u(\gamma(t)) = \langle \nabla u, \dot{\gamma}(t)\rangle_{\gamma(t)} = \langle \nabla u, X_{\gamma(t)}\rangle_{\gamma(t)} = \langle \nabla u, A(\gamma(t), u, \nabla u)\rangle_{\gamma(t)} \ge 0 ,$$

we have $u(\gamma(t)) \ge u(\gamma(0)) = u(x) > s_\varepsilon$ for every $t \in [0, T_x)$. Thus, $\gamma(t)$ lies in Ω_ε for every $t \in [0, T_x)$. Suppose by contradiction that $T_x < +\infty$. Since $|X| \le k$, the curve γ has length $\ell(\gamma) \le kT_x < +\infty$. As M is complete, γ has a limit end point $\tilde{x} = \lim_{t \to T_x} \gamma(t) \in M$. Since $\gamma([0, T_x)) \subseteq \Omega_\varepsilon$, we have $\tilde{x} \in \overline{\Omega_\varepsilon} \subseteq \Omega$. Thus, $u(\tilde{x})$ is defined and by continuity $u(\tilde{x}) = \lim_{t \to T_x} u(\gamma(t)) \ge u(x) > s_\varepsilon$. But then $\tilde{x} \in \Omega_\varepsilon$, and thus, T_x is not the maximal existence time for the solution of (33), contradiction. That is, $T_x = +\infty$.

Once we have guaranteed that $\Psi(t, \Omega_\varepsilon) \subseteq \Omega_\varepsilon$ for every $t \ge 0$, the proof follows exactly the same lines as the proof of Theorem 1 (including the choice of $\beta^* \in \mathbb{R}$ satisfying $\beta > \beta^* > \frac{c^*}{c^*+\varepsilon}\beta$) until inequality (28). From that point on, since $U_t \subseteq B_{\delta+kt}(p) \subseteq B_{R_1+kt}$ and also $U_t \subseteq \Omega_\varepsilon \subseteq \Omega$, we have $g(t) \le \mathrm{vol}(B_{R_1+kt} \cap \Omega)$, and proceeding as in the proof of Theorem 1, we obtain

$$\log \mathrm{vol}(B_R \cap \Omega) \ge \log(g(0)) + \frac{(c^* + \varepsilon)\beta^*}{k} \begin{cases} \left(\dfrac{R^{1-\mu}}{1-\mu} - \dfrac{R_1^{1-\mu}}{1-\mu}\right) & \text{if } 0 \le \mu < 1 \\ (\log R - \log R_1) & \text{if } \mu = 1 . \end{cases}$$

for every $R \ge R_1$. Hence, by (29), we conclude from here that $c_\mu(\Omega) \ge (c^* + \varepsilon)\beta^*/k > c^*\beta/k \ge c_\mu(\Omega)$, which is a contradiction. □

In the next theorems, we state the weighted version of our previous results, where by this we mean that we are working with operators L of the form

$$Lu = w^{-1}\mathrm{div}(w\,A(x,u,\nabla u)) \tag{34}$$

where A is C^1 regular and w is a strictly positive C^1 function, acting as a "weight." Given $h \in C^1(M)$, it is standard to denote by div_h the "weighted" or "drift" divergence operator defined by

$$\mathrm{div}_h X = \mathrm{div} X - \langle \nabla h, X\rangle$$

for vector fields $X : M \to TM$ and by vol_h the "weighted" volume defined by

$$\mathrm{vol}_h(U) = \int_U e^{-h}\mathrm{d}v$$

for measurable subsets $U \subseteq M$. With this notation, the definition (34) restates as

$$Lu = \mathrm{div}_h A(x,u,\nabla u) \qquad \text{for } h = -\log w\,.$$

The next two theorems are stated in terms of weigthed versions of the parameters $c_\mu(M)$ and $c_\mu(\Omega)$: fixing as usual an origin $o \in M$ and denoting $B_R = B_R(o)$, if $\Omega \subseteq M$ is open and non-empty, then for $w \in C^1(\Omega)$ such that $w > 0$ in Ω and for $\mu \in [0,1]$, we define

$$c^w_\mu(\Omega) = \begin{cases} \liminf\limits_{R\to+\infty} \dfrac{(1-\mu)\log\int_{B_R\cap\Omega} w}{R^{1-\mu}} & \text{if } 0 \le \mu < 1 \\[2ex] \liminf\limits_{R\to+\infty} \dfrac{\log\int_{B_R\cap\Omega} w}{\log R} & \text{if } \mu = 1\,. \end{cases} \tag{35}$$

Theorem 3 *Let $(M, \langle\,,\,\rangle)$ and A be as in Theorem 1, let $w \in C^1(M)$ be strictly positive, and let L be the corresponding operator given by* (34). *Let $f \in C(\mathbb{R})$ be nondecreasing and $b \in C(M)$ be positive and satisfy condition* (B) *for given parameters $\mu \in [0,1]$ and $\beta > 0$. If $u \in C^2(M)$ satisfies*

$$Lu \ge b(x)f(u) \qquad \text{on } M \tag{36}$$

then

$$f(u) \le \frac{k}{\beta}c^w_\mu(M) \qquad \text{on } M\,. \tag{37}$$

In particular, if $c^w_\mu(M) = 0$*, then*

$$f(u) \leq 0 \qquad on\ M.$$

Proof Setting $h = -\log w$, the proof goes exactly as that of Theorem 1, after replacing the function $g(t)$ with the function

$$g_h(t) = \mathrm{vol}_h(U_t) = \int_{\Psi_t(B_\delta(p))} e^{-h}\mathrm{d}v = \int_{B_\delta(p)} \Psi_t^*(e^{-h}\mathrm{d}v)\,,$$

by observing that

$$\begin{aligned} g_h'(t) &= \left.\frac{\mathrm{d}}{\mathrm{d}\tau}\right|_{\tau=0} \int_{\Psi_{\tau+t}(B_\delta(p))} e^{-h}\mathrm{d}v = \left.\frac{\mathrm{d}}{\mathrm{d}\tau}\right|_{\tau=0} \int_{\Psi_t(B_\delta(p))} \Psi_\tau^*(e^{-h}\mathrm{d}v) \\ &= \int_{\Psi_t(B_\delta(p))} \left.\frac{\mathrm{d}}{\mathrm{d}\tau}\right|_{\tau=0} \Psi_\tau^*(e^{-h}\mathrm{d}v) = \int_{\Psi_t(B)} \left.\frac{\mathrm{d}}{\mathrm{d}\tau}\right|_{\tau=0} \left(e^{-h\circ\Psi_\tau}\Psi_\tau^*(\mathrm{d}v)\right) \\ &= \int_{\Psi_t(B_\delta(p))} e^{-h}(\mathrm{div}X - \langle\nabla h, X\rangle)\mathrm{d}v = \int_{\Psi_t(B_\delta(p))} e^{-h}\mathrm{div}_h X\mathrm{d}v. \end{aligned}$$

□

Similarly, reasoning as in the proof of Theorem 2, we also have

Theorem 4 *Let* $(M, \langle\ ,\ \rangle)$*,* Ω *and* A *be as in Theorem 2, let* $w \in C^1(\Omega)$ *be strictly positive, and let* L *be the corresponding operator given by* (34)*. Let* $f \in C(\mathbb{R})$ *be nondecreasing and* $b \in C(M)$ *be positive and satisfy condition* (B) *for given parameters* $\mu \in [0, 1]$ *and* $\beta > 0$*. If* $u \in C(\overline{\Omega}) \cap C^2(\Omega)$ *satisfies*

$$Lu \geq b(x)f(u) \quad on\ \Omega \tag{38}$$

then

$$f(u) \leq \max\left\{\frac{k}{\beta}c^w_\mu(\Omega), \sup_{\partial\Omega} f(u)\right\} \quad on\ \Omega. \tag{39}$$

In particular, if $c^w_\mu(\Omega) = 0$*, then*

$$f(u) \leq \max\left\{0, \sup_{\partial\Omega} f(u)\right\} \quad on\ \Omega. \tag{40}$$

More generally, the same line of reasoning can be applied to first-order differential inequalities of the form $\mathrm{div}_h(X) \geq b(x)\psi$ with X a bounded C^1 vector field, and ψ a C^1 function only related by the condition $\langle\nabla\psi, X\rangle \geq 0$.

Theorem 5 *Let $(M, \langle\, , \rangle)$ be a complete, non-compact manifold, $\Omega \subseteq M$ a non-empty open set and $X : \Omega \to TM$ a C^1 vector field satisfying $|X| \leq k$ in Ω for some constant $k > 0$. Let $b \in C(M)$ be positive and satisfy condition* (B) *for some $\mu \in [0, 1]$ and $\beta > 0$. Let $\psi \in C^1(\Omega) \cap C(\overline{\Omega})$ be such that*

$$\langle \nabla \psi, X \rangle \geq 0 \qquad on\ \Omega \tag{41}$$

and for some $h \in C^1(\Omega)$

$$\mathrm{div}_h(X) \geq b(x)\, \psi \quad on\ \Omega. \tag{42}$$

Then

$$\sup_{\Omega} \psi \leq \max \left\{ \frac{k}{\beta} c_\mu^w(\Omega), \sup_{\partial\Omega} \psi \right\}$$

where $w = e^{-h}$.

4 Applications to Capillary Graphs and Prescribed Mean Curvature Graphs

Here, we introduce the mean curvature operator:

$$\mathrm{div}\left(\frac{\nabla u}{\sqrt{1 + |\nabla u|^2}} \right) \tag{43}$$

as an example of weakly-1-coercive operator with coercivity constant $k = 1$.

In our next result, we assume that $c_\mu(M) = 0$ for some $\mu \in [0, 1]$. Recall from Remark 1 that this is the case if $(M, \langle\, , \rangle)$ has, for instance, at most polynomial volume growth and $\mu \in [0, 1)$.

Theorem 6 *Let $(M, \langle\, , \rangle)$ be a complete non-compact Riemannian manifold with $c_\mu(M) = 0$ for some $\mu \in [0, 1]$ and let $0 < b \in C(M)$ satisfy*

$$\begin{cases} b(x) \geq \beta & \forall x \in M \quad\quad if\ \mu = 0 \\ b(x) \geq \beta r(x)^{-\mu} & for\ r(x) \gg 1 \quad if\ 0 < \mu \leq 1 \end{cases}$$

for some $\beta > 0$. Then, the only globally defined C^2 solution $u : M \to \mathbb{R}$ of the capillary surface equation

$$\operatorname{div}\left(\frac{\nabla u}{\sqrt{1+|\nabla u|^2}}\right) = b(x)u \tag{44}$$

is the trivial solution $u \equiv 0$.

Proof Assume that $u \in C^2(M)$ is a global solution of (44). Then, from the last assertion in Theorem 1, we deduce $u \leq 0$ on M. Consider now the function $v = -u$, which clearly satisfies

$$\operatorname{div}\left(\frac{\nabla v}{\sqrt{1+|\nabla v|^2}}\right) = b(x)v.$$

Then, we also obtain that $v \leq 0$ on M; in other words, $u \geq 0$. Summing up $u \equiv 0$ on M. □

A "localized" version of the previous result holds as well.

Theorem 7 *Let $(M, \langle\ ,\ \rangle)$ be a complete non-compact Riemannian manifold and let $\Omega \subsetneq M$ be an open set. Let $0 < b \in C(M)$ satisfy for some parameters $\mu \in [0, 1]$ and $\beta > 0$*

$$\begin{cases} b(x) \geq \beta & \forall x \in M \qquad \text{if } \mu = 0 \\ b(x) \geq \beta r(x)^{-\mu} & \text{for } r(x) \gg 1 \qquad \text{if } 0 < \mu \leq 1\,. \end{cases}$$

Let $u \in C(\overline{\Omega}) \cap C^2(\Omega)$ be a solution of the capillary surface equation on Ω:

$$\operatorname{div}\left(\frac{\nabla u}{\sqrt{1+|\nabla u|^2}}\right) = b(x)u. \tag{45}$$

Then

$$\sup_{\Omega} u \leq \max\left\{\frac{c_\mu(\Omega)}{\beta}, \sup_{\partial\Omega} u\right\}, \qquad \inf_{\Omega} u \geq \min\left\{-\frac{c_\mu(\Omega)}{\beta}, \inf_{\partial\Omega} u\right\}$$

In particular

$$\sup_{\Omega} |u| \leq \max\left\{\frac{c_\mu(\Omega)}{\beta}, \sup_{\partial\Omega} |u|\right\}.$$

Proof Assume that $u \in C(\overline{\Omega}) \cap C^2(\Omega)$ is a solution of the capillary surface equation on Ω. Then, from Theorem 2, we obtain that $u \le \max\left\{\frac{c_\mu(\Omega)}{\beta}, \sup_{\partial\Omega} u\right\}$ on Ω. Consider now the function $v = -u$, which clearly satisfies

$$\operatorname{div}\left(\frac{\nabla v}{\sqrt{1+|\nabla v|^2}}\right) = b(x)v.$$

with $\sup_{\partial\Omega} v = -\inf_{\partial\Omega} u$. Then, we also obtain that $v = -u \le \max\left\{\frac{c_\mu(\Omega)}{\beta}, -\inf_{\partial\Omega} u\right\}$ on Ω, that is, $u \ge -\max\left\{\frac{c_\mu(\Omega)}{\beta}, -\inf_{\partial\Omega} u\right\} = \min\left\{-\frac{c_\mu(\Omega)}{\beta}, \inf_{\partial\Omega} u\right\}$. Summing up, we have

$$\sup_{\Omega} u \le \max\left\{\frac{c_\mu(\Omega)}{\beta}, \sup_{\partial\Omega} u\right\}, \qquad \inf_{\Omega} u \ge \min\left\{-\frac{c_\mu(\Omega)}{\beta}, \inf_{\partial\Omega} u\right\}$$

□

Of course, when $c_\mu(\Omega) = 0$ and $\partial\Omega \neq \emptyset$ from the above theorem, we deduce that every solution of (45) on Ω with $u|_{\partial\Omega} \equiv c =$ constant has constant sign and satisfies $|u| \le |c|$ on Ω. In particular, in that case, the only solution with $u|_{\partial\Omega} \equiv 0$ is $u \equiv 0$ on Ω.

We now consider the case of a graph with prescribed mean curvature $H(x)$. If the graph Γ_u is given by $\Gamma_u(x) = (x, u(x))$, then u has to satisfy

$$\operatorname{div}\left(\frac{\nabla u}{\sqrt{1+|\nabla u|^2}}\right) = mH(x) \qquad \text{on } M\,. \tag{46}$$

Theorem 8 *Let $(M, \langle\,,\,\rangle)$ be an m-dimensional complete non-compact Riemannian manifold and let $u : M \to \mathbb{R}$ define an entire graph with positive mean curvature $H(x)$ satisfying, for some parameters $\mu \in [0, 1]$ and $\beta > 0$*

$$\begin{cases} H(x) \ge \dfrac{\beta}{m} & \forall x \in M \qquad \text{if } \mu = 0 \\ H(x) \ge \dfrac{\beta}{m} r(x)^{-\mu} & \text{for } r(x) \gg 1 \qquad \text{if } 0 < \mu \le 1\,. \end{cases} \tag{47}$$

Then

$$\beta \le c_\mu(M)$$

where $c_\mu(M)$ is defined in (16).

The proof of the theorem is a straightforward application of Theorem 1 using the mean curvature equation (46), with the choices $f(u) = 1$ and $b(x) = mH(x)$.

Corollary 1 *Let $(M, \langle\, , \rangle)$ be an m-dimensional complete non-compact Riemannian manifold. If $\beta > c_\mu(M)$ for some $\mu \in [0, 1]$, then there is no global solution $u : M \to \mathbb{R}$ of the prescribed mean curvature equation (46) for a positive function $H \in C(M)$ satisfying* (47).

Remark 2 In the special case, H is a positive constant we can take $\mu = 0$ and $\beta = mH$. Then, Corollary 1 tells us that there exists no entire graph of constant mean curvature H over M if

$$H > \frac{1}{m} \liminf_{R\to+\infty} \frac{\log \mathrm{vol}(B_R)}{R} = \frac{c_0(M)}{m}.$$

In particular, there exists no such graph for any $H \neq 0$ if $c_0(M) = 0$, which is the case, for instance, if M has at most polynomial volume growth in view of Remark 1. Thus, if $(M, \langle\, , \rangle) = (\mathbb{R}^m, \langle\, , \rangle_{can})$, we recover a well-known result of Chern, [7], and Flanders, [15], that is, there are no entire graphs on $\mathbb{R}^m$ with positive constant mean curvature H; while for $(M, \langle\, , \rangle) = (\mathbb{H}^m, \langle\, , \rangle_{\mathbb{H}^m})$, there are no entire graphs on $\mathbb{H}^m$ with constant mean curvature H satisfying

$$H > \frac{m-1}{m}$$

On the other hand, it is well known, see [23] and [4], that there exist entire graphs of constant positive mean curvature H for any

$$H \leq \frac{m-1}{m}.$$

This shows that Corollary 1 is sharp. We also observe that the necessary condition

$$H \leq \frac{1}{m} c_0(M)$$

for the existence of a graph of constant positive mean curvature H is somehow related to that of Salavessa, [23], that is,

$$H \leq \frac{1}{m} \mathfrak{h}(M)$$

where $\mathfrak{h}(M)$ is the Cheeger constant of M.

We shall now discuss an application of Theorem 3. Let $(\mathcal{P}, \langle\, , \rangle_{\mathcal{P}})$ be an m-dimensional Riemannian manifold and let $h \in C^\infty(\mathcal{P})$, $h > 0$ on $\mathcal{P}$. We can then consider $\overline{M} = \mathbb{R}_h \times \mathcal{P}$, that is, the product manifold $\mathbb{R} \times \mathcal{P}$ with the warped product metric $\overline{\langle\, , \rangle}$ given by

$$\overline{\langle\, , \rangle} = h(x)^2 \mathrm{d}t^2 + \langle\, , \rangle_{\mathcal{P}} \tag{48}$$

where t is the coordinate on $\mathbb{R}$. For instance, $\mathbb{H}^{m+1}$, the hyperbolic space of constant sectional curvature -1 and dimension $m+1$, admits the description $\mathbb{H}^{m+1} = \mathbb{R}_h \times \mathbb{H}^m$ with $h(x) = \cosh(r(x))$, where $r(x) = \mathrm{dist}(x, o)$ in $\mathbb{H}^m$ for some fixed origin $o \in \mathbb{H}^m$. Thus, the metric in $\mathbb{H}^{m+1}$ is represented in the form

$$\overline{\langle\, ,\, \rangle} = \cosh^2(r(x))\mathrm{d}t^2 + \langle\, ,\, \rangle_{\mathbb{H}^m}\,.$$

Given a smooth function $u : \mathcal{P} \to \mathbb{R}$, we consider the graph ψ of u on $\mathcal{P}$, that is, the imbedding

$$\psi : \mathcal{P} \to \overline{M}$$

given by

$$\psi(x) = (u(x), x)\,, \qquad x \in \mathcal{P}\,. \tag{49}$$

Graphs of this kind are called equidistant graphs. For more details and results, see, for instance, [6] and the references therein.

It is well known that $\frac{\partial}{\partial t}$ is a Killing vector field on $\overline{M}$ and we set ϕ_t to indicate its flow. Letting D denote the gradient on $(\mathcal{P}, \langle\, ,\, \rangle_{\mathcal{P}})$, the vector field

$$\nu = \frac{1}{h\sqrt{1 + h^2|Du|^2}}\left(\frac{\partial}{\partial t} - h^2(\phi_u)_*(Du)\right) \tag{50}$$

is a unit normal vector field to the graph ψ. We set H to denote the (normalized) mean curvature of ψ in the direction of ν. A computation shows the validity of the formula:

$$\mathrm{div}_{\mathcal{P}}\left(\frac{hDu}{\sqrt{1 + h^2|Du|^2}}\right) + \left\langle \frac{hDu}{\sqrt{1 + h^2|Du|^2}}, \frac{Dh}{h}\right\rangle = mH \tag{51}$$

on $\mathcal{P}$, that we can put into the form

$$\mathrm{div}_{-\log h}\left(\frac{hDu}{\sqrt{1 + h^2|Du|^2}}\right) = mH \tag{52}$$

where the $\mathrm{div}_{-\log h} = (\mathrm{div}_{\mathcal{P}})_{-\log h}$; we have omitted the reference to $\mathcal{P}$ for the sake of simplicity.

In case of (52)

$$A(x, s, \xi) = \frac{h(x)\xi}{\sqrt{1 + h^2(x)|\xi|^2}}$$

and therefore the operator in (52) is weakly 1-coercive with coercivity constant $k = 1$. We recall the definition:

$$c_{\mu}^{h}(\mathcal{P}) = \begin{cases} \liminf_{R\to+\infty} \dfrac{(1-\mu)\log\int_{B_R} h}{R^{1-\mu}} & \text{if } 0 \leq \mu < 1 \\ \liminf_{R\to+\infty} \dfrac{\log\int_{B_R} h}{\log R} & \text{if } \mu = 1 . \end{cases} \tag{53}$$

Applying Theorem 3, we have

Theorem 9 *Let* $(\mathcal{P}, \langle\ ,\ \rangle_{\mathcal{P}})$ *be a complete manifold of dimension* m, $h \in C^{\infty}(\mathcal{P})$, $h > 0$ *on* $\mathcal{P}$. *Let* ψ *as in* (49) *be an equidistant graph in* $\overline{M} = \mathbb{R}_h \times \mathcal{P}$ *defined via the smooth function* $u : \mathcal{P} \to \mathbb{R}$ *with mean curvature* H *in the direction of* ν *as in* (50). *Suppose that* $H > 0$ *and that* (47) *is satisfied for some parameters* $\mu \in [0, 1]$ *and* $\beta > 0$ *on* $\mathcal{P}$. *Then*

$$c_{\mu}^{h}(\mathcal{P}) \geq \beta . \tag{54}$$

In particular, we have

Corollary 2 *In the assumptions of Theorem 9, suppose that* H *is a positive constant. Then*

$$mH \leq c_0^h(\mathcal{P}) = \liminf_{R\to+\infty} \frac{\log\int_{B_R} h}{R} .$$

It is immediate to deduce some nonexistence results from Corollary 2.

We now give some applications of Theorem 5. It gives an extended version of Theorem 3.1 of [1]. Toward this aim, we let $(\overline{M}, \langle\ ,\ \rangle)$ be an oriented Riemannian manifold of dimension $m+1$ endowed with a Killing vector field Y. We let $\psi : M \to \overline{M}$ be an isometric immersion of a connected, orientable, complete non-compact Riemannian manifold M of dimension m into $\overline{M}$. We choose a globally defined unit normal vector field ν and let $v = \langle \nu, Y \rangle$. Let $\{\theta^a\}$, $\{\theta^a_b\}$, $1 \leq a, b, \dots \leq m+1$ be a local orthonormal oriented Darboux coframe along ψ with respective Levi-Civita connection forms. The dual frame is therefore given by $\{e_i\}$, $i = 1, \dots, m$ tangent to M and $e_{m+1} = \nu$. Then

$$v = Y^{m+1}$$

and

$$dv = dY^{m+1} = Y_i^{m+1}\theta^i - Y^t\theta_t^{m+1} = (Y_i^{m+1} - Y^t A_{ti})\theta^i$$

where $A = A_{ij}\theta^i \otimes \theta^j$ is the second fundamental form in the direction of ν. Note that, since Y is Killing, we have

$$v_i = -(Y^i_{m+1} + Y^t A_{ti}) \,. \tag{55}$$

Set $X = -\nabla v$. We have

$$\langle \nabla v, X \rangle = -|X|^2 \le 0 \,. \tag{56}$$

From equation (2.8) in Proposition 2.2 of [19] (see also Proposition 1 in [16]), using the fact that Y is Killing, we deduce

$$\Delta v = -[|A|^2 + \overline{\mathrm{Ric}}(\nu, \nu)]v - m\langle Y, \nabla H \rangle \tag{57}$$

where $\overline{\mathrm{Ric}}$ is the Ricci tensor of $\overline{M}$; we refer the reader to Proposition 6 in [3] for an extension of (57) to the case where Y is conformal Killing. Observe that, setting Z for the vector field of components

$$Z_i = Y^t A_{ti}$$

an immediate computation gives

$$|X|^2 = |\nabla_\nu Y|^2 + 2\langle Z, \nabla_\nu Y \rangle + |Z|^2$$

so that X is bounded anytime Y, $\nabla_\nu Y$, and A are bounded.

Theorem 10 *Let $(M, \langle\ ,\ \rangle)$ be a complete, non-compact manifold of dimension m isometrically immersed into a manifold $(\overline{M}, \langle\ ,\ \rangle)$ of dimension $m+1$ with constant mean curvature. Let M and $\overline{M}$ be orientable and let Y a bounded Killing vector field on $\overline{M}$. Assume that there exists $\alpha > 0$ such that*

$$\sup_M |Y| \le \frac{1}{\alpha} \qquad \text{and} \qquad \overline{\mathrm{Ric}}(\nu, \nu) > -\alpha \,. \tag{58}$$

where ν is a chosen globally defined unit normal on M and $\overline{\mathrm{Ric}}$ is the Ricci tensor of $\overline{M}$. We let A be the second fundamental form of ψ in the direction of ν. Let $v = \langle Y, \nu \rangle$ satisfy

$$v \ge \frac{1}{|A|^2 + \overline{\mathrm{Ric}}(\nu, \nu) + \alpha} \qquad \text{on } M \tag{59}$$

and assume that the vector field

$$X = \overline{\nabla}_\nu Y + A(Y^\top, \ )^\sharp = \overline{\nabla}_\nu Y + Z \tag{60}$$

has bounded length. If $c_0(M) = 0$*, then* $|A|^2 = -\overline{\mathrm{Ric}}(\nu, \nu)$*. In particular, if* $\overline{\mathrm{Ric}}(\nu, \nu) \geq 0$ *then* $\psi(M)$ *is totally geodesic.*

Proof Since $H = \frac{1}{m}\mathrm{tr}A$ is constant, (57) yields

$$\Delta v = -[|A|^2 + \overline{\mathrm{Ric}}(\nu, \nu)]v \qquad \text{on } M\,, \tag{61}$$

where $v = \langle \nu, Y\rangle \leq |Y| \leq \frac{1}{\alpha}$ since $v^2 = |Y|^2 - |Y^\top|^2 \leq |Y|^2$. Define $u = \frac{1}{\alpha} - v \geq 0$ on M and observe that because of (56)

$$\langle \nabla u, X\rangle \geq 0 \qquad \text{on } M\,. \tag{62}$$

From (58), (59), and (61), we get

$$\mathrm{div}X = -\Delta v = [|A|^2 + \overline{\mathrm{Ric}}(\nu, \nu) + \alpha]v - \alpha v \geq 1 - \alpha v = \alpha u\,.$$

Then, we apply Theorem 5 with the choices $\Omega = M$, $h = 0$, or, equivalently, $w = 1$, $\mu = 0$, $\beta = \alpha > 0$, $k = 1$, and $f(t) = t$ to obtain that

$$u \leq \frac{c_0(M)}{\alpha}\,.$$

If $c_0(M) = 0$, then $u \equiv 0$; hence, $v \equiv \frac{1}{\alpha} > 0$, and substituting into (61), we get

$$|A|^2 + \overline{\mathrm{Ric}}(\nu, \nu) \equiv 0$$

and if further $\overline{\mathrm{Ric}}(\nu, \nu) \geq 0$, then $A \equiv 0$, so that ψ is totally geodesic. □

The next result is in the same vein. Let $\psi : M \to \mathbb{R}^{m+1}$ be an oriented, m-dimensional mean curvature flow soliton with respect to the position vector field Z of $\mathbb{R}^{m+1}$ with soliton constant $c \in \mathbb{R}$. This means that the isometric immersion ψ satisfies the equation:

$$m\mathbf{H} = cZ_\psi^\perp = c\psi^\perp \tag{63}$$

where $\mathbf{H}$ is the mean curvature vector of the immersion and $Z_\psi^\perp = \psi^\perp$ is the normal part of the vector field Z along ψ. We let

$$\nu = \nu_\psi : M \to \mathbb{S}^m \subseteq \mathbb{R}^{m+1} \tag{64}$$

be the spherical Gauss map of the immersion and B a Killing vector field on $\mathbb{R}^{m+1}$ and define

$$\eta(x) = \frac{1}{2}|\psi|^2(x)\,.$$

The following computation can be found in Proposition 1 of [8].

Lemma 1 *In the above setting, let $v = \langle B, \nu\rangle$. Then, v satisfies the equation:*

$$\Delta_{-cZ^\top} v + |A|^2 v = -\langle [B, cZ], \nu\rangle + cv \tag{65}$$

where A is the second fundamental form in the direction of ν and $\Delta_{-cZ^\top}$ is the operator $\Delta + \langle cZ^\top, \nabla\ \rangle$ and $[\ ,\ ]$ the Lie bracket.

Theorem 11 *Let $\psi : M \to \mathbb{R}^{m+1}$ be an oriented, complete, non-compact, m-dimensional mean curvature flow soliton with respect to the position vector field Z of $\mathbb{R}^{m+1}$ and with soliton constant $c \in \mathbb{R}$. Assume that $c_0^w(M) = 0$, where $w = e^{c\eta}$, that $|A|$ is bounded and that the spherical Gauss map $\nu : M \to \mathbb{S}^m$ of the immersion satisfies*

$$v = \langle b, \nu\rangle \geq \frac{1}{1 + |A|^2} \tag{66}$$

for some fixed vector $b \in \mathbb{S}^m$. Denote by B the parallel extension of b to all of $\mathbb{R}^{m+1}$. Then, $\psi(M)$ is a hyperplane orthogonal to B, and if $c \neq 0$, we have $0_{\mathbb{R}^{m+1}} \in \psi(M)$.

Remark 3 Note that (66) implies that the image $\nu(M)$ of the spherical Gauss map of the immersion is contained in the compact subset:

$$\left\{ y \in \mathbb{S}^m : \operatorname{dist}(b, y) \leq \cos^{-1}\left(\frac{1}{1 + \sup_M |A|^2}\right)\right\}$$

of the open hemisphere of $\mathbb{S}^m$ centered at b.

Proof First, observe that we can apply Lemma 1 since B is parallel, hence Killing, on $\mathbb{R}^{m+1}$. In this case, we have $[B, Z] = B$, as it can be seen by fixing Cartesian coordinates (x^i) on $\mathbb{R}^{m+1}$ such that $B = \partial_{x^1}$ and computing

$$[B, Z] = \left[\partial_{x^1}, \sum_{i=1}^{m+1} x^i \partial_{x^i}\right] = \sum_{i=1}^{m+1} \frac{\partial x^i}{\partial x^1}\partial_{x^i} = \partial_{x^1} = B\,,$$

so that formula (65) reduces to

$$\Delta_{-cZ^\top} v + |A|^2 v = 0\,. \tag{67}$$

Furthermore, setting $\bar\eta(x) = \frac{1}{2}|x|^2$ for $x \in \mathbb{R}^{m+1}$, we have $Z = \nabla\bar\eta$ in $\mathbb{R}^{m+1}$, and since $\eta = \bar\eta \circ \psi$, it follows that $Z^\top = \nabla\eta$. Hence, $\Delta_{-cZ^\top}$ is in fact a weighted Laplacian

$$\Delta_{-cZ^\top} = \Delta_{-c\eta} = \Delta + c\langle\nabla\eta,\ \rangle$$

and using (67), we get

$$\Delta_{-c\eta}v = -|A|^2 v\,.$$

As in Theorem 10, and since B is parallel, we have $\nabla v = -AB^\top$. Setting $X = AB^\top$, we can further restate the above relation as

$$\mathrm{div}_{-c\eta}X = |A|^2 v\,. \tag{68}$$

Thus, for $u = 1 - v \geq 0$ (note that $v \leq |B| = 1$), we have, using (66)

$$\mathrm{div}_{-c\eta}X \geq u\,. \tag{69}$$

We note that X is bounded since and $|B^\top| \leq |B| = 1$. Also, $\langle \nabla u, X\rangle = -\langle \nabla v, X\rangle = |X|^2 \geq 0$. We can then apply Theorem 5 as in the proof of Theorem 10 to deduce

$$u \leq c_0^w(M) = 0$$

and since $u \geq 0$, we conclude $u \equiv 0$, that is, $v \equiv 1$. Since $0 < v = \langle B, \nu\rangle = |B^\perp| \leq |B| \equiv 1$, we must have $B^\perp = B$ along $\psi(M)$, that is, $\nu = B$. So, $\psi(M) \subseteq \mathbb{R}^{m+1}$ is a hyperplane perpendicular to B. Moreover, if $c \neq 0$, then by (63) it must be $Z_\psi^\perp \equiv \psi^\perp \equiv 0$, so that $\psi(M)$ is a hyperplane passing through the origin of $\mathbb{R}^{m+1}$. □

We observe that the condition $c_0^w(M) = 0$, with $w = e^{c\eta}$, means

$$\liminf_{R\to+\infty} \frac{1}{R} \log \int_{B_R} e^{\frac{c}{2}|x|^2} = 0\,.$$

Note that when $c < 0$ the above condition allows a growth of the volume of geodesic balls which is faster than polynomial. Thus, the assumptions of the above theorem do not imply those of Theorem 3.1 of [11]. See also Theorem 2.2 of [8].

5 Comparison for Strictly Monotone 1-Coercive Operators

Let $(M, \langle\,,\,\rangle)$ be a complete Riemannian manifold. Let $A : \mathbb{R} \times TM \to TM$ be a weakly-1-coercive generalized bundle map and let L be the differential operator weakly defined by

$$Lu(x) = \mathrm{div}A(x, u, \nabla u)\,.$$

In this section, we assume that A is also *strictly monotone*, that is, that

$$\langle A(x,t,\xi) - A(x,s,\eta), \xi - \eta\rangle \geq 0 \tag{70}$$

for every $x \in M$, $\xi, \eta \in T_xM$ and $t, s \in \mathbb{R}$ with $t \geq s$, and we require strict inequality in case $t > s$ and $\xi \neq \eta$.

As an example, let

$$A(x,s,\xi) = \frac{\xi}{\sqrt{1+|\xi|^2}}\,.$$

Then, by the Mīkljukov inequality in (see [20], page 165)

$$\begin{aligned}
&\langle A(x,t,\xi) - A(x,s,\eta), \xi - \eta\rangle \\
&= \left\langle \frac{\xi}{\sqrt{1+|\xi|^2}} - \frac{\eta}{\sqrt{1+|\eta|^2}}, \xi - \eta \right\rangle \\
&\geq \frac{1}{2}[\sqrt{1+|\xi|^2} + \sqrt{1+|\eta|^2}] \left| \frac{\xi}{\sqrt{1+|\xi|^2}} - \frac{\eta}{\sqrt{1+|\eta|^2}} \right|^2 .
\end{aligned}$$

Hence, (70) holds, and if $\xi \neq \eta$, the inequality is clearly strict. Thus, in particular, our subsequent results can be applied to the mean curvature operator.

Theorem 12 *Let $(M, \langle\, ,\, \rangle)$ be a complete manifold, $A : \mathbb{R} \times TM \to TM$ a C^1 strictly monotone weakly-1-coercive generalized bundle map with coercivity constant $k > 0$ and let L be the corresponding weakly-1-coercive operator.*

Let $b \in C(M)$ be positive and satisfy condition (B) *for two given parameters $\mu \in [0, 1]$ and $\beta > 0$, and let $f \in C(\mathbb{R})$ be such that*

$$f(s) - f(t) \geq \alpha(s-t) \qquad \forall s, t \in \mathbb{R} \textit{ with } s > t \tag{71}$$

for some constant $\alpha > 0$. If $u, v \in C^2(M)$ satisfy

$$Lu \geq b(x) f(u)\,, \qquad Lv \leq b(x) f(v) \qquad \textit{in } M \tag{72}$$

then

$$u \leq v + 2\frac{k}{\alpha\beta} c_\mu(M) \qquad \textit{in } M\,. \tag{73}$$

Proof We let $w = u - v$ and we suppose by contradiction that (73) is false so that, having set $c^* = 2\frac{k}{\alpha\beta} c_\mu(M)$, we have $\{x \in M : w(x) > c^*\} \neq \emptyset$. Thus, choose $\varepsilon > 0$ small enough so that the set

$$\Omega_\varepsilon = \{x \in M : w(x) > c^* + \varepsilon\} \neq \emptyset\,.$$

Ω_ε is open by continuity of w. For short, write $A_u = A(x, u, \nabla u)$, $A_v = A(x, v, \nabla v)$, and consider the vector field $X : M \to TM$ given by

$$X = A_u - A_v \,.$$

From monotone 1-coercivity (70) of A, we have

$$\langle X, \nabla w \rangle \geq 0 \qquad \text{on } \{w \geq 0\}\,. \tag{74}$$

The assumptions on A, u, and v ensure that X is of class C^1 and that

$$|X| = |A_u - A_v| \leq |A_u| + |A_v| \leq 2k \quad \text{ on } M\,.$$

We observe that X has a well-defined flow

$$\Psi : [0, +\infty) \times \Omega_\varepsilon \to \Omega_\varepsilon. \tag{75}$$

Indeed, since M is complete and X is bounded on M, for any $x \in \Omega_\varepsilon$, the initial value problem

$$\begin{cases} \gamma(0) = x \\ \dot{\gamma}(t) = X_{\gamma(t)} \end{cases} \tag{76}$$

has a (unique) solution $\gamma : [0, +\infty) \to M$. For all $t \in [0, +\infty)$, we have

$$\frac{\mathrm{d}}{\mathrm{d}t} w(\gamma(t)) = \langle \nabla w, \dot{\gamma}(t) \rangle_{\gamma(t)} = \langle \nabla w, X_{\gamma(t)} \rangle_{\gamma(t)}\,. \tag{77}$$

We show that $\gamma(t) \in \Omega_\varepsilon$ for all $t \in [0, +\infty)$. Since $w(\gamma(0)) > 0$, we also have $w(\gamma(t)) > 0$ for all sufficiently small $t > 0$. Let

$$\tilde{T}_x = \sup\{T \in (0, +\infty] : w(\gamma(t)) > 0 \text{ for all } t \in [0, T)\}$$

We know that $0 < \tilde{T}_x \leq +\infty$. Suppose by contradiction that $\tilde{T}_x < +\infty$. Then, it must be $w(\gamma(\tilde{T}_x)) \leq 0$. But then $\gamma(t) \in \{w > 0\}$ for all $t \in [0, \tilde{T}_x)$, so by (77) and (74), we have

$$\frac{\mathrm{d}}{\mathrm{d}t} w(\gamma(t)) \geq 0 \qquad \forall t \in [0, \tilde{T}_x)$$

so that $[0, \tilde{T}_x) \ni t \mapsto w(\gamma(t))$ is nondecreasing and therefore by continuity

$$w(\gamma(\tilde{T}_x)) = \lim_{t \to \tilde{T}_x} w(\gamma(t)) \geq w(\gamma(0)) > 0\,,$$

contradiction. So, $\tilde{T}_x = +\infty$ and $\gamma([0, +\infty)) \subseteq \{w > 0\}$. We just observed that $[0, +\infty) \ni t \mapsto w(\gamma(t))$ is nondecreasing, so in fact $w(\gamma(t)) \geq w(\gamma(0)) = w(x) > c^* + \varepsilon$ for every $t \in [0, +\infty)$. Thus, $\gamma(t)$ lies in Ω_ε for every $t \in [0, +\infty)$.

Now, fix $p \in \Omega_\varepsilon$ and $\delta > 0$ such that $B_\delta(p) \subseteq \Omega_\varepsilon$ and $\partial B_\delta(p)$ is smooth. For every $t \in [0, +\infty)$, let

$$U_t = \Psi(t, B_\delta(p)) \qquad \text{and} \qquad g(t) = \mathrm{vol}(U_t)\,.$$

Using that $\mathcal{L}_X(\mathrm{d}v) = \mathrm{div}X\mathrm{d}v$, with $\mathcal{L}_X$ the Lie derivative with respect to X, we have

$$\begin{aligned} g'(t) = \int_{U_t} \mathrm{div}X\mathrm{d}v = \int_{U_t} (Lu - Lv)\mathrm{d}v &\geq \int_{U_t} b(x)[f(u) - f(v)]\mathrm{d}v \\ &= \int_{U_t} b(x)\frac{f(u) - f(v)}{u - v}(u - v)\mathrm{d}v \\ &> (c^* + \varepsilon)\alpha \int_{U_t} b(x)\mathrm{d}v\,, \end{aligned} \tag{78}$$

where have used that $u - v = w > c^* + \varepsilon > 0$ in $\Omega_\varepsilon \supseteq U_t$ and that

$$\frac{f(u) - f(v)}{u - v} \geq \alpha$$

by (71). Reasoning as in the proof of Theorem 1 (see (25)), fixing $\beta^* \in \mathbb{R}$ such that

$$\beta > \beta^* > \frac{c^*}{c^* + \varepsilon}\beta$$

there exists $R_0 > 0$ such that

$$\inf_{B_R} b \geq \beta^* R^{-\mu} \qquad \forall\, R \geq R_0\,. \tag{79}$$

As a consequence of $|X| \leq 2k$ and of the triangle inequality, we have

$$U_t \subseteq B_{\delta+2kt}(p) \subseteq B_{r(p)+\delta+2kt} \qquad \forall\, t \geq 0\,. \tag{80}$$

Setting $R_1 = \max\{R_0, r(p) + \delta\}$, from (78), (79), and (80), we get

$$g'(t) \geq (c^* + \varepsilon)\alpha\beta^*(R_1 + 2kt)^{-\mu} g(t) \qquad \forall\, t \geq 0\,.$$

Integrating over $[0, t]$

$$\log(g(t)) \geq \log(g(0)) + (c^* + \varepsilon)\alpha\beta^* \int_0^t (R_1 + 2ks)^{-\mu}\, ds\,. \tag{81}$$

Note that $U_t \subseteq B_{\delta+2kt}(p) \subseteq B_{R_1+2kt}$ and also $U_t \subseteq \Omega_\varepsilon$, we have $g(t) \leq \mathrm{vol}(B_{R_1+2kt})$, so with a change of integration variable $y = R_1 + 2ks$ in (81), we get

$$\log \mathrm{vol}(B_R) \geq \log(g(0)) + \frac{1}{2}(c^* + \varepsilon)\alpha\beta^* \int_{R_1}^R y^{-\mu}\, dy \qquad \forall R \geq R_1$$

and therefore

$$\log \mathrm{vol} B_R(o) \geq \log g(0) + (c^* + \varepsilon)\frac{\alpha\beta^*}{2k}\frac{R^{1-\mu} - R_1^{1-\mu}}{1-\mu} \qquad \text{if } 0 \leq \mu < 1$$

$$\log \mathrm{vol} B_R(o) \geq \log g(0) + (c^* + \varepsilon)\frac{\alpha\beta^*}{2k}(\log R - \log R_1) \qquad \text{if } \mu = 1\,.$$

This yields

$$c_\mu(M) \geq (c^* + \varepsilon)\frac{\alpha\beta^*}{2k} > c^*\frac{\alpha\beta}{2k} = c_\mu(M)\,,$$

contradiction. □

Similarly, one proves the local version by adapting the proof of Theorem 12 in the same way as we proved Theorem 2 adapting the proof of Theorem 1.

Theorem 13 *Let $(M, \langle\, ,\, \rangle)$ be complete, $\Omega \subseteq M$ a non-empty open set, $A : \mathbb{R} \times TM|_\Omega \to TM|_\Omega$ a C^1 strictly monotone weakly-1-coercive generalized bundle map with coercivity constant $k > 0$ and let L be the corresponding operator.*

Let $b \in C(M)$ be positive and satisfy (B) *for some $\mu \in [0, 1]$ and $\beta > 0$, and let $f \in C(\mathbb{R})$ be such that*

$$f(s) - f(t) \geq \alpha(s - t) \qquad \forall s, t \in \mathbb{R} \text{ with } s > t$$

for some constant $\alpha > 0$. If $u, v \in C(\overline{\Omega}) \cap C^2(\Omega)$ satisfy

$$\begin{cases} Lu \geq b(x) f(u) & \text{in } \Omega \\ Lv \leq b(x) f(v) & \text{in } \Omega \end{cases}$$

then

$$u \leq v + \max\left\{\frac{2k}{\alpha\beta}c_\mu(\Omega), \sup_{\partial\Omega}(u - v)\right\} \qquad \text{in } \Omega. \tag{82}$$

Remark 4 Theorem 13 generalizes a comparison principle for the capillarity equation in the Euclidean space $M = \mathbb{R}^m$ first proved (in case $\Omega \subseteq \mathbb{R}^m$ is bounded) by Concus and Finn [10] and later extended (to possibly unbounded Ω) in Finn and Hwang [14]. The case of the capillarity equation corresponds to $A(\xi) = (1 + |\xi|^2)^{-1/2}\xi$, $f(t) = \alpha t$, $b(x) \equiv 1$. Note that in this setting, one can take $\mu = 0$, and since $c_0(\Omega) = 0$ for any $\Omega \subseteq \mathbb{R}^m$, the conclusion (82) reduces to

$$\sup_{\Omega}(u - v) \leq \sup_{\partial\Omega}(u - v)_+ .$$

See also [21] Theorems 1.3, 1.4, and 1.5 for further comparison results in the case where M is a general Riemannian manifold.

Remark 5 Theorems 12 and 13 can be also deduced directly as applications of Theorem 5 with the choices $X = A(x, u, \nabla u) - A(x, v, \nabla v)$ and $\psi = w = u - v$. Indeed, we already observed in the proof of Theorem 12 that $|X| \leq 2k$ and $\langle \nabla\psi, X\rangle \geq 0$ due to weak coercivity and strict monotonicity of A, and by the assumptions on f, we have

$$\text{div} X = Lu - Lv \geq b(x)[f(u) - f(v)] \geq \alpha b(x)(u - v) = \alpha b(x)\psi \qquad \text{on } \Omega_+$$

where $\Omega_+ = \{x \in \Omega : u(x) > v(x)\} \equiv \{x \in \Omega : \psi(x) > 0\}$. Since $c_\mu(\Omega) \geq 0$, if $\Omega_+ = \emptyset$, then $u \leq v$ everywhere in Ω and (82) immediately follows; otherwise, $\sup_\Omega \psi = \sup_{\Omega_+} \psi > 0$, and from Theorem 5, we obtain

$$\sup_{\Omega} \psi \leq \sup_{\Omega_+} \psi \leq \max\left\{\frac{2k}{\alpha\beta} c_\mu(\Omega_+), \sup_{\partial\Omega_+} \psi\right\}.$$

Since $c_\mu(\Omega_+) \leq c_\mu(\Omega)$ and $\partial\Omega_+ \subseteq (\partial\Omega) \cup \{u = v\}$, we further obtain

$$\sup_{\Omega} \psi \leq \max\left\{\frac{2k}{\alpha\beta} c_\mu(\Omega), \sup_{\partial\Omega} \psi\right\},$$

that is, (82).

Applying Theorem 13 to the capillarity equation

$$\text{div}\left(\frac{\nabla u}{\sqrt{1 + |\nabla u|^2}}\right) = b(x)u \qquad \text{on } \Omega \subseteq M \tag{83}$$

we have

Theorem 14 *Let $(M, \langle\, ,\, \rangle)$ be complete and $\Omega \subsetneq M$ a non-empty open set such that $c_\mu(\Omega) = 0$ for a given $\mu \in [0, 1]$. Let $0 < b \in C(M)$ satisfy*

$$\begin{cases} b(x) \geq \beta & \forall x \in M & \text{if } \mu = 0 \\ b(x) \geq \beta r(x)^{-\mu} & \text{for } r(x) \gg 1 & \text{if } 0 < \mu \leq 1 \end{cases}$$

for some $\beta > 0$. If u, v are solutions of equation (83) *in Ω such that $u = v$ on $\partial\Omega$, then $u \equiv v$ in Ω.*

Corollary 3 *Let $(M, \langle\, ,\, \rangle)$ be a complete manifold of dimension m satisfying*

$$\mathrm{Ric}(\nabla r, \nabla r) \geq -(m-1)\frac{B^2}{(1+r^2)^{\alpha/2}}$$

for some $0 < \alpha \leq 2$ and $B > 0$. Let $\mu \in [0, \alpha/2)$ and let $0 < b \in C(M)$ satisfy

$$\begin{cases} b(x) \geq \beta & \forall x \in M & \text{if } \mu = 0 \\ b(x) \geq \beta r(x)^{-\mu} & \text{for } r(x) \gg 1 & \text{if } 0 < \mu < \frac{\alpha}{2} \end{cases}$$

for some $\beta > 0$. Let $\Omega \subsetneq M$ be a non-empty open set and let $u, v \in C^2(\Omega)$ be solutions of (83) *such that $u = v$ on $\partial\Omega$. Then, $u \equiv v$ in Ω.*

Proof Use Theorem 14 and Remark 1 to guarantee that in the above assumptions $c_\mu(\Omega) = 0$. □

6 Extension to the General Setting

The aim of this last section is to show how the results proved in Sect. 3 can be extended to the general setting of weakly-1-coercive operators:

$$Lu = w^{-1}\mathrm{div}(wA(x, u, \nabla u)) \tag{84}$$

with A just a Carathéodory-type weakly-1-coercive function, w is a positive, measurable and locally bounded weight, and u is just continuous and locally $W^{1,1}$ regular. To do so, we first prove the following generalization of Theorem 5.

Theorem 15 *Let $(M, \langle\, ,\, \rangle)$ be a complete, non-compact manifold, $\Omega \subseteq M$ a non-empty open set. Let $w \in L^\infty_{\mathrm{loc}}(\Omega)$ satisfy $w > 0$ a.e. in Ω and let $0 < b \in C(M)$ satisfy condition* (B) *for some $\mu \in [0, 1]$ and $\beta > 0$. Let $X \in L^\infty(\Omega; TM)$ and*

$\psi \in W^{1,1}_{\mathrm{loc}}(\Omega) \cap C(\overline{\Omega})$ be such that

$$-\int_{\Omega} w\,\langle X, \nabla\varphi\rangle \geq \int_{\Omega} w\,b\,\psi\,\varphi \qquad \forall\,\varphi \in W^{1,1}_{c}(\Omega),\ \varphi \geq 0 \tag{85}$$

and

$$\langle \nabla\psi, X\rangle \geq 0 \qquad \text{a.e. in } \Omega\,. \tag{86}$$

Then

$$\sup_{\Omega} \psi \leq \max\left\{\frac{\|X\|_{L^\infty(\Omega)}}{\beta} c^w_\mu(\Omega),\, \sup_{\partial\Omega} \psi\right\}. \tag{87}$$

In particular, in case $\Omega = M$, we have

$$\sup_{M} \psi \leq \frac{\|X\|_{L^\infty(\Omega)}}{\beta} c^w_\mu(M)\,.$$

Remark 6 In case $\|X\|_{L^\infty(\Omega)} = 0$ and $c^w_\mu(\Omega) = +\infty$, we adopt the convention $0 \cdot (+\infty) = 0$ to make sense of the first term in brackets on the RHS of (87). Note that if $\|X\|_{L^\infty(\Omega)} = 0$, then the validity of (85) forces $wb\psi \leq 0$ a.e. in Ω, and thus, $\psi \leq 0$ a.e. in Ω since w and b are a.e. strictly positive. Hence, in case $\|X\|_{L^\infty(\Omega)} = 0$, we necessarily have $\sup_\Omega \psi \leq 0$, and then, even in case $c^w_\mu(\Omega) = +\infty$, inequality (87) certainly holds because its RHS is 0.

Proof Suppose, by contradiction, that (87) is not satisfied. Then, we can find $\gamma \in \mathbb{R}$ such that

$$\sup_{\Omega} \psi > \gamma > \max\left\{\frac{\|X\|_{L^\infty(\Omega)}}{\beta} c^w_\mu(\Omega),\, \sup_{\partial\Omega} \psi\right\}.$$

Since the RHS of (87) is nonnegative, necessarily $\gamma > 0$. We define

$$\Omega_\gamma = \{x \in \Omega : \psi(x) > \gamma\}$$

and $G : (0, +\infty) \to [0, +\infty)$ by setting

$$G(t) = \mathrm{vol}_{-\log w}(B_t \cap \Omega_\gamma) = \int_{B_t \cap \Omega_\gamma} w \qquad \forall\, t > 0\,.$$

The set Ω_γ is open and non-empty and satisfies $\overline{\Omega_\gamma} \subseteq \Omega$ because $\gamma > \sup_{\partial\Omega} \psi$. The function G is well defined, nondecreasing, and absolutely continuous on any compact interval contained in $(0, +\infty)$. In particular, it is differentiable a.e. in $(0, +\infty)$. Moreover, by monotone convergence together with the fact that $w > 0$

a.e. and $\mathrm{vol}(\Omega_\gamma) > 0$, we have

$$\lim_{R\to+\infty} G(R) = \int_{\Omega_\gamma} w > 0\,. \tag{88}$$

As in the proof of Theorem 1, fix $\beta^* \in \mathbb{R}$ such that

$$\beta > \beta^* > \frac{\|X\|_{L^\infty(\Omega)}}{\gamma} c^w_\mu(\Omega)$$

and let $R_0 > 0$ be large enough so that

$$\inf_{B_R} b(x) \geq \beta^* R^{-\mu} \qquad \forall\, R \geq R_0\,.$$

Similarly to what we did in the proof of Theorem 1, we aim at showing that

$$\frac{\|X\|_{L^\infty(\Omega)}}{\beta^*} G'(t) \geq \frac{\gamma}{t^\mu} G(t) \qquad \text{for a.e. } t \in (R_0, +\infty)\,. \tag{89}$$

Provided (89) holds, if we choose $R_1 = R_1(\gamma) > R_0$ such that $G(R_1) > 0$ (the existence of such R_1 is guaranteed by (88)), then integrating the differential inequality (89) on $[R_1, R]$ for any $R > R_1$ (and using the fact that $\log G$ is absolutely continuous on $[R_1, R]$ because G is absolutely continuous, positive, and bounded away from 0 on that interval), we get

$$\frac{\|X\|_{L^\infty(\Omega)}}{\beta^*}[\log G(R) - \log G(R_1)] \geq \gamma \begin{cases} \left(\dfrac{R^{1-\mu}}{1-\mu} - \dfrac{R_1^{1-\mu}}{1-\mu}\right) & \text{if } 0 \leq \mu < 1 \\ (\log R - \log R_1) & \text{if } \mu = 1 \end{cases}$$

for all $R > R_1$, from which we obtain, letting $R \to +\infty$

$$\frac{\|X\|_{L^\infty(\Omega)}}{\beta^*} c^w_\mu(\Omega) \geq \frac{\|X\|_{L^\infty(\Omega)}}{\beta^*} c^w_\mu(\Omega_\gamma) \geq \gamma$$

thus reaching the desired contradiction. Hence, we are only left to prove validity of (89) under the aformentioned assumptions on γ and R_0. Let $t > R_0$ be a value for which $G'(t)$ exists. For any $0 < \delta < t$, choose $\eta_\delta \in C^\infty(M)$ satisfying

$$\begin{cases} \eta_\delta \equiv 1 & \text{on } B_{t-\delta} \\ \eta_\delta \equiv 0 & \text{on } M \setminus B_t \\ 0 \leq \eta_\delta \leq 1 & \text{on } B_t \setminus B_{t-\delta} \\ |\nabla \eta_\delta| \leq \dfrac{1}{\delta} + 1 & \text{on } M\,. \end{cases}$$

Let $\lambda \in C^\infty(\mathbb{R})$ be such that

$$\lambda(s) = 0 \quad \text{if } s \leq 1\,, \qquad \lambda(s) = 1 \quad \text{if } s \geq 2\,, \qquad \lambda' \geq 0 \quad \text{on } \mathbb{R}$$

and for any $\varepsilon > 0$ define $\lambda_\varepsilon \in C^\infty(\mathbb{R})$ by

$$\lambda_\varepsilon(s) = \lambda(s/\varepsilon)\,.$$

We have

$$0 \leq \lambda_\varepsilon \leq \mathbf{1}_{(0,+\infty)} \quad \forall\, \varepsilon > 0 \qquad \text{and} \qquad \lambda_\varepsilon \nearrow \mathbf{1}_{(0,+\infty)} \quad \text{as } \varepsilon \to 0^+\,,$$

where $\mathbf{1}$ stands for the indicator function and $\nearrow$ denotes monotone convergence from below. Now, for any $0 < \delta < t$ and $\varepsilon > 0$, we consider the test function $\varphi_{\delta,\varepsilon} \in W^{1,1}_c(\Omega)$ defined by

$$\varphi_{\delta,\varepsilon} = \eta_\delta \cdot \lambda_\varepsilon(\psi - \gamma)\,.$$

By construction, $\varphi_{\delta,\varepsilon}$ is nonnegative and compactly supported inside $\Omega_\gamma \subseteq \Omega$, so it is an admissible test function for the differential inequality (85), yielding

$$-\int_{\Omega_\gamma} w\langle X, \nabla\varphi_{\delta,\varepsilon}\rangle \geq \int_{\Omega_\gamma} w\, b\, \psi \varphi_{\delta,\varepsilon}\,.$$

We have

$$\begin{aligned} -\, w\langle X, \nabla\varphi_{\delta,\varepsilon}\rangle &= -w\lambda_\varepsilon(\psi - \gamma)\langle X, \nabla\eta_\delta\rangle - w\eta_\delta\lambda'_\varepsilon(\psi - \gamma)\langle X, \nabla\psi\rangle \\ &\leq -w\lambda_\varepsilon(\psi - \gamma)\langle X, \nabla\eta_\delta\rangle \end{aligned}$$

where the inequality holds since $w\,\eta_\delta \geq 0$ and $\lambda'_\varepsilon \geq 0$ by construction and $\langle X, \nabla\psi\rangle \geq 0$ by assumption. Thus, we have

$$-\int_{\Omega_\gamma} w\,\lambda_\varepsilon(\psi - \gamma)\langle X, \nabla\eta_\delta\rangle \geq \int_{\Omega_\gamma} w\, b\, \psi\lambda_\varepsilon(\psi - \gamma)\eta_\delta\,.$$

Letting $\varepsilon \to 0^+$ and applying the dominated convergence theorem and the monotone convergence theorem to the left and right sides of this inequality, respectively, we obtain

$$-\int_{\Omega_\gamma} w\langle X, \nabla\eta_\delta\rangle \geq \int_{\Omega_\gamma} w\, b\, \psi\eta_\delta = \int_{\Omega_\gamma \cap B_t} w\, b\, \psi\eta_\delta \geq \frac{\beta^*\gamma}{t^\mu}\int_{\Omega_\gamma} w\eta_\delta$$

where in the middle equality we used the fact that $\eta_\delta = 0$ outside B_t, and in the last inequality, we used that $\psi \geq \gamma$ on Ω_γ and that $b \geq \beta^* t^{-\mu}$ on B_t. By Cauchy-Schwarz inequality, we further estimate (using that $\nabla\eta_\delta$ is supported in $B_t \setminus B_{t-\delta}$ and that $\eta_\delta = 1$ on $B_{t-\delta}$)

$$\|X\|_{L^\infty(\Omega)} \|\nabla\eta_\delta\|_{L^\infty(M)} \int_{\Omega\gamma\cap B_t\setminus B_{t-\delta}} w \geq \frac{\beta^*\gamma}{t^\mu} \int_{\Omega\gamma\cap B_{t-\delta}} w\,,$$

that is, using the upper bound $|\nabla\eta_\delta| \leq 1 + 1/\delta$ and the definition of G

$$\frac{\|X\|_{L^\infty(\Omega)}}{\beta^*}(1+\delta)\frac{G(t) - G(t-\delta)}{\delta} \geq \frac{\gamma}{t^\mu} G(t-\delta)\,.$$

Letting $\delta \to 0^+$ and using that t is a differentiability point for G, we obtain

$$\frac{\|X\|_{L^\infty(\Omega)}}{\beta^*} G'(t) \geq \frac{\gamma}{t^\mu} G(t)\,.$$

So, we proved that this inequality is satisfied for every $t \in (R_0, +\infty)$ such that $G'(t)$ exists, so in particular (89) is proved and this concludes the argument. □

From Theorem 15, we deduce the following extensions of Theorem 4 to the general aforementioned setting.

Theorem 16 *Let $(M, \langle\,,\,\rangle)$ be a complete Riemannian manifold, $\Omega \subseteq M$ an open set. Let L be a weakly-1-coercive differential operator weakly defined as in* (84) *for a weakly-1-coercive Carathéodory-type function $A : \mathbb{R} \times TM|_\Omega \to TM|_\Omega$ with coercivity constant $k > 0$ and a weight $w \in L^\infty_{\mathrm{loc}}(\Omega)$ satisfying $w > 0$ a.e. in Ω.*

Let $0 < b \in C(M)$ satisfy condition (B) *for some $\mu \in [0, 1]$ and $\beta > 0$ and let $f \in C^1(\mathbb{R})$ be nondecreasing. If $u \in W^{1,1}_{\mathrm{loc}}(\Omega) \cap C(\overline{\Omega})$ satisfies*

$$Lu \geq b(x)f(u) \quad \text{in } \Omega$$

then

$$\sup_\Omega f(u) \leq \max\left\{\frac{k}{\beta} c^w_\mu(\Omega), \sup_{\partial\Omega} f(u)\right\}.$$

In particular, in case $\Omega = M$, we have $f(u) \leq kc^w_\mu(M)/\beta$ on M.

Proof The thesis follows by direct application of Theorem 15 with the choice $X = A(x, u, \nabla u)$ and $\psi = f(u)$. □

Remark 7 The regularity condition $f \in C^1(\mathbb{R})$ ensures that $f(u) \in W^{1,1}_{\mathrm{loc}}(\Omega)$ as long as $u \in W^{1,1}_{\mathrm{loc}}(M) \cap C(\overline{\Omega})$, so that we can directly apply Theorem 15 with the choice $\psi = f(u)$. In fact, an analogue of the conclusion of Theorem 16 still holds also in case f is only assumed to be measurable and nondecreasing. Namely, in this case, one can show that

$$\lim_{s \nearrow u^*} f(s) \leq \max \left\{ \frac{k}{\beta} c_\mu^w(\Omega), \sup_{\partial\Omega} f(u) \right\} \tag{90}$$

where $u^* = \sup_\Omega u$ and by $\lim_{s \nearrow u^*} f(s)$ we intend the limit of $f(s)$ as s approaches u^* from the left. This limit exists by monotonicity of f, and it is always less or equal than $\sup_\Omega f(u)$. The two values concide if either $u < u^*$ everywhere in Ω or if u^* is actually attained at some point $x_0 \in \Omega$ and f is continuous at u^*. Inequality (90) can be proved by following the same reasoning used in the proof of Theorem 15, with $\lambda_\varepsilon(\psi - \gamma)$ replaced by $\lambda_\varepsilon(u - u_0)$ for a suitable choice of $u_0 \in \mathbb{R}$ (viz., if by contradiction there exists $\gamma \in \mathbb{R}$ such that

$$\lim_{s \nearrow u^*} f(s) > \gamma > \max \left\{ \frac{k}{\beta} c_\mu^w(\Omega), \sup_{\partial\Omega} f(u) \right\}$$

then one can choose any $u_0 < u^*$ such that $f(u_0) > \gamma$, so that on the open, non-empty set $\Omega_0 = \{x \in \Omega : u(x) > u_0\}$ it holds $f(u) \geq f(u_0) > \gamma$; by continuity of u, the inequality $u \geq u_0$ and therefore $f(u) > \gamma$ holds on $\overline{\Omega_0}$ as well, implying that $\overline{\Omega_0} \subseteq \Omega$.)

Similarly, one can prove the following extension of Theorem 13.

Theorem 17 *Let $(M, \langle \, , \, \rangle)$ be a complete Riemannian manifold, $\Omega \subseteq M$ an open set. Let L be the differential operator defined as in* (84) *for a strictly monotone weakly-1-coercive Carathéodory-type function $A : \mathbb{R} \times TM|_\Omega \to TM|_\Omega$ with coercivity constant $k > 0$ and a weight $w \in L^\infty_{\mathrm{loc}}(\Omega)$ satisfying $w > 0$ a.e. in Ω.*

Let $b \in C(M)$ be positive and satisfy (B) *for some $\mu \in [0, 1]$ and $\beta > 0$, and let $f \in C^1(\mathbb{R})$ be such that*

$$f(s) - f(t) \geq \alpha(s - t) \qquad \forall s, t \in \mathbb{R} \text{ with } s > t$$

for some constant $\alpha > 0$. If $u, v \in W^{1,1}_{\mathrm{loc}}(\Omega) \cap C(\overline{\Omega})$ satisfy

$$\begin{cases} Lu \geq b(x) f(u) & \text{in } \Omega \\ Lv \leq b(x) f(v) & \text{in } \Omega \end{cases}$$

then

$$u \leq v + \max \left\{ \frac{2k}{\alpha\beta} c_\mu(\Omega), \sup_{\partial\Omega}(u - v) \right\} \qquad \text{in } \Omega. \tag{91}$$

Acknowledgments The authors would like to thank the anonymous referee for reading the manuscript in great detail and for giving several valuable suggestions and useful corrections to improve the paper.

Luis J. Alías and Marco Rigoli are partially supported by the grant PID2021-124157NB-I00 funded by MCIN/AEI/10.13039/501100011033/ "ERDF A way of making Europe," Spain. Marco Rigoli is also partially supported by the MUR grant for the PRIN project 20225J97H5.

References

1. Alías, L.J., Caminha, A., do Nascimento, Y.: A maximum principle related to volume growth and applications, Ann. Mat. Pura Appl. **200**, 1637–1650 (2021).
2. Alías, L.J., Colombo, G., Rigoli,M.: Growth of subsolutions of $\Delta_p u = V|u|^{p-2}u$ and of a general class of quasilinear equations, J. Geom. Anal. **34**, 44 (2024).
3. Alías, L.J., Dajczer, M., Ripoll, J.: A Bernstein-type theorem for Riemannian manifolds with a Killing field, Ann. Global Anal. Geom. **31**, 363–373 (2007).
4. Alías, L.J., de Lira, J., Rigoli, M.: Remarks on mean curvature flow solitons in the presence of conformal vector fields, J. Geom. Anal. **30**, 1466–1529 (2020).
5. Alías, L.J., Mastrolia, P., Rigoli, M.: Maximum principles and geometric applications. Springer Monographs in Mathematics, Springer, Cham (2016).
6. Bianchini, B., Mari, L., Pucci, P., Rigoli, M.: Geometric Analysis of Quasilinear Inequalities on Complete Manifolds. Birkhäuser/Springer, Cham (2021).
7. Chern, S.S.: On the curvatures of a piece of hypersurface in Euclidean space, Abh. Math. Sem. Univ. Hamburg **29**, 77–91 (1965).
8. Colombo, G., Mari, L., Rigoli, M.: Remarks on mean curvature flow solitons in warped products, Discrete and Continuous Dyn. Systems, Series S, **13**, 1957–1991 (2020).
9. Concus, P., Finn, R.: On capillary free surfaces in the absence of gravity, Acta Math., **132**, 177–198 (1974).
10. Concus, P., Finn, R.: On capillary free surfaces in a gravitational field, Acta Math., **132**, 207–223 (1974).
11. Ding, Q., Xin, Y.L., Yang, L.: The rigidity theorems of self-shrinkers via Gauss maps, Adv. Math. **303** 151–174 (2016).
12. Finn, R.: Isolated singularities of solutions of non-linear partial differential equations, Trans. Amer. Math. Soc., **75**, 385–404 (1953).
13. Finn, R.: A note on the capillary problem, Acta Math., **132**, 199–205 (1974).
14. Finn, R., Hwang, J.-F.: On the comparison principle for capillary surfaces, J. Fac. Sci. Univ. Tokyo, Sect. IA, Math., **36**, 131–134 (1989).
15. Flanders, H.: Remark on mean curvature, J. London Math. Soc. **41**, 364–366 (1966).
16. Fornari S., Ripoll J.: Killing fields, mean curvature, translations maps, Illinois J. of Math. **48**, 1385–1403 (2004).
17. Grigor'yan, A.: Analytic and geometric background of recurrence and non-explosion of the Brownian motion on Riemannian manifolds, Bull. Amer. Math. Soc. (N.S.) **36** 135–249 (1999).
18. Hwang, J.-F.: Comparison principles and Liouville theorems for prescribed mean curvature equations in unbounded domains, Ann. Scuola Norm. Sup. Pisa Cl. Sci. (4), **15** 341–355 (1988).
19. Mari, L., Mastrolia, P., Rigoli, M.: A note on Killing fields and CMC hypersurfaces, J. Math. Anal. Appl. **431**, 919–934 (2015).
20. Mĭkljukov, V.M.: A new approach to the Bernšteĭn theorem and to related questions of equations of minimal surface type. (Russian) Mat. Sb. (N.S.) **108(150)**, no. 2, 268–289, 304 (1979).
21. Pigola, S., Rigoli, M., Setti, A.G.: Some remarks on the prescribed mean curvature equation on complete manifolds, Pacific J. Math. **206** 195–210 (2002).

22. Pigola, S., Rigoli, M., Setti, A.G.: A remark on the maximum principle and stochastic completeness, Proc. Amer. Math. Soc. **131** 1283–1288 (2003).
23. Salavessa, I.: Graphs with parallel mean curvature, Proc. Amer. Math. Soc. **107**, 449–458 (1989).

The Intrinsic Metric of Constant Mean Curvature Surfaces and Minimal Hypersurfaces with Free Boundary

Lucas Ambrozio

Dedicated to Professor Marcos Dajczer on his 75th birthday.

Abstract Ricci-Curbastro established necessary and sufficient conditions for a Riemannian metric on a surface to be the first fundamental form of a minimal immersion of that surface into the Euclidean space. We revisit certain developments arising from his theorem and propose new versions of these results in the context of the theory of constant mean curvature surfaces in three-dimensional space forms that meet umbilical surfaces orthogonally along their boundary components. Higher-dimensional generalisations, inspired by a theorem of do Carmo and Dajczer, are discussed as well.

In 1981, Marcos Dajczer and Manfredo do Carmo published a short article on the *Boletim da Sociedade Brasileira de Matemática* [4]. The main result of that work was a higher-dimensional generalisation, for minimal hypersurfaces, of classical theorems about minimal surfaces in three-dimensional space forms. All these results trace back to a theorem of Ricci-Curbastro about the first fundamental form of minimal surfaces in $\mathbb{R}^3$ [16].

We want to revisit these theorems with the aim of connecting them to a topic in the theory of minimal surfaces that has attracted great attention in recent years, namely, free boundary minimal surfaces in Euclidean balls [11]. On the way, we establish a correspondence between constant mean curvature surfaces in three-dimensional space forms that meet umbilical surfaces in right angles along their boundary components, and we prove the analogue of do Carmo-Dajczer main theorem in [4] under this sort of boundary condition.

L. Ambrozio (✉)
IMPA, Rio de Janeiro, RJ, Brazil
e-mail: l.ambrozio@impa.br

© The Author(s), under exclusive license to Springer Nature Switzerland AG 2025

R. Tojeiro (ed.), *Advances in Submanifold Theory and Related Topics*,
Latin American Mathematics Series – UFSCar subseries,
https://doi.org/10.1007/978-3-032-00918-0_3

We have not attempted to exhaust the subject. Rather, we decided to focus on explaining a couple of key ideas and on discussing how they fit in current research. In any case, our expectation is that the systematic study of the ideas sketched here only briefly may bear fruits in the future.

It is a pleasure to congratulate Professor Marcos Dajczer on this special occasion!

1 The First Fundamental Form of CMC Immersions

1.1 Motivation

Let (Σ^2, g) be a connected, oriented Riemannian surface without boundary, not necessarily complete. The basic problem in the classical theory of isometric immersions is to identify properties of g that can determine whether or not (Σ^2, g) admits an isometric immersion into the Euclidean space $\mathbb{R}^3$ or, more generally, into one of the (simply connected) space forms $\mathbb{Q}^3(c)$ of constant sectional curvature $c \in \mathbb{R}$.

By the fundamental theorem of isometric immersions (see Theorem 1.11 in [5]), passing to the universal cover of Σ^2, the problem boils down to determine conditions which allow a symmetric bilinear form A to be constructed on Σ^2, out of g and invariant by deck transformations, in such way that the pair (g, A) satisfies the two fundamental equations:

$$(\textit{Gauss equation}): \quad K = c + det(A), \quad \text{and}$$

$$(\textit{Coddazi equation}): \quad (\nabla_X A)(Y, Z) = (\nabla_Y A)(X, Z) \text{ for all } X, Y, Z \in \mathcal{X}(\Sigma),$$

and moreover the right period conditions are met. In the above formulas, K denotes the Gaussian curvature, and ∇ denotes the Levi-Civita connection of (Σ^2, g), while $det(A)$ denotes the product of the two eigenvalues of A.

If such tensor A is found, then A is the second fundamental form of an isometric immersion of (Σ^2, g) into $\mathbb{Q}^3(c)$, which is unique up to orientation preserving isometries of $\mathbb{Q}^3(c)$. (Given the chosen orientations of Σ^2 and $\mathbb{Q}^3(c)$, a unique unit normal vector field N is determined and, up to sign, $A = \langle D_{-}N, -\rangle$, where D is the Levi-Civita connection of $\mathbb{Q}^3(c)$.)

However, the two fundamental equations form a difficult and subtle geometric PDE system for A, which is tricky to address directly.

A more approachable problem is, perhaps, to look for conditions that are necessary and sufficient for the metric g to arise from an isometric immersion of (Σ^2, g) into $\mathbb{Q}^3(c)$ that satisfies, moreover, further meaningful geometric restrictions. For instance, one could naively ask the following basic question:

What is special about the metric g if it is the first fundamental form of a constant mean curvature immersion of Σ^2 into $\mathbb{Q}^3(c)$?

Recall that an isometric immersion has constant mean curvature when H—the *average* of the trace of the second fundamental form A with respect to g—is constant on Σ^2.

The same question can be formulated, of course, also in higher dimensions and codimensions. But in that degree of generality, interesting necessary conditions are less easy to find and even more so sufficient conditions. See [2] for a comprehensive discussion on the case of minimal submanifolds of the Euclidean space $\mathbb{R}^n$.

Since in this article we want to discuss only the case of hypersurfaces, we therefore return to the formulation we started with, turning our attention back to the case of immersed surfaces in $\mathbb{Q}^3(c)$.

1.2 Dimension Two

A rather satisfactory answer to the basic question has been known in the case of *minimal* ($H = 0$) surfaces in $\mathbb{R}^3$ since at least the eighteen hundreds. The original theorem of Ricci-Curbastro [16] was generalised by Lawson in the case of constant mean curvature (CMC) surfaces in the Euclidean sphere $\mathbb{S}^3 = \mathbb{Q}^3(1)$ (see [9], Theorem 8). The generalisation for constant mean curvature surfaces in all three-dimensional space forms is actually straightforward and well known. It reads as follows:

Theorem A (The Ricci Condition) *Assume that (Σ^2, g) with Gaussian curvature K is isometrically immersed into $\mathbb{Q}^3(c)$ with constant mean curvature H. Then, $c+H^2-K$ is non-negative, vanishes either everywhere or on a discrete set of points, and on the complement of its zero set the Riemannian metric $\sqrt{c+H^2-K}\,g$ is flat.*

Conversely, given (Σ^2, g), suppose that there exists a pair of constants (c, H) such that $c + H^2 - K$ is positive everywhere and the Riemannian metric $\sqrt{c+H^2-K}\,g$ is flat. Then, (Σ^2, g) admits, locally, a constant mean curvature H isometric immersion into $\mathbb{Q}^3(c)$.

In the converse statement, we remark that the universal cover of (Σ^2, g) always admits a constant mean curvature H isometric immersion into $\mathbb{Q}^3(c)$.

We want to sketch the proof of the above result, emphasising the role played by the traceless part of the second fundamental form:

$$\mathring{A} = A - Hg,$$

which is conformally invariant, and by a Simons-type identity for the Laplacian of the squared-norm of $\mathring{A}$ on (Σ^2, g), see (3) below.

Proof (⇒) The Gauss equation can be rewritten as

$$K = c + H^2 - \frac{1}{2}|\mathring{A}|^2. \tag{1}$$

Thus, $c + H^2 - K = |\mathring{A}|^2/2$ is non-negative on Σ^2.

The Codazzi equation implies that, in local isothermal coordinates $z = x + iy$ compatible with the Riemann surface structure of (Σ^2, g), the complex function

$$\phi(z) = 2\mathring{A}\left(\frac{\partial}{\partial z}, \frac{\partial}{\partial z}\right) = \frac{1}{2}\left(\mathring{A}\left(\frac{\partial}{\partial x}, \frac{\partial}{\partial x}\right) - \mathring{A}\left(\frac{\partial}{\partial y}, \frac{\partial}{\partial y}\right)\right) - i\mathring{A}\left(\frac{\partial}{\partial x}, \frac{\partial}{\partial y}\right) \tag{2}$$

is holomorphic. Since $\phi(z) = 0$ if, and only if, the traceless tensor $\mathring{A}$ vanishes at z, we conclude that the function $c + H^2 - K = |\mathring{A}|^2/2$ either vanishes everywhere or vanishes on a discrete set of points of Σ^2.

Finally, combining Gauss and Codazzi equations, a straightforward computation of the Laplacian of $|\mathring{A}|^2$ in (Σ^2, g) leads to the identity:

$$\frac{1}{2}\Delta|\mathring{A}|^2 = |\nabla\mathring{A}|^2 + 2K|\mathring{A}|^2. \tag{3}$$

Since, in dimension $n = 2$, any traceless tensor $\mathring{A}$ satisfies

$$|\nabla\mathring{A}|^2 = 2|\nabla|\mathring{A}||^2$$

at points where $|\mathring{A}| \neq 0$, equation (3) can be rewritten as

$$\Delta\log(|\mathring{A}|) = 2K$$

on the set of points where $|\mathring{A}| \neq 0$.

Recalling the equation that relates the Gaussian curvature of conformal metrics $h = e^{2u}g$, namely,

$$-\Delta_g u + K_g = K_h e^{2u}, \tag{4}$$

we conclude that the conformal factor $e^{2u} = |\mathring{A}|$ turns g into a flat metric. The same can be said about the conformal factor $\sqrt{c + H^2 - K} = |\mathring{A}|/\sqrt{2}$.

(⇐) Consider local isothermal coordinates $z = x + iy$ on a disc in (Σ^2, g), compatible with the underlying Riemann surface structure, so that $g = e^{2u}(dx^2 + dy^2)$ for some smooth function u. By hypothesis, $\sqrt{c + H^2 - K}g$ is a flat Riemannian metric. By equation (4), the assumption is therefore equivalent to the equation:

$$\Delta(\log(c + H^2 - K)e^{4u}) = 0,$$

where $\Delta = \partial_x^2 + \partial_y^2$. By Lemma 3 (see the Appendix), there exists a holomorphic function $\phi(z)$ on this coordinate neighbourhood such that

$$|\phi|^2 = (c + H^2 - K)e^{4u}.$$

Next, define a two-tensor A so that, on this coordinate neighbourhood

$$A(\partial_x, \partial_x) = Re(\phi) + He^{2u},$$

$$A(\partial_y, \partial_y) = -Re(\phi) + He^{2u},$$

$$A(\partial_x, \partial_y) = -Im(\phi) = A(\partial_y, \partial_x).$$

Then, A defines a symmetric two-tensor, with average trace H, whose traceless part has squared-norm $2|\phi|^2 e^{-4u}$ with respect to the metric $g = e^{2u}(dx^2 + dy^2)$. (We may use $\{e^{-u}\partial_x, e^{-u}\partial_y\}$ as g-orthonormal frame to perform these computations.)

A long but straightforward computation then shows that (g, A) satisfy both the Gauss equation (rewritten as in (1)) and the Codazzi equation for a constant mean curvature H isometric immersion into $\mathbb{Q}^3(c)$. By the fundamental theorem of isometric immersions, this simply connected coordinate neighbourhood of (Σ^2, g) isometrically immerses into $\mathbb{Q}^3(c)$ with constant mean curvature H. □

Remark 1 Lawson gave an example of Riemannian metric g on $\mathbb{R}^2$ such that its Gaussian curvature is negative everywhere except at the origin, such that the Riemannian metric $\sqrt{-K}g$ is flat where it is defined, and yet $(\mathbb{R}^2, g)$ does not isometrically immerse as a minimal surface into $\mathbb{R}^3$ (see [9], Remark 12.1). Thus, in a sense, Theorem A is sharp. Nevertheless, it is not difficult to see what fails in his example. As soon as one understand the special structure of the zeroes of K—namely, that they are the zeroes of $|\phi|^2$ for some holomorphic function ϕ in local isothermal coordinates—one can figure out the right extra necessary condition that rules out Lawson's examples. This was done by Jost-Hinrich Eschenburg and Renato Tribuzy in [7], Theorem 0. This observation has been explored also by Andrei Moroianu and Sergiu Moroianu [14], who introduced recently the very apt concept of *Ricci surfaces*. We will discuss this notion further in Sect. 3.1.

1.3 The Lawson Correspondence

There is an intriguing observation regarding the converse statement of Theorem A. If different pairs of constants (c, H) and $(\tilde{c}, \tilde{H})$ satisfy $c + H^2 = \tilde{c} + \tilde{H}^2$, then the sufficient condition that needs to be verified is satisfied by the pair (c, H) if, and only if, it is satisfied by the pair $(\tilde{c}, \tilde{H})$. This means that every simply connected CMC surface in a space form is isometric to a CMC "cousin" surface with different constant mean curvature in a different space form.

This correspondence is known as the *Lawson correspondence*. It has motivated many interesting developments, for instance, in the theory of CMC $H = 1$ surfaces in the hyperbolic space (also known as Bryant surfaces after [1]), which are simply the cousins to minimal surfaces in $\mathbb{R}^3$.

Example 1 Complete minimal surfaces of revolution in the Euclidean space $\mathbb{R}^3$ have been classified long ago: they are the *catenoids*. Up to isometries and dilations of $\mathbb{R}^3$, there exists only one catenoid. Robert Bryant showed that catenoids are in correspondence with complete CMC $H = 1$ surfaces of revolution in the hyperbolic space $\mathbb{H}^3 = \mathbb{Q}^3(-1)$, called the *catenoid cousins* [1]. Some catenoid cousins are embedded, while others are immersed but not embedded. It is interesting to notice that dilated Euclidean catenoids correspond to very different hyperbolic catenoid cousins.

1.4 Higher Dimensions

Let (Σ^n, g) be a connected, oriented Riemannian manifold without boundary, not necessarily complete, that is isometrically immersed into the space form $\mathbb{Q}^{n+1}(c)$. In dimensions $n \geq 3$, while the Codazzi equation for the second fundamental form A is the same, the Gauss equation for the Riemman curvature tensor of g and its traces read as

$$Rm = \frac{1}{2} c\, g \odot g + \frac{1}{2} A \odot A, \tag{5}$$

$$Ric = c(n-1)g + nHA - A \circ A, \tag{6}$$

$$R = cn(n-1) + n^2H^2 - |A|^2.$$

In (5), $\odot$ denotes the Kulkarni-Nomizu product of symmetric two-tensors, while in (6), $A \circ A$ denotes the symmetric two-tensor $(A \circ A)(X, Y) = \sum_{i=1}^{n} A(X, e_i)A(e_i, Y)$ for every $X, Y \in \mathcal{X}(\Sigma)$, where $\{e_i\}$ is any local g-orthonormal frame.

It will be convenient to have the following notation at hand. Let $\hat{A}$ be the (fiberwise) endomorphism of the tangent bundle defined by

$$g(\hat{A}(X), Y) = A(X, Y) \quad \text{for every} \quad X, Y \in \mathcal{X}(\Sigma).$$

Then, $\hat{A}$ is symmetric with respect to g and

$$(A \circ A)(X, Y) = g(\hat{A}(X), \hat{A}(Y)) \quad \text{for every} \quad X, Y \in \mathcal{X}(\Sigma).$$

Also, by construction

$$(\nabla_X A)(Y, Z) = g((\nabla_X \hat{A})(Y), Z) \quad \text{for every} \quad X, Y, Z \in \mathcal{X}(\Sigma).$$

Rewriting the equation (6) in terms of the traceless part $\mathring{A} = A - Hg$ of $\mathring{A}$, we obtain

$$Ric = (n-1)(c+H^2)g + (n-2)H^2\mathring{A} - \mathring{A} \circ \mathring{A}.$$

Thus, in contrast with the case $n = 2$ (where $Ric = Kg$), except when H vanishes identically, it is not clear whether the symmetric two-tensor $(n-1)(c+H^2)g - Ric$ is non-negative or not. (At least, it does not need to be so from a purely *algebraic* point of view, even when assuming H is constant.)

In order to continue the analysis, we therefore restrict our attention to the case $H = 0$. Chern and Osserman [2] analysed the case $c = 0$ in detail, but the final statement, for arbitrary values of c, is precisely the theorem proven by Dajczer and do Carmo in [4]. We state it as follows:

Theorem B (The Osserman-do Carmo-Dajczer Condition) *Suppose that (Σ^n, g) is isometrically immersed into $\mathbb{Q}^{n+1}(c)$ as a minimal hypersurface, and denote by A its second fundamental form. Then, the symmetric two-tensor $\overline{g} = c(n-1)g - Ric_g$ is non-negative, and wherever it is positive definite, it is a Riemannian metric such that:*

i) $\overline{g}(\overline{\nabla}_X Y, Z) = g(\nabla_X(\hat{A}(Y)), \hat{A}(Z))$ *for every* $X, Y, Z \in \mathcal{X}(\Sigma)$.
ii) $Rm_{\overline{g}} = \frac{1}{2}\overline{g} \odot \overline{g} + \frac{c}{2} A \odot A$.

Conversely, assume that (Σ^n, g) is such that $\overline{g} = c(n-1)g - Ric_g$ is a Riemannian metric on Σ^n and that there exists a smooth symmetric two-tensor A on (Σ^n, g) such that $\overline{g} = A \circ A$ satisfies properties $i)$ and $ii)$ above. Then, (Σ^n, g) admits, locally, a minimal isometric immersion into $\mathbb{Q}^{n+1}(c)$.

When $c = 0$, the condition $ii)$ above holds if, and only if, $\overline{g}$ is a metric of constant sectional curvature 1. In fact, in this case, $\overline{g}$ is nothing but the pulled-back metric from the sphere $\mathbb{S}^n(1) \subset \mathbb{R}^{n+1}$ by the Gauss map N.

The original proof in [4] was formulated in the language of Cartan's *repère mobile* method. We sketch a proof below using common tensor language, as in [5], Section 3.6.

Proof ($\Rightarrow$) The second fundamental form A has average trace $H = 0$ and, by the Gauss equation (6)

$$(n-1)cg - Ric_g = A \circ A$$

is a non-negative symmetric tensor, because $(A \circ A)(X, X) = |\hat{A}(X)|^2$ for all tangent vector fields X on (Σ^n, g).

Consider the set where $\overline{g} = (n-1)cg - Ric_g = A \circ A$ is positive definite. It is the same set as the set of points where $\hat{A}$ is invertible. By a straightforward computation using the Koszul equation, the Levi-Civita connection $\overline{\nabla}$ of $\overline{g}$ satisfies

$$2\overline{g}(\overline{\nabla}_X Y, Z) = 2g(\nabla_X(\hat{A}(Y)), \hat{A}(Z)) \tag{7}$$

$$
\begin{aligned}
&+ \langle(\nabla_X \hat{A})(Z) - (\nabla_Z \hat{A})(X), \hat{A}(Y)\rangle \\
&+ \langle(\nabla_Y \hat{A})(Z) - (\nabla_Z \hat{A})(Y), \hat{A}(X)\rangle \\
&- \langle(\nabla_X \hat{A})(Y) - (\nabla_Y \hat{A})(X), \hat{A}(Z)\rangle.
\end{aligned}
$$

Thus, the Codazzi equation implies i).

Another straightforward computation using i) and the definition of the Riemann curvature tensor $\overline{Rm}$ of $\overline{g}$ gives

$$
\overline{Rm}(X, Y, X, Y) = Rm(X, Y, \hat{A}(X), \hat{A}(Y)) \text{ for every } X, Y \in \mathcal{X}(\Sigma). \tag{8}
$$

By the Gauss equation (5)

$$
\begin{aligned}
\overline{Rm}(X, Y, X, Y) &= c(g(X, \hat{A}(X))g(Y, \hat{A}(Y)) - g(X, \hat{A}(Y))g(Y, \hat{A}(X))) \\
&\quad + A(X, \hat{A}(X))A(Y, \hat{A}(Y)) - A(X, \hat{A}(Y))A(Y, \hat{A}(X)). \\
&= c(A(X, X)A(X, Y) - A(X, Y)^2) \\
&\quad + \overline{g}(X, X)\overline{g}(Y, Y) - \overline{g}(X, Y)^2.
\end{aligned}
$$

Since X and Y are arbitrary (and a four-tensor T with the same symmetries of the Riemann curvature tensor is uniquely determined by the values $T(X, Y, X, Y)$), ii) follows.

($\Leftarrow$) By definition of A, $|\hat{A}(X)|^2 = (A \circ A)(X, X) = \overline{g}(X, X)$ for every tangent vector X. Since, by assumption, $\overline{g}$ is positive definite, $\hat{A}$ is invertible.

Using the symmetry of the Levi-Civita connection $\overline{\nabla}$ of $\overline{g}$, equation (7) together with i) implies that for every tangent vectors X, Y, and Z on Σ^n

$$
\begin{aligned}
0 &= \overline{g}(\overline{\nabla}_X Y - \overline{\nabla}_Y X - [X, Y], \hat{A}^{-1}(Z)) \\
&= g(\nabla_X(\hat{A}(Y)) - \nabla_Y(\hat{A}(X)) - \hat{A}([X, Y]), Z) \\
&= g((\nabla_X \hat{A})(Y) - (\nabla_Y \hat{A})(X), Z) \\
&= (\nabla_X A)(Y, Z) - (\nabla_Y A)(X, Z).
\end{aligned}
$$

Hence, A satisfies the Codazzi equation on (Σ^n, g).

As seen above, equation (8) is a consequence of the definition of $\overline{g}$ and A and together with i). Since $\hat{A}$ is invertible, (8) and ii) imply that

$$
\begin{aligned}
Rm(X, Y, X, Y) =& \overline{Rm}(X, Y, \hat{A}^{-1}(X), \hat{A}^{-1}(Y)) \\
=& \overline{g}(X, \hat{A}^{-1}(X))\overline{g}(Y, \hat{A}^{-1}(Y)) - \overline{g}(X, \hat{A}^{-1}(Y))\overline{g}(Y, \hat{A}^{-1}(X)) \\
& + cA(X, \hat{A}^{-1}(X))A(Y, \hat{A}^{-1}(Y))
\end{aligned}
$$

$$- cA(X, \hat{A}^{-1}(Y))A(Y, \hat{A}^{-1}(X))$$
$$= g(\hat{A}(X), X)g(\hat{A}(Y), Y) - g(\hat{A}(X), Y)g(\hat{A}(Y), X)$$
$$+ cg(X, X)g(Y, Y) - cg(X, Y)^2$$

for all $X, Y \in \mathcal{X}(\Sigma)$. In other words, A satisfies the Gauss equation (5) for an isometric immersion of (Σ^n, g) into $\mathbb{Q}^{n+1}(c)$.

Therefore, the fundamental theorem of isometric immersions applies for the pair (g, A) on any simply connected neighbourhood in Σ and provides local isometric immersions of (Σ^n, g) into $\mathbb{Q}^{n+1}(c)$ with second fundamental form A. *A posteriori*, since $Ric_g = c(n-1)g - A \circ A$ (by assumption) as well as $Ric_g = c(n-1)g + nHA - A \circ A$ (by (6)), and since $\hat{A}$ is invertible, we must have $H = 0$. In other words, these local isometric immersions are minimal. □

Remark 2 For minimal immersed hypersurfaces in $\mathbb{Q}^{n+1}(c)$, $n \geq 3$, the structure of the set where $c(n-1)g - Ric_g = A \circ A$ ceases to be positive definite has a more complicated structure. In three dimensions, $A \circ A$ is not positive definite at a point p if, and only if, either $\hat{A}$ has a one-dimensional kernel or vanishes at p. In higher dimensions, the situation is of course much wilder. (See, however, the discussions in [2] about the specific case $c = 0$.) We think that it could be interesting to understand this behaviour somehow, specially in three dimensions, because this understanding could lead to the formulation of a meaningful higher-dimensional generalisation of the concept of Ricci surfaces in the sense of [14], *cf.* Sect. 3.1.

2 The Free Boundary Condition

After this overview, let us consider isometric immersions of a connected, oriented Riemannian manifold (Σ^n, g) with boundary $\partial\Sigma \neq \emptyset$, not necessarily complete. Our aim is to find analogues of Theorems A and B under geometrically reasonable boundary conditions on the pair (g, A), added to the Gauss and Codazzi equations.

Before we formulate these boundary conditions, let us fix some notation once and for all. The unit outward-pointing conormal of $\partial\Sigma$ inside Σ^n will be denoted by ν. If $\partial_i\Sigma$ is a boundary component of Σ^n, we denote by $B_i = g(\nabla_{-}\nu, -)$ its second fundamental form in (Σ^n, g).

2.1 Umbilical Hypersurfaces and Capillary Boundary Conditions

Inside any space form $\mathbb{Q}^{n+1}(c)$, there are distinguished hypersurfaces called *umbilical*, which are defined by the property that their second fundamental form is a multiple of the metric at every point. Connected umbilical hypersurfaces in $\mathbb{Q}^{n+1}(c)$ have therefore constant mean curvature, and they are easily classified. In fact, connected umbilical hypersurfaces are just the standard embeddings of space forms $\mathbb{Q}^n(b)$ in $\mathbb{Q}^{n+1}(c)$, which exist only when $b \geq c$. (See Proposition 1.20 in [5].)

Denote by $\overline{N}$ the unit normal to an umbilical hypersurface $\mathbb{Q}^n(b)$ of $\mathbb{Q}^{n+1}(c)$, chosen in such way that its mean curvature $\overline{H}$ is non-negative. The relation between the ambient sectional curvature c, the intrinsic sectional curvature b, and the mean curvature $\overline{H}$ of the umbilical immersion of $\mathbb{Q}^n(b)$ into $\mathbb{Q}^{n+1}(c)$ is

$$b = c + \overline{H}^2.$$

(Notice that $\overline{H} = 0$ if, and only if, $b = c$, in which case the second fundamental form vanishes identically, by virtue of the Gauss equation (5), the choice of $\overline{N}$ is ambiguous in this case and can be determined only by a further choice of orientation.)

The geometric situation that interests us is when a hypersurface Σ^n of $\mathbb{Q}^{n+1}(c)$ intersects an umbilical hypersurface $\mathbb{Q}^n(b_i)$, transversally along a boundary component $\partial_i\Sigma$. Moreover, we want to suppose that the angle between ν and the unit normal $\overline{N}$ to $\mathbb{Q}^n(b_i)$ is constant along $\partial_i\Sigma$. This boundary condition is known as the *capillary condition*. In the particular case where $\overline{N} = \nu$ or $\overline{N} = -\nu$ on $\partial_i\Sigma$, that is, when Σ^n meets $\mathbb{Q}^n(b_i)$ orthogonally along $\partial_i\Sigma$, we say that Σ^n is *free boundary* with respect to $\mathbb{Q}^n(b_i)$ along $\partial_i\Sigma$.

Since $\partial_i\Sigma$ is contained in $\mathbb{Q}^n(b_i)$, there exists a unique unit vector field $\overline{\nu}$ along $\partial_i\Sigma$, tangent to $\mathbb{Q}^n(b_i)$ but normal to $\partial_i\Sigma$, in such way that $\overline{N}$ and $\overline{\nu}$ (in that order) form together another orthonormal normal frame along $\partial_i\Sigma$ that induces the same orientation on the normal bundle of $\partial_i\Sigma$ as N and ν (in that order). Assuming the capillary condition, there exists moreover a constant $\alpha_i \in (0, \pi) \cup (\pi, 2\pi)$ such that

$$\overline{N} = \cos(\alpha_i)N + \sin(\alpha_i)\nu,$$
$$\overline{\nu} = -\sin(\alpha_i)N + \cos(\alpha_i)\nu.$$

Equations (9) below will be key to the results we are searching for.

Lemma 1 *Under the capillary conditions, for every tangent vector fields X and Y on $\partial_i\Sigma$, we have:*

i) $A(X, \nu) = 0$.
ii) $\overline{H}g(X, Y) = \cos(\alpha_i)A(X, Y) + \sin(\alpha_i)B_i(X, Y)$.

In particular, in the free boundary case

$$A(X, \nu) = 0 \quad \text{and} \quad B_i = \pm\sqrt{b_i - c}\,g, \tag{9}$$

where the sign of the trace of B_i depends on whether $\nu = \overline{N}$ or $\nu = -\overline{N}$.

Proof Let X be a vector field tangent to $\partial_i \Sigma$. The two vector fields X and $\overline{\nu}$ are tangent to $\mathbb{Q}^n(b_i)$ and orthogonal between themselves. Since $\mathbb{Q}^n(b_i)$ is an umbilical hypersurface

$$\langle D_X \overline{N}, \overline{\nu} \rangle = \overline{H} \langle X, \overline{\nu} \rangle = 0.$$

On the other hand, substituting the expressions of $\overline{N}$ and $\overline{\nu}$ in terms of the orthonormal frame N and ν and using that α_i is constant

$$\begin{aligned}\langle D_X \overline{N}, \overline{\nu} \rangle &= \cos(\alpha_i)^2 \langle D_X N, \nu \rangle - \sin(\alpha_i)^2 \langle D_X \nu, N \rangle \\ &= \langle D_X N, \nu \rangle = A(X, \nu),\end{aligned}$$

by definition of A. Item $i)$ follows.

Item $ii)$ is also a consequence of a direct computation. For every vector fields X and Y tangent to $\partial_i \Sigma$

$$\begin{aligned}\langle D_X \overline{N}, Y \rangle &= \cos(\alpha_i) \langle D_X N, Y \rangle + \sin(\alpha_i) \langle D_X \nu, Y \rangle \\ &= \cos(\alpha_i) A(X, Y) + \sin(\alpha_i) g(\nabla_X \nu, Y).\end{aligned}$$

In particular, when $\alpha_i = \pi/2$ or $\alpha_i = 3\pi/2$, we have

$$B_i(X, Y) = g(\nabla_X \nu, Y) = \pm \langle D_X \overline{N}, Y \rangle = \pm \overline{H} g(X, Y),$$

because $\mathbb{Q}^n(b_i)$ is umbilical in $\mathbb{Q}^{n+1}(c)$. Since $b_i = c + \overline{H}^2$ and we have chosen orientations in such way that $\overline{H} \geq 0$, the result follows. □

In dimension $n = 2$, we can proceed our analysis of the free boundary condition a bit further and derive properties of the squared-norm of $\mathring{A}$ along the boundary that will be crucial for our reasoning.

Denote by T a unit tangent vector field along the boundary component $\partial_i \Sigma$, so that T and ν form an orthonormal frame along $\partial_i \Sigma$. Since $\mathring{A}$ has zero trace with respect to g by definition

$$\mathring{A}(T, T) + \mathring{A}(\nu, \nu) = 0. \tag{10}$$

Moreover, by Lemma 1

$$\mathring{A}(T, \nu) = A(T, \nu) + Hg(T, \nu) = 0. \tag{11}$$

Therefore

$$|\mathring{A}|^2 = \mathring{A}(T, T)^2 + \mathring{A}(\nu, \nu)^2 + 2\mathring{A}(T, \nu)^2 = 2\mathring{A}(T, T)^2. \tag{12}$$

Since $\nabla_T T$ is ortogonal to T, $\nabla_T \nu$ is orthogonal to ν, and these vector fields satisfy

$$g(\nabla_T T, \nu) = -g(T, \nabla_T \nu) = -B_i(T, T) = \mp\sqrt{b_i - c}$$

because of Lemma 1, we have

$$\nabla_T \nu = \pm\sqrt{b_i - c}\, T \quad \text{and} \quad \nabla_T T = \mp\sqrt{b_i - c}\, \nu. \tag{13}$$

Lemma 2 *Under the free boundary condition along $\partial_i \Sigma \subset \mathbb{Q}^2(b_i)$*

$$-\frac{\partial}{\partial \nu}|\mathring{A}|^2 = \pm 4\sqrt{b_i - c}|\mathring{A}|^2 \quad on \quad \partial_i \Sigma, \tag{14}$$

where the sign depends on whether $\nu = \overline{N}$ or $\nu = -\overline{N}$.

Proof Using (10) and (11)

$$\begin{aligned}\frac{1}{2}\frac{\partial}{\partial \nu}|\mathring{A}|^2 &= g(\nabla_\nu \mathring{A}, \mathring{A})\\ &= (\nabla_\nu \mathring{A})(T, T)\mathring{A}(T, T) + (\nabla_\nu \mathring{A})(\nu, \nu)\mathring{A}(\nu, \nu)\\ &\quad + 2(\nabla_\nu \mathring{A})(T, \nu)\mathring{A}(T, \nu)\\ &= 2(\nabla_\nu \mathring{A})(T, T)\mathring{A}(T, T) = 2(\nabla_T \mathring{A})(\nu, T)\mathring{A}(T, T),\end{aligned}$$

by the Codazzi equation. (Since H is constant, $\nabla A = \nabla \mathring{A}$.)

By definition

$$(\nabla_T \mathring{A})(\nu, T) = T\mathring{A}(\nu, T) - \mathring{A}(\nabla_T \nu, T) - \mathring{A}(\nu, \nabla_T T).$$

Substituting (11), (12), (10), and (13) successively, we obtain

$$\begin{aligned}-\frac{\partial}{\partial \nu}|\mathring{A}|^2 &= \pm 4\sqrt{b_i - c}\left(\mathring{A}(T, T) - \mathring{A}(\nu, \nu)\right)\mathring{A}(T, T)\\ &= \pm 8\sqrt{b_i - c}\mathring{A}(T, T)^2 = \pm 4\sqrt{b_i - c}|\mathring{A}|^2.\end{aligned}$$

□

2.2 Dimension Two

Let (Σ^2, g) be a connected, oriented Riemannian surface with non-empty boundary $\partial\Sigma$, not necessarily complete. Our generalisation of Theorem A under the free boundary condition reads as follows:

Theorem C *Assume that (Σ^2, g) with Gaussian curvature K is isometrically immersed into $\mathbb{Q}^3(c)$ with constant mean curvature H and in such way that it intersects a collection of umbilical $\mathbb{Q}^2(b_i)$ orthogonally along the boundary components $\partial_i\Sigma$. Then, $c + H^2 - K$ is non-negative, vanishes on a discrete set of points, and on the complement of its zero set the Riemannian metric $\sqrt{c + H^2 - K}\,g$ is flat and such that each $\partial_i\Sigma$ has zero geodesic curvature with respect to it.*

Conversely, let (Σ^2, g) be such that its boundary components $\partial_i\Sigma$ have constant geodesic curvature $k_i = \pm\sqrt{b_i - c}$ for some constants $b_i \geq c$, and assume that there exists a constant H such that the function $c + H^2 - K$ is positive everywhere, that the Riemannian metric $\sqrt{c + H^2 - K}\,g$ is flat, and that the boundary components are geodesics with respect to that metric. Then, (Σ^2, g) admits, locally, constant mean curvature H isometric immersions into $\mathbb{Q}^3(c)$ that meet a collection of umbilical $\mathbb{Q}^2(b_i)$ orthogonally on points of the boundary component $\partial_i\Sigma$.

Proof ($\Rightarrow$) In the proof of Theorem A, we have already analysed the neighbourhoods of points in $\Sigma \setminus \partial\Sigma$. Consider isothermal coordinates $z = x + iy$ in a neighbourhood of a boundary point in $\partial_i\Sigma$, compatible with the underlying Riemann surface structure of (Σ^2, g), in such way that the points with $y > 0$ are interior points and $\{y = 0\}$ parametrises boundary points.

In this coordinate system, at each point, ν is parallel to $-\partial_y$ and ∂_x is tangent to $\partial_i\Sigma$ on $y = 0$. By Lemma 1, $A(\partial_x, \partial_y) = 0$ on $y = 0$. Hence, the holomorphic function ϕ defined in (2) is such that its imaginary part vanishes at $y = 0$. By Schwartz reflection, ϕ extends as a holomorphic function in the reflected domain across $\{y = 0\}$. In particular, ϕ vanishes at isolated boundary points or vanishes identically. Since ϕ vanishes precisely at points where $\mathring{A} = 0$, the same can be said about the zero set of $\sqrt{c + H^2 - K} = |\mathring{A}|^2/2$.

At boundary points where $\mathring{A} \neq 0$, the conformal factor $e^{2u} = |\mathring{A}|$, which turns $\overline{g} = e^{2u}g$ into a flat metric as in the proof of Theorem A, also satisfies

$$\frac{\partial}{\partial\nu}\log|\mathring{A}| \pm 2\sqrt{b_i - c^2} = 0,$$

by virtue of Lemma 2. Recalling the equation that relates the geodesic curvature of $\partial_i\Sigma$ in Σ^2 (with respect to ν) for conformal metrics $h = e^{2u}g$, namely,

$$\frac{\partial u}{\partial\nu_g} + k_g = k_h e^u, \tag{15}$$

and since the geodesic curvature of $\partial_i\Sigma$ in (Σ^2, g) is precisely $\pm\sqrt{b_i - c}$ by virtue of Lemma 1, the geodesic curvature of $\partial_i\Sigma$ with respect to the conformal metric $|\mathring{A}|g$ is zero. The same can be said about $\partial_i\Sigma$ with respect to the conformal metric $\overline{g} = \sqrt{c + H^2 - K}g = (|\mathring{A}|/\sqrt{2})g$.

($\Leftarrow$) For simply connected neighbourhoods of interior points of Σ^2, we argue as in the proof of Theorem A. Consider then local isothermal coordinates $z = x+iy$ on a half-disc on (Σ^2, g), compatible with the underlying Riemann surface structure, so that interior points correspond to $y > 0$ and boundary points of $\partial_i\Sigma$ correspond to $y = 0$. We have $g = e^{2u}(dx^2 + dy^2)$ for a smooth function u.

By assumption, $(\Sigma^2, \sqrt{c + H^2 - K}g)$ has zero Gaussian curvature and its boundary has zero geodesic curvature. Equivalently, in view of (4) and (15)

$$\Delta \log((c + H^2 - K)e^{4u}) = 0 \quad \text{and} \quad \frac{\partial}{\partial y}\log((c + H^2 - K)e^{4u}) = 0,$$

where $\Delta = \partial_x^2 + \partial_y^2$. By Lemma 3 (see Appendix), there exists a holomorphic function $\phi(z)$ on this coordinate neighbourhood that has zero imaginary part on $\{y = 0\}$ and is such that

$$|\phi|^2 = (c + H^2 - K)e^{4u}.$$

As in the proof of Theorem A, let A be the two-tensor defined in this coordinates system as follows:

$$A(\partial_x, \partial_x) = Re(\phi) + He^{2u},$$
$$A(\partial_y, \partial_y) = -Re(\phi) + He^{2u},$$
$$A(\partial_x, \partial_y) = -Im(\phi) = A(\partial_y, \partial_x).$$

Then, A is a symmetric two-tensor, with average trace H, whose traceless part has squared-norm $2|\phi|^2e^{-4u}$ with respect to the metric $g = e^{2u}(dx^2 + dy^2)$. Moreover, since ϕ has zero imaginary part on $\{y = 0\}$, $A(\partial_x, \partial_y) = 0$ on this set.

As in the proof of Theorem C, the pair (g, A) satisfies the Gauss and Codazzi equations for a constant mean curvature H isometric immersion into $\mathbb{Q}^3(c)$ on this neighbourhood. By the fundamental theorem of isometric immersions, this simply connected region isometrically immerses into $\mathbb{Q}^3(c)$ with second fundamental form A and constant mean curvature H. Denote by N the unit normal with $A = \langle D_- N, -\rangle$, and by T a unit vector field tangent to $\partial_i\Sigma$ at points of this region.

It remains to compute the image of the points of $\partial_i\Sigma$ under this immersion. First of all, notice that $A(T, \nu) = 0$ along points of $\partial\Sigma$ in this neighbourhood. In fact, $\nu = -e^{-u}\partial_y$ and $T = \pm e^{-u}\partial_x$ at each boundary point, so that the vanishing of $A(T, \nu)$ follows from the definition of A and the vanishing of the imaginary part of ϕ.

It follows that ν is a parallel section of the normal bundle of $\partial_i \Sigma$ in $\mathbb{Q}^3(c)$. In fact, $D_T \nu$ has no normal component because $D_T \nu$ is orthogonal to ν, and $\langle D_T \nu, N\rangle = -\langle D_T N, \nu\rangle = -A(T, \nu) = 0$.

Moreover, $\langle D_T \nu, T\rangle = g(\nabla_T \nu, T)$ is the geodesic curvature of $\partial_i \Sigma$ in (Σ^2, g), which is constant and equal to $\pm\sqrt{b_i - c}$ by assumption.

Hence, standard results imply that the boundary points in question lie inside an umbilical $\mathbb{Q}^2(x_i)$ in $\mathbb{Q}^{n+1}(c)$, whose Gaussian curvature is $x_i = c + (\pm\sqrt{b_i - c})^2 = b_i$, and moreover, ν is normal to $\mathbb{Q}^2(x_i)$ (see Exercise 2.9 of [5] or Theorem 1 of [18]). In other words, Σ meets an umbilical $\mathbb{Q}^2(b_i)$ orthogonally along the boundary points of $\partial_i \Sigma$ in this region, as we wanted to prove. □

2.3 *The Free Boundary Correspondence*

Theorem A implies a version of Lawson's correspondence for CMC surfaces in three-dimensional space forms meeting umbilical surfaces orthogonally.

The Free Boundary Correspondence Let $(c, H, \{b_i\})$, $(\tilde{c}, \tilde{H}, \{\tilde{b}_i\})$ be ordered real numbers such that $c + H^2 = \tilde{c} + \tilde{H}^2$ and $b_i - c = \tilde{b}_i - \tilde{c} \geq 0$ for all i. If a simply connected surface with boundary (Σ^2, g) admits a constant mean curvature H isometric immersion into $\mathbb{Q}^3(c)$ meeting a collection of umbilical surface $\mathbb{Q}^2(b_i)$ orthogonally along a boundary component $\partial_i \Sigma$, then it also admits a constant mean curvature $\tilde{H}$ isometric immersion into $\mathbb{Q}^3(\tilde{c})$ meeting a collection of umbilical $\mathbb{Q}^2(\tilde{b}_i)$ orthogonally along the boundary component $\partial_i \Sigma$.

Example 2 Free boundary minimal surfaces of revolution with free boundary in the Euclidean sphere of radius one centred at the origin have also been classified: up to rotations, there is only one piece of catenoid with this property. This piece is called the *critical catenoid*. According to the above correspondence, the critical catenoid is also isometrically immersed into the hyperbolic space as a piece of a catenoid cousin that meets (possibly different) *horospheres* orthogonally along each boundary component.

2.4 *Higher Dimensions*

Let (Σ^n, g) denote a connected, oriented Riemannian surface with non-empty boundary $\partial\Sigma$, not necessarily complete. The higher-dimensional generalisation of Theorem B reads as follows:

Theorem D *Assume (Σ^n, g) is isometrically immersed into $\mathbb{Q}^{n+1}(c)$ as a minimal hypersurface that intersects a collection of umbilical $\mathbb{Q}^n(b_i)$ orthogonally along the boundary components $\partial_i \Sigma$, and denote by A its second fundamental form. Then, each boundary component $\partial_i \Sigma$ is umbilical in (Σ^n, g) with second fundamental* *form* $B_i = \pm\sqrt{b_i - c}\, g$ *with respect to the outward-pointing unit conormal* ν. *Moreover,* $A(X, \nu) = 0$ *for all vector fields* X *tangent to* $\partial\Sigma$. *Finally, the symmetric two-tensor* $\overline{g} = c(n-1)g - Ric_g$ *is non-negative and, wherever it is positive definite, it is a Riemannian metric such that:*

(i) $\overline{g}(\overline{\nabla}_X Y, Z) = g(\nabla_X(\hat{A}(Y)), \hat{A}(Z))$ *for every* $X, Y, Z \in \mathcal{X}(\Sigma)$.
(ii) $Rm_{\overline{g}} = \frac{1}{2}\overline{g} \odot \overline{g} + \frac{c}{2} A \odot A$.

Conversely, assume that (Σ^n, g) *has boundary components* $\partial_i\Sigma$ *that are umbilical in* (Σ^n, g) *with* $B_i = \pm\sqrt{b_i - c}\, g$ *for some constants* $b_i \geq c$, *that the symmetric tensor* $\overline{g} = c(n-1)g - Ric_g$ *is a Riemannian metric on* Σ^n, *and that there exists a smooth symmetric two-tensor* A *on* Σ^n, *with* $A(X, \nu) = 0$ *for every vector field* X *tangent to* $\partial\Sigma$, *such that* $\overline{g} = A \circ A$ *satisfies conditions* (i) *and* (ii). *Then,* (Σ^n, g) *admits, locally, minimal isometric immersions into* $\mathbb{Q}^{n+1}(c)$ *that intersect a collection of umbilical* $\mathbb{Q}^n(b_i)$ *in right angles along boundary points in* $\partial_i\Sigma$.

Proof ($\Rightarrow$) The necessity of all these properties is clear from the proof of the boundary-less case (Theorem B) and Lemma 1.

($\Leftarrow$) Given the hypotheses, the proof of Theorem B guarantees that the pair (g, A) satisfies the Gauss and Codazzi equations for a minimal isometric immersion into $\mathbb{Q}^{n+1}(c)$. Thus, every point of Σ^n, including boundary points, belongs to a simply connected neighbourhood that admits a minimal isometric immersion into $\mathbb{Q}^n(c)$. It remains only to understand how the boundary components $\partial_i\Sigma$ are mapped inside $\mathbb{Q}^{n+1}(c)$ by these immersions.

Denote by N the unit normal so that $A = \langle D_- N, - \rangle$. The unit conormal ν of $\partial_i\Sigma$ in Σ^n and the unit normal N of Σ^n along $\partial_i\Sigma$ form an orthonormal frame of the normal bundle of $\partial_i\Sigma$ in $\mathbb{Q}^{n+1}(c)$. The section ν satisfies two properties:

(a) ν is parallel with respect to the normal connection. In fact, by assumption

$$\langle D_X \nu, N \rangle = -\langle D_X N, \nu \rangle = -A(X, \nu) = 0$$

for all vector fields X tangent to $\partial_i\Sigma$. Also, $\langle \nabla_X \nu, \nu \rangle = X|\nu|^2/2 = 0$, so that there is no component of $\nabla_X \nu$ in the direction of ν either. Hence, the normal component of $\nabla_X \nu$ is zero for all such vector fields X.

(b) By the assumption that $\partial_i\Sigma$ is an umbilical hypersurface of (Σ^n, g), for all vector fields X and Y tangent to $\partial_i\Sigma$

$$\langle D_X \nu, Y \rangle = g(\nabla^{\Sigma}_X \nu, Y) = \pm\sqrt{b_i - c}\, g(X, Y) = \pm\sqrt{b_i - c}\langle X, Y \rangle.$$

Standard results now imply that $\partial_i\Sigma$ lies inside an umbilical $\mathbb{Q}^n(x_i)$ in $\mathbb{Q}^{n+1}(c)$, whose sectional curvature is $x_i = c + (\pm\sqrt{b_i - c})^2 = b_i$, and moreover, ν is normal to $\mathbb{Q}^n(x_i)$ (see again Exercise 2.9 of [5] and Theorem 1 of [18]). In other words, under this local isometric immersion, Σ^n meets an umbilical $\mathbb{Q}^n(b_i)$ orthogonally along the points corresponding to the boundary component $\partial_i\Sigma$. □

3 Other Intrinsic Properties

To conclude this article, we would like to mention briefly two notions related to intrinsic properties of minimal surfaces in $\mathbb{R}^3$ meeting spheres of radius one orthogonally along their boundary components, from their inside.

3.1 Ricci Surfaces with Boundary

Let (Σ^2, g) be a surface with negative Gaussian curvature K. It is straighforward to check that the equation that says $\sqrt{-K}g$ is a flat metric, namely,

$$-\Delta \log(-K) + 4K = 0,$$

is equivalent to the equation

$$-K\Delta K + |\nabla K|^2 + 4K^3 = 0. \tag{16}$$

But there is a difference between the two equations: the latter makes sense even where $K = 0$. In [14], Andrei Moroianu and Sergiu Moroianu noticed that, somehow, equation (16) contains more information than the original Ricci condition, in the sense that it forces a specific behaviour of its solutions near a zero. (Beware their different convention for the Laplacian.) As a consequence, they were able to prove a very satisfactory equivalence:

Theorem 1 (*cf.* Theorem 2 in [14]) *The following assertions about a Riemannian surface* (Σ^2, g) *(without boundary) are equivalent:*

(i) (Σ^2, g) *admits local minimal isometric immersions into* $\mathbb{R}^3$.
(ii) (Σ^2, g) *has non-positive Gaussian curvature* K*, which is a solution to equation* (16).

While $(i) \Rightarrow (ii)$ follows directly from the proof of Theorem A using Gauss equation and equation (3), the converse, as alluded before, requires a careful study of the zero set of solutions to equation (16), which ruled out behaviours like that of Lawson's example (*Cf.* Remark 1).

Because of this and other results, Moroianu and Moroianu proposed to call *Ricci surfaces* the Riemannian surfaces whose Gaussian curvature satisfies equation (16) and initiated the systematic investigation of these objects.

The works of Yiming Zang [17] and Benoît Daniel and Yiming Zang [3] developed the concept further, the first by imposing conditions on the ends of complete non-compact Ricci surfaces that match the intrinsic properties of the ends of finite total curvature minimal surfaces and the second by investigating a generalised notion of Ricci surface which captures, among other things, the key properties of the intrinsic geometry of other types of immersed CMC surfaces in space forms.

In the case of surfaces with boundary, one would like to combine equation (16) with a geometrically meaningful boundary condition. The ideas we have been exploring in this article suggest clearly that the following boundary condition fulfils such requirement.

Definition Let (Σ^2, g) be a Riemannian surface with boundary. We say that (Σ^2, g) is a *Ricci surface with boundary* when its boundary $\partial\Sigma$ has constant geodesic curvature one and its Gaussian curvature K satisfies:

$$\begin{cases} -K\Delta K + |\nabla K|^2 + 4K^3 = 0 & \text{in} \quad \Sigma, \\ \frac{\partial K}{\partial \nu} = -4K & \text{in} \quad \partial\Sigma. \end{cases}$$

For instance, consider a Riemannian surface (Σ^2, g) with boundary that is isometrically immersed as a minimal surface in $\mathbb{R}^3$ in such way that it meets a collection of unit spheres orthogonally along their boundary components, from their inside. According to Lemma 1 and 2 (recall in particular equation (14) for $|A|^2 = -2K$), (Σ^2, g) is a *Ricci surface with boundary* in the sense of the definition above.

We believe that a statement similar to Theorem 1 should also be true. But to establish it would lead us much further than we would like in this article, and so we leave it here.

In the absence of an Enneper-Weierstrass representation for free boundary minimal surfaces inside the unit ball, we think that the study of compact Ricci surfaces with boundary in the sense of Definition 3.1 could be fruitful. (For instance, Jaehoon Lee and Eungbeom Yeon adopted a similar but distinct perspective in [10].)

Remark 3 In [6], the authors classified surfaces of revolution in $\mathbb{R}^3$ that are Ricci surfaces in the sense of [14]. They also studied the intersection of these surfaces with the Euclidean sphere of radius one centred at the origin and found a 1-parameter family of such surfaces interpolating between the critical catenoid and its axis of rotation. None of them, except the critical catenoid, are Ricci surfaces with boundary in the sense of Definition 3.1.

3.2 Steklov Eigenvalues

If Σ^2 is a compact minimal surface with boundary in $\mathbb{R}^3$ meeting the unit sphere centred at the origin orthogonally, then its coordinate functions x^1, x^2, x^3 satisfy

$$\begin{cases} \Delta x^i = 0 & \text{on} \quad \Sigma, \\ \frac{\partial x^i}{\partial \nu} = x^i & \text{in} \quad \partial\Sigma. \end{cases}$$

This means that the three coordinate functions are *Steklov eigenfunctions* of (Σ^2, g), associated to the eigenvalue 1. This observation was a starting point for the work of Ailana Fraser and Rick Schoen [8], who sparked a renovated interest in the theory of compact free boundary minimal surfaces, by connecting it to a spectral maximisation problem for the Steklov (or Dirichlet-to-Neumann) operator.

The relation between the spectral theory of the Steklov operator and compact minimal surfaces with free boundary is not restricted to the Euclidean case. Vanderson Lima and Ana Menezes [12] identified the right spectral problem that describes minimal surfaces with free boundary in geodesic spheres of the sphere $\mathbb{S}^3$. An interesting new feature is that this problem involves a combination of the first two Steklov eigenvalues. Later, Vladimir Medvedev [13], among other things, extended the work done in [12] to the case of minimal surfaces with free boundary in geodesic balls of the hyperbolic space $\mathbb{H}^3$. (See also the more general setup proposed Romain Petrides [15].)

The free boundary correspondence discussed in Sect. 2.3 suggests that such spectral relationships—which are, after all, part of the intrinsic geometry of these immersed surfaces with boundary—are even more widespread than earlier thought. They could be useful, for instance, in the study of compact constant mean curvature $H = 1$ surfaces with boundary in $\mathbb{H}^3$ that meet horospheres orthogonally.

Acknowledgments L.A. is supported by CNPq (Conselho Nacional de Desenvolvimento Científico e Tecnológico (309908/2021-3, Bolsa PQ and 406666/2023-7, Universal)) and by FAPERJ (Fundação Carlos Chagas Filho de Amparo à Pesquisa do Estado do Rio de Janeiro (grant SEI-260003/000534/2023, BOLSA E-26/200.175/2023 and grant SEI-260003/001527/2023, APQ1 E-26/210.319/2023)).

I thank Ruy Tojeiro, who had the excellent idea of commemorating Professor Marcos Dajczer seventy-fifth birthday with an Festschrift volume, for the kind invitation to contribute to it with an article. I also thank the reviewer for pointing out [7] and for their other comments. Finally, I would also like to express my appreciation of my collaborators Alessandro Carlotto and Ben Sharp, with whom I enjoyed, over already eight years of scientific collaboration and friendship, so many interesting conversations about free boundary minimal surfaces. Several ideas that I exposed in this article were certainly born out of these conversations.

Appendix

Consider the following subsets of the complex plane $\mathbb{C}$:

$$\mathbb{D} = \{z = x + iy \in \mathbb{C} \mid |z| < 1\} \quad \text{and} \quad \mathbb{D}_+ = \{z = x + iy \in \mathbb{D} \mid y > 0\},$$

and let Γ denote the horizontal boundary of $\mathbb{D}_+$:

$$\Gamma = \{z = x + iy \in \overline{\mathbb{D}}_+ \mid y = 0\}.$$

The following elementary lemma was an important ingredient of the proof of Theorem A and Theorem C (*Cf.* Lemma 4.2 in [14]).

Lemma 3

(i) Let $F : \mathbb{D} \to \mathbb{R}$ be a positive smooth function such that $\log(F)$ *is harmonic on* $\mathbb{D}$*. Then, there exists a holomorphic function $h : D \to \mathbb{C}$ such that $F = |h|^2$.*
(ii) Let $F : \mathbb{D}_+ \cup \Gamma \to \mathbb{R}$ be a positive smooth function such that $\log(F)$ *is harmonic on $\mathbb{D}_+$ and satisfies the Neumann boundary condition on Γ. Then, there exists a holomorphic function $h : \mathbb{D} \to \mathbb{C}$ that has zero imaginary part on Γ and is such that $F = |h|^2$ on $\mathbb{D}_+ \cup \Gamma$.*

Proof
(i) Recall that $\Delta = 4\partial_{\bar{z}}\partial_z$. Thus, the assumption is that $\partial_z \log(F)$ is a holomorphic function on $\mathbb{D}$.

Integrating over paths based at 0, we obtain a holomorphic function $g_1 : \mathbb{D} \to \mathbb{C}$ such that $\partial_z g_1 = \partial_z \log(F)$. Then, the function $g_2 = \overline{\log(F) - g_1}$ is holomorphic, because

$$\partial_{\bar{z}} g_2 = \overline{\partial_z(\log(F) - g_1)} = 0.$$

Hence, $\log(F) = g_1 + \overline{g}_2$. The imaginary part of g_1 must be equal to the imaginary part of g_2, because $\log(F)$ is real. Furthermore, $\log(F)$ equals the real part of the sum $g_1 + g_2$ as well. Therefore, $h = \exp((g_1 + g_2)/2)$ is a holomorphic function such that $|h|^2 = F$.
(ii) By virtue of the boundary condition, it is possible to extend F to $\mathbb{D}$ by reflection across Γ, so that we may assume that $\log(F)$ is a harmonic function on $\mathbb{D}$ invariant by the reflection across Γ. Applying the same reasoning as in case (i), we obtain holomorphic functions g_1 and g_2 with the same imaginary part in such way that $\partial_z g_1 = \partial_z \log(F)$ and $\log(F) = g_1 + \overline{g}_2$.

By assumption, $\partial_y \log(F) = 0$ on Γ. Hence

$$2\partial_z \log(F) = \partial_x \log(F) - i\partial_y \log(F) = \partial_x \log(F)$$

is real on Γ. If we write $g_1 = a + ib$ according to its real and imaginary part

$$2\partial_z g_1 = (\partial_x a + \partial_y b) + i(\partial_x b - \partial_y a).$$

Thus, the vanishing of the imaginary part of $\partial_z g_1 = \partial_z \log(F)$ on Γ together with the Cauchy-Riemann equations satisfied by g_1 implies, in particular, that

$$\partial_x b = \partial_y a = -\partial_x b \quad \text{on} \quad \Gamma \quad \Rightarrow \quad \partial_x b = 0 \quad \text{on} \quad \Gamma.$$

Hence, g_1 has constant imaginary part on Γ. But then the imaginary part of g_2 is constant as well.

In conclusion, the holomorphic function $g_1 + g_2$ has constant imaginary part on Γ, say $k \in \mathbb{R}$. The function $h = \exp((g_1 + g_2)/2)\exp(-ki/2)$ is therefore a holomorphic function, with squared-norm equal to F, which moreover has zero imaginary part on Γ. And this is all we wanted to show. □

References

1. Robert Bryant, *Surfaces of mean curvature one in hyperbolic space*, Théorie des variétés minimales et applications (Palaiseau, 1983-1984) Astérisque (1987), no.154-155, 12, 321-347, 353.
2. S. S. Chern and Robert Osserman, *Remarks on the Riemannian metric of a minimal submanifold*, Geometry Symposium, Utrecht 1980 (Utrecht, 1980), pp. 49-90 Lecture Notes in Math., 894 Springer-Verlag, Berlin-New York, 1981.
3. Benoît Daniel and Yiming Zang, *Generalized Ricci surfaces*, J. Geom. Anal. 34 (2024), no. 8, Paper No. 255, 35 pp.
4. Manfredo do Carmo and Marcos Dajczer, *Necessary and sufficient conditions for existence of minimal hypersurfaces in spaces of constant curvature*, Bol. Soc. Brasil. Mat. 12 (1981), no. 2, 113-121.
5. Marcos Dajczer and Ruy Tojeiro, *Submanifold theory. Beyond an introduction*, Universitext, Springer, New York, 2019. xx+628 pp.
6. Iury Domingos, Roney Santos and Feliciano Vitório, *Rotational Ricci surfaces*, Ann. Mat. Pura Appl. (4) 203 (2024), no. 5, 2075–2093.
7. Jost-Hinrich Eschenburg and Renato Tribuzy, *Constant mean curvature surfaces in 4-space forms*, Rend. Semin. Mat. Univ. Padova 79 (1988), 185-202.
8. Ailana Fraser and Richard Schoen, *Minimal surfaces and eigenvalue problems*, in *Geometric analysis, mathematical relativity, and nonlinear partial differential equations*, 105-121, Contemp. Math., 599 American Mathematical Society, Providence, RI, 2013.
9. H. Blaine Lawson, Jr., *Complete minimal surfaces in S^3*, Ann. of Math. (2) 92 (1970), 335-374.
10. Jaehoon Lee and Eungbeom Yeon, *A new approach to the Fraser-Li conjecture with the Weierstrass representation formula*, Proc. Amer. Math. Soc. 149 (2021), no. 12, 5331-5345.
11. Martin Li, *Free boundary minimal surfaces in the unit ball: recent advances and open questions*, Proceedings of the International Consortium of Chinese Mathematicians 2017, 401-435. International Press, Boston, MA, 2020.
12. Vanderson Lima and Ana Menezes, *Eigenvalue problems and free boundary minimal surfaces in spherical caps*, arXiv:2307.13556.
13. Vladimir Medvedev, *On free boundary minimal submanifolds in geodesic balls in $\mathbb{H}^n$ and $\mathbb{S}^n$*, Math. Z. 310 (2025), no. 1, Paper No. 10, 32 pp.
14. Andrei Moroianu and Sergiu Moroianu, *Ricci surfaces*, Ann. Sc. Norm. Super. Pisa Cl. Sci. (5) 14 (2015), no. 4, 1093-1118.
15. Romain Petrides, *Laplace eigenvalues and non-planar minimal spheres into 3-dimensional ellipsoids*, J. Funct. Anal. 285 (2023), no.10, Paper No. 110087, 75 pp.
16. Gregorio Ricci-Curbastro, *Sulla teoria intrinseca delle superficie ed in ispecie di quelle di secondo grado*, Atti R. Ist. Ven. di Lett, ed Arti 6 (1895), 445–488.
17. Yiming Zang, *Non-positively curved Ricci surfaces with catenoidal ends*, Manuscripta Math. 172 (2023), no.1-2, 531-565.
18. Shing-Tung Yau, *Submanifolds with constant mean curvature. I, II.*, Amer. J. Math. 96 (1974), 346-366.

Conformally Einstein Left-Invariant Metrics on Four-Dimensional Lie Groups

Esteban Calviño-Louzao, Eduardo García-Río, Ixchel Gutiérrez-Rodríguez, and Ramón Vázquez-Lorenzo

Dedicated to Professor Marcos Dajczer on his 75th birthday.

Abstract Conformally Einstein left-invariant metrics on four-dimensional Lie groups are reviewed, providing a complete description in the Riemannian and Lorentzian settings.

1 Introduction

Einstein metrics, being critical for the Einstein-Hilbert functional, are central in geometry and physics. Three-dimensional Einstein metrics are of constant sectional curvature. Hence, dimension four is not only the first nontrivial case for consideration but also of special relevance due to its physical significance. Four-dimensional Einstein Lorentzian metrics are not yet well understood although much

E. Calviño-Louzao
I.E.S Eduardo Pondal, Santiago de Compostela, Spain
e-mail: estebcl@edu.xunta.gal

E. García-Río (✉)
Department of Mathematics, CITMAga, University of Santiago de Compostela, Santiago de Compostela, Spain
e-mail: eduardo.garcia.rio@usc.es

I. Gutiérrez-Rodríguez
Department of Mathematics, CITMAga, Facultad de Ciencias de la Educación y del Deporte, University of Vigo, Pontevedra, Spain
e-mail: ixchel.dzohara.gutierrez.rodriguez@uvigo.gal

R. Vázquez-Lorenzo
IES Dionisio Gamallo, Ribadeo, Spain
e-mail: ravazlor@edu.xunta.gal

© The Author(s), under exclusive license to Springer Nature Switzerland AG 2025
R. Tojeiro (ed.), *Advances in Submanifold Theory and Related Topics*,
Latin American Mathematics Series – UFSCar subseries,
https://doi.org/10.1007/978-3-032-00918-0_4

progress has been made, specially by assuming the existence of symmetries on the underlying manifold structure.

Lie groups with left-invariant Lorentzian metrics, being homogeneous spaces, play a distinguished role in the analysis since the Einstein field equations reduce to a purely algebraic system. Therefore, it is not surprising that left-invariant Lorentzian Einstein metrics on four-dimensional Lie groups are well described (see [14, 36]).

In this paper, we analyze a conformally related question by focusing on conformally Einstein manifolds (i.e., pseudo-Riemannian manifolds (M, g) with an Einstein representative of the conformal class $[g]$). Our purpose is to present a complete description of conformally Einstein left-invariant Riemannian and Lorentzian metrics on four-dimensional Lie groups. As it will be pointed out in Sect. 2.1, this condition is closely related to Bach-flat metrics, which are the underlying background in conformal gravity [32–34]. Hence, we also pay special attention to the description of strictly Bach-flat Lorentzian Lie groups.

Nontrivial conformally Einstein Riemannian Lie groups are semi-direct extensions of the Heisenberg and Abelian Lie groups, as described in Theorem 1. Conformally Einstein Lorentzian Lie groups are solvable and correspond to semi-direct extensions of the Euclidean, Poincaré, Heisenberg, and Abelian three-dimensional Lie groups. They are described in Theorem 4, Theorem 6, and Theorem 8. The class of Bach-flat metrics is much larger, with many non-solvable examples. Their description is summarized in Theorems 5 and 7, together with the results in Sects. 8.2 and 9. More information and detailed proofs are given in [17–19].

2 Preliminaries

It was shown by Jensen in [26] that any Einstein homogeneous Riemannian manifold is symmetric in dimension four. Hence, a real space form (locally isometric to $\mathbb{S}^4$, $\mathbb{R}^4$, or $\mathbb{H}^4$), a complex space form (locally isometric to $\mathbb{C}P^2$ or $\mathbb{C}H^2$), or a product of two surfaces of the same constant Gauss curvature $N_1^2(c) \times N_2^2(c)$. It follows from the work of Bérard-Bergery [5] that any simply connected Riemannian homogeneous four-manifold is either symmetric or isometric to a Lie group with a left-invariant metric. Hence, Jensen's result follows by showing that left-invariant Riemannian Einstein metrics in dimension four are necessarily symmetric.

Four-dimensional Einstein Lorentzian Lie groups constitute a larger class described in [14] (see also [36]). They split into three categories: symmetric spaces, plane waves, and left-invariant metrics which do not correspond to any of these. Einstein Lorentzian symmetric spaces are of constant sectional curvature in the irreducible case [11]. In the non-irreducible case, either they are products $N_1^2(c) \times N_2^2(c)$ of two surfaces with the same constant sectional curvature, or they correspond to Ricci-flat Cahen-Wallach symmetric spaces [11] (see Sect. 2.2). The remaining situations are locally homothetic to the semi-direct extensions $\mathbb{R}^3 \rtimes \mathbb{R}$ of the Abelian group with left-invariant Lorentzian metric determined by the following

Lie algebra structures (see [36]):

$$\begin{array}{llll} (i)\ [e_1,e_4]=-2e_1, & [e_2,e_4]=e_2+\sqrt{3}e_3, & [e_3,e_4]=-\sqrt{3}e_2+e_3, & \\ (ii)\ [u_1,u_4]=-u_1+\delta u_2, & [u_2,u_4]=5u_2, & [u_3,u_4]=2u_3, & \delta\neq 0, \\ (iii)\ [u_1,u_4]=4u_1, & [u_2,u_4]=-2u_2+\delta u_3, & [u_3,u_4]=\delta u_1+u_3, & \delta\neq 0, \end{array}$$

where $\{e_i\}$ is an orthonormal basis with e_3 timelike in case (i) and $\{u_j\}$ is a pseudo-orthonormal basis with $\langle u_1,u_2\rangle=\langle u_3,u_3\rangle=\langle u_4,u_4\rangle=\ 1$ in cases (ii)–(iii).

2.1 Conformally Einstein Metrics

Although the Einstein condition is not conformally invariant, a pseudo-Riemannian manifold (M^n,g) is *conformally Einstein* if there is an Einstein representative of the conformal class $[g]$. A conformally related metric $\overline{g}=\varphi^{-2}g$ is Einstein if and only if there exists a nowhere zero solution of the overdetermined PDE:

$$(n-2)\operatorname{Hes}_\varphi+\varphi\rho=\tfrac{1}{n}\{(n-2)\Delta\varphi+\varphi\tau\}g, \tag{1}$$

where $\operatorname{Hes}_\varphi$ is the Hessian tensor of φ and $\Delta\varphi=\operatorname{tr}_g\operatorname{Hes}_\varphi$ is the Laplacian of the conformal factor φ (see [8]). Any two-dimensional pseudo-Riemannian manifold is conformally Einstein, and three-dimensional conformally Einstein manifolds reduce to locally conformally flat ones. Despite its apparent simplicity, the integrability of Eq. (1) is surprisingly difficult (see, e.g., [29]), and only few results are available in higher dimensions.

A necessary condition to be conformally Einstein is derived from the fact that the Weyl tensor of any Einstein metric is divergence-free. Hence, if a conformal metric $\overline{g}=e^{-2\sigma}g$ is Einstein, then $\overline{\operatorname{div}}\,\overline{W}=0$, and a straightforward calculation now shows that $\operatorname{div}_4 W-W(\cdot,\cdot,\cdot,\nabla\sigma)=0$. A pseudo-Riemannian manifold is *conformally Cotton-flat* if there is a conformal metric $\overline{g}=e^{-2\sigma}g$ which is Cotton-flat. Equivalently, there is a function σ so that

$$\operatorname{div}_4 W-W(\cdot,\cdot,\cdot,\nabla\sigma)=0. \tag{2}$$

More generally, one says that (M,g) is a *conformal C-space* if there is a (not necessarily gradient) vector field X so that $\operatorname{div}_4 W-W(\cdot,\cdot,\cdot,X)=0$ (see [25]).

A special feature of dimension four is that the quadratic curvature functional given by the L^2-norm of the Weyl conformal curvature tensor, $g\mapsto\int_M\|W\|^2\,\mathrm{dvol}_g$, is conformally invariant. Hence, so is its gradient, given by the *Bach tensor* [4]:

$$\mathfrak{B}=\operatorname{div}_2\operatorname{div}_4 W+\tfrac{1}{2}W[\rho], \tag{3}$$

where $W[\rho]$ denotes the contraction of the Weyl tensor and the Ricci tensor, $W[\rho]_{ij} = W_{iajb}\rho^{ab}$. Locally conformally flat metrics are trivially Bach-flat, and so are Einstein metrics since the Weyl tensor is trace-free. Due to the conformal invariance, being Bach-flat is a necessary condition in order to be conformally Einstein in dimension four.

The special significance of the conformally Cotton-flat property was given in [28], where it is shown that a weakly generic Bach-flat manifold is conformally Einstein if and only if it is conformally Cotton-flat, where *weakly genericity* means that the Weyl tensor, viewed as a map $W : TM \to \otimes^3 TM$, is injective. It must be remarked that there are many important space-times which are not weakly generic and thus the above criterion does not apply directly. This is, for instance, the case of pp-waves, where the Weyl curvature operator acting on the space of two-forms is two-step nilpotent.

2.2 Four-Dimensional PP-Waves and Plane Waves

A *pp-wave* is a transversally flat space-time (M, g) admitting a null parallel vector field ℓ (i.e., the curvature endomorphism satisfies $R(\ell^\perp, \ell^\perp) = 0$). In the special case that the covariant derivative of the curvature is also transversally flat (i.e., $\nabla_{\ell^\perp} R = 0$), (M, g) is referred to as a *plane wave* (see [30]).

The metric tensor of a pp-wave is locally described in adapted coordinates (u, v, x^1, x^2) so that it takes the form

$$g = 2du \circ dv + H(v, x^1, x^2)dv \circ dv + dx^1 \circ dx^1 + dx^2 \circ dx^2 , \qquad (4)$$

and the parallel vector field is given by $\ell = \partial_u$. Furthermore, plane waves correspond to the special case where the function $H(v, x^1, x^2)$ is a quadratic polynomial in the spacelike coordinates $\mathbf{x} = (x^1, x^2)$. The Ricci operator of any pp-wave is two-step nilpotent. Moreover, a metric (4) is Einstein (indeed Ricci-flat) if and only if the spacelike Laplacian $\Delta_{\mathbf{x}} H = 0$.

Plane waves can be described in suitable coordinates by (4) with $H(v, x^1, x^2) = a_{ij}(v)x^i x^j$, for some functions $a_{ij}(v)$. It now follows that a plane wave is Ricci-flat if and only if the matrix $(a_{ij}(v))$ is trace-free, and it is locally conformally flat if and only if $a_{ij}(v) = a(v)$ for all $i, j = 1, 2$. *Cahen-Wallach symmetric spaces* are a special class of plane waves determined by constant functions $a_{ij}(v) = a_{ij} \in \mathbb{R}$.

Four-dimensional pp-waves are conformally Einstein if and only if $\operatorname{div} W = 0$, as shown in [9]. Therefore, one has that four-dimensional plane waves are conformally Einstein. As a matter of notation, we say that a conformally Einstein metric is *nontrivial* if it is neither Einstein, nor locally conformally flat, nor a plane wave. Moreover, a Bach-flat metric is said to be *strict* if it is not conformally Einstein. We refer to [31] for a description of four-dimensional Bach-flat pp-waves.

It was shown by Globke and Leistner [24] that any four-dimensional Ricci-flat locally homogeneous pp-wave is a plane wave. Our analysis shows that the previous result holds true also for left-invariant Bach-flat Lorentzian metrics on four-dimensional Lie groups. Homogeneous plane waves in dimension four are described in terms of a 2×2 skew-symmetric matrix F and a 2×2 symmetric matrix A_0 so that the defining function $H(v, x^1, x^2)$ takes the form $H = \mathbf{x}^T A(v)\mathbf{x}$, where the matrix $A(v)$ is given by (see [6])

$$A(v) = e^{vF} A_0 e^{-vF}, \quad \text{or} \quad A(v) = \frac{1}{(v+b)^2} e^{\log(v+b)F} A_0 e^{-\log(v+b)F}.$$

Furthermore, homogeneous plane wave metrics are Ricci-flat if and only if A_0 is trace-free. Locally conformally flat homogeneous plane waves reduce to the case $A(v) = e^{vF} A_0 e^{-vF}$ in the previous description where A_0 is a multiple of the identity.

3 Four-Dimensional Lorentzian Lie Groups

Any real Lie algebra is a semi-direct product of the radical (i.e., the maximal solvable ideal) and a semi-simple subalgebra (the Levi factor) [27]. Hence, the classification of four-dimensional Lie algebras reduces to classifying low-dimensional semi-simple Lie algebras, solvable Lie algebras, and semi-direct products of semi-simple Lie algebras and solvable ones. Semi-simple Lie algebras can be decomposed into a direct sum of simple subalgebras which are orthogonal with respect to the Killing-Cartan form. Solvable Lie algebras are also classified in low dimensions (up to six). We refer, for example, to [37] for a description of all four-dimensional Lie algebras. Simply connected four-dimensional Lie groups are isomorphic to the products $\widetilde{SL}(2, \mathbb{R}) \times \mathbb{R}$ or $SU(2) \times \mathbb{R}$ or, otherwise, they are solvable Lie groups which correspond to semi-direct extensions of the Euclidean and Poincaré groups $\widetilde{E}(2) \rtimes \mathbb{R}$ and $E(1,1) \rtimes \mathbb{R}$, the Heisenberg group $\mathcal{H}^3 \rtimes \mathbb{R}$, or the Abelian group $\mathbb{R}^3 \rtimes \mathbb{R}$ [3].

The description of left-invariant metrics is more subtle in the Lorentzian case than in the Riemannian one, and we follow the approach in [12]. Since the restriction of the metric to the unimodular three-dimensional Lie group plays an essential role, we firstly review the description of those metrics as in [35, 38].

Three-Dimensional Lorentzian Unimodular Lie Groups

Let $(\mathfrak{g}, \langle \cdot, \cdot \rangle)$ be a three-dimensional Lie algebra equipped with a positive definite inner product, and let L be the structure operator $[x, y] = L(x \times y)$, where "$\times$" denotes the vector cross product on $(\mathfrak{g}, \langle \cdot, \cdot \rangle)$. Then, the associated Lie group G is unimodular if and only if L is self-adjoint [35], in which case there is an orthonormal basis $\{e_1, e_2, e_3\}$ of L-eigenvectors so that

$$[e_2, e_3] = \lambda_1 e_1, \qquad [e_3, e_1] = \lambda_2 e_2, \qquad [e_1, e_2] = \lambda_3 e_3,$$

where $\lambda_1, \lambda_2, \lambda_3$ are the corresponding eigenvalues of L. The Lie algebra is $\mathfrak{su}(2)$ (resp., $\mathfrak{sl}(2,\mathbb{R})$) if L is non-singular and eigenvalues λ_k do not change sign (resp., change sign). If $\ker L$ is one dimensional, then the Lie algebra is $\mathfrak{e}(2)$ (resp., $\mathfrak{e}(1,1)$) if the nonzero eigenvalues of L have the same sign (resp., opposite sign). The Heisenberg Lie algebra $\mathfrak{h}^3$ corresponds to the case when $\ker L$ is two dimensional, and the Abelian Lie algebra $\mathfrak{r}^3$ corresponds to $L = 0$.

Rahmani extended in [38] the work of Milnor to the Lorentzian situation. As in the previous case, one considers the structure operator $L(x \times y) = [x, y]$, where the vector-cross product $\langle x \times y, z\rangle = \det(x, y, z)$ is now defined with respect to a Lorentzian inner product on $\mathfrak{g}_3$, where $\mathfrak{g} = \mathfrak{g}_3 \rtimes \mathfrak{r}$. One must now consider the possible Jordan normal forms of L as in [38]:

Ia. *L is diagonalizable*. Hence, there exists an orthonormal basis $\{e_1, e_2, e_3\}$, where we assume e_3 to be timelike, so that $L(e_i) = \lambda_i e_i$.

Ib. *L has complex eigenvalues*. Then, there exists an orthonormal basis $\{e_1, e_2, e_3\}$, where we assume e_3 to be timelike, so that

$$L = \begin{pmatrix} \lambda & 0 & 0 \\ 0 & \alpha & \beta \\ 0 & -\beta & \alpha \end{pmatrix}, \qquad \beta \neq 0 .$$

II. *L has a double root of its minimal polynomial*. Then, there exists a pseudo-orthonormal basis $\{u_1, u_2, u_3\}$ so that

$$L = \begin{pmatrix} \lambda_1 & 0 & 0 \\ \varepsilon & \lambda_1 & 0 \\ 0 & 0 & \lambda_2 \end{pmatrix}, \qquad \varepsilon = \pm 1, \qquad \text{where} \quad \langle u_1, u_2\rangle = \langle u_3, u_3\rangle = 1 .$$

III. *L has a triple root of its minimal polynomial*. Then, there exists a pseudo-orthonormal basis $\{u_1, u_2, u_3\}$ so that

$$L = \begin{pmatrix} \lambda & 0 & 1 \\ 0 & \lambda & 0 \\ 0 & 1 & \lambda \end{pmatrix}, \qquad \text{where} \quad \langle u_1, u_2\rangle = \langle u_3, u_3\rangle = 1 .$$

Attending to the Jordan normal form of L, the unimodular Lorentzian Lie algebras $(\mathfrak{g}, \langle \cdot, \cdot \rangle)$ are described by one of the four types given below, where $\{e_1, e_2, e_3\}$ and $\{u_1, u_2, u_3\}$ are orthonormal and pseudo-orthonormal bases as specified in each case.

$$\begin{array}{lll}
\text{Ia} & \begin{array}{c}\langle e_1, e_1\rangle = \langle e_2, e_2\rangle = -\langle e_3, e_3\rangle = 1,\\ [e_1, e_2] = -\lambda_3\, e_3, \quad [e_1, e_3] = -\lambda_2\, e_2, \quad [e_2, e_3] = \lambda_1\, e_1,\end{array} & \\
\text{Ib} & \begin{array}{c}\langle e_1, e_1\rangle = \langle e_2, e_2\rangle = -\langle e_3, e_3\rangle = 1,\\ [e_1, e_2] = -\beta\, e_2 - \alpha\, e_3, \quad [e_1, e_3] = -\alpha\, e_2 + \beta\, e_3, \quad [e_2, e_3] = \lambda\, e_1,\end{array} & \beta \neq 0, \\
\text{II} & \begin{array}{c}\langle u_1, u_2\rangle = \langle u_3, u_3\rangle = 1,\\ [u_1, u_2] = \lambda_2\, u_3, \quad [u_1, u_3] = -\lambda_1\, u_1 - \varepsilon\, u_2, \quad [u_2, u_3] = \lambda_1\, u_2,\end{array} & \varepsilon = \pm 1, \\
\text{III} & \begin{array}{c}\langle u_1, u_2\rangle = \langle u_3, u_3\rangle = 1,\\ [u_1, u_2] = u_1 + \lambda\, u_3, \quad [u_1, u_3] = -\lambda\, u_1, \quad [u_2, u_3] = \lambda\, u_2 + u_3.\end{array} &
\end{array}$$

It is important to emphasize that, although all unimodular Lie groups admit left-invariant Lorentzian metrics of Type Ia, metrics of other types are not always realizable. Indeed, metrics of Type Ib only exist on $\widetilde{SL}(2, \mathbb{R})$ or $E(1, 1)$, as well as metrics of Type III. Metrics of Type II exist on $\widetilde{SL}(2, \mathbb{R})$, $E(1, 1)$, $\widetilde{E}(2)$ and on the Heisenberg group $\mathcal{H}^3$ (see [38]).

Left-Invariant Lorentzian Metrics on Four-Dimensional Lie Groups
For a given three-dimensional unimodular Lie group G_3, Lorentzian left-invariant metrics on semi-direct extensions $G_3 \rtimes \mathbb{R}$ may induce left-invariant Riemannian, Lorentzian, or degenerate metrics on G_3. Hence, all these possibilities must be considered separately. The case when the induced metric on G_3 is positive definite resembles the Riemannian situation, although many new conformally Einstein and Bach-flat metrics appear in this situation. The case when the induced metric on G_3 is Lorentzian is more involved, and one has to take into consideration the different possibilities corresponding to types Ia, Ib, II, and III previously discussed.

Furthermore, the induced metric on G_3 may be degenerate. In this case, we follow the approach in [12] based on the dimension of the derived algebra $\mathfrak{g}_3' = [\mathfrak{g}_3, \mathfrak{g}_3]$. The case of $\dim \mathfrak{g}_3' = 3$ may only occur on the product Lie algebras $\mathfrak{sl}(2, \mathbb{R}) \times \mathfrak{r}$ or $\mathfrak{su}(2) \times \mathfrak{r}$. One has that $\dim \mathfrak{g}_3' = 2$ is only possible on Lie algebras $\mathfrak{e}(2) \rtimes \mathfrak{r}$ and $\mathfrak{e}(1, 1) \rtimes \mathfrak{r}$. The case $\dim \mathfrak{g}_3' = 1$ may only occur on semi-direct extensions of the Heisenberg group, while $\dim \mathfrak{g}_3' = 0$ only occurs in semi-direct extensions of the Abelian group.

Bach-Flat Equations for Lorentzian Lie Groups
A program to describe all conformally Einstein left-invariant Lorentzian metrics in dimension four was recently developed by the authors in [17–19], which solves a question posed in [2]. While the Bach-flat and the conformally Einstein conditions are very restrictive in the Riemannian setting (see Theorems 1 and 2), there are plenty of Lorentzian examples. We consider the different families of solvable and non-solvable four-dimensional Lie groups separately in order to describe nontrivial conformally Einstein and strictly Bach-flat Lorentzian Lie groups. The classification of nontrivial conformally Einstein Lorentzian Lie groups is obtained as a combination of Theorems 4, 6, and 8. The description of strictly Bach-flat metrics is more involved. It is summarized in Theorems 5 and 7, together with the results in Sects. 8.2 and 9.

The Bach-flatness condition $\mathfrak{B} = 0$ is expressed as a system of polynomial equations $\mathcal{S}$ (given by the components $\mathfrak{B}_{ij}$ of the Bach tensor) on the structure constants which one has to solve in order to obtain a complete classification. Since two sets of polynomials generating the same ideal have the same zeroes, **a** is a solution of $\mathcal{S}$ if and only if it is a solution of $\mathcal{I} = \langle \mathfrak{B}_{ij} \rangle$, the ideal generated by the $\mathfrak{B}_{ij}$'s. The theory of Gröbner bases provides a well-known strategy to solve rather large polynomial systems obtaining "better" polynomials that belong to the ideal generated by the initial polynomial system [21]. The calculations of Gröbner bases in [17–19] have been done using MATHEMATICA. We refer to those papers for more information on the orderings used for the different calculations.

4 Conformally Einstein Homogeneous Riemannian Metrics

Four-dimensional Einstein metrics, as well as locally conformally flat ones, are symmetric in the Riemannian homogeneous setting [26, 39]. Moreover, simply connected four-dimensional homogeneous Riemannian manifolds are symmetric or isometric to a Lie group with a left-invariant metric (see [5]). A four-dimensional locally symmetric Riemannian manifold is Einstein, or it locally splits as a product of spaces of constant sectional curvature $N^3(c)\times\mathbb{R}$ or $N_1^2(c_1)\times N_2^2(c_2)$ with $c_1 \neq c_2$. The manifold is locally conformally flat in the former case and in the second one whenever $c_1 = -c_2$. Since a product $N_1^2(c_1) \times N_2^2(c_2)$ is Bach-flat if and only if $c_1^2 = c_2^2$, a non-Einstein locally symmetric four-manifold is conformally Einstein if and only if it is locally conformally flat. Hence, the classification of nontrivial conformally Einstein Riemannian homogeneous four-manifolds reduces to the case of Lie groups, and they are described as follows.

Theorem 1 ([15]) *Let (M, g) be a four-dimensional simply connected conformally Einstein homogeneous Riemannian manifold. Then, (M, g) is symmetric or otherwise it is homothetic to one of the following:*

(i) The left-invariant metric on $\mathbb{R}^3 \rtimes \mathbb{R}$ determined by

$$[e_1, e_4] = e_1, \quad [e_2, e_4] = (\sqrt{p} - 1)^2 e_2, \quad [e_3, e_4] = p\, e_3, \qquad \tfrac{1}{4} \leq p < 1\,.$$

(ii) The left-invariant metric on $\mathcal{H}^3 \rtimes \mathbb{R}$ determined by

$$[e_1, e_2] = e_3, \quad [e_1, e_4] = e_1, \quad [e_2, e_4] = e_2, \quad [e_3, e_4] = 2e_3,$$

where $\{e_1, \dots, e_4\}$ is an orthonormal basis.

We note that the families in Theorem 1 are simplified with respect to those in [15] in such a way that any two of them are non-homothetic (see also [10]). Moreover, the left-invariant metric on $\mathcal{H}^3 \rtimes \mathbb{R}$ determined by (ii) is the only nonsymmetric half-conformally flat homogeneous four-manifold [22].

The only homogeneous strictly Bach-flat metrics were originally discovered in [1], and it was shown in [15] that no other possibility may exist, as follows.

Theorem 2 ([15]) *Let (M, g) be a four-dimensional simply connected strictly Bach-flat homogeneous Riemannian manifold. Then, (M, g) is homothetic to one of the following:*

(i) The left-invariant metric on $E(1,1) \rtimes \mathbb{R}$ determined by

$$[e_2, e_3] = e_1, \qquad [e_1, e_3] = (2+\sqrt{3})\, e_2,$$
$$[e_4, e_1] = \sqrt{6+3\sqrt{3}}\, e_1, \qquad [e_4, e_2] = \sqrt{6+3\sqrt{3}}\, e_2.$$

(ii) The left-invariant metric on $\mathcal{H}^3 \rtimes \mathbb{R}$ determined by

$$[e_1, e_2] = e_3, \qquad [e_4, e_1] = \tfrac{1}{4}\sqrt{7-3\sqrt{5}}\, e_1,$$
$$[e_2, e_4] = \tfrac{1}{4}\sqrt{7+3\sqrt{5}}\, e_2, \qquad [e_3, e_4] = \tfrac{\sqrt{5}}{2\sqrt{2}}\, e_3,$$

where $\{e_1, \ldots, e_4\}$ is an orthonormal basis.

5 Conformally Einstein Nonreductive Homogeneous Lorentzian Four-Manifolds

A specific feature of pseudo-Riemannian geometry is the existence of nonreductive homogeneous manifolds. While homogeneous pseudo-Riemannian manifolds are reductive in the three-dimensional case, dimension four supports nonreductive homogeneous Lorentzian manifolds. Nonreductive homogeneous four-manifolds were classified by Fels and Renner in [23]. Moreover, they determined all Einstein nonreductive homogeneous four-manifolds.

In the Lorentzian signature, Bach-flatness is a quite restrictive condition as shown in [16]. Only three families of such metrics exist which are neither Einstein nor locally conformally flat. Moreover, they all are conformally Einstein, which shows the nonexistence of strictly Bach-flat nonreductive homogeneous Lorentzian four-manifolds. They are described as follows, where we use the labelling (A2) and ($A3^{\pm}$) to agree with the notation in [23].

Theorem 3 ([16]) *Let (M, g) be a four-dimensional nonreductive homogeneous Lorentzian manifold. If (M, g) is Bach-flat but neither locally conformally flat nor Einstein, then it is locally isometric to:*

(A2) The space $\mathbb{R}^4$ with coordinates (x, y, z, t) and metric tensor

$$g = -2ae^{2t}dx \circ dz + ae^{2t}dy \circ dy + bdz \circ dz + 2cdz \circ dt + qdt \circ dt,$$

where a, b, c, and q are constants with $aq > 0$ and $b \neq 0$.

(A3$^+$) An open subset $\mathfrak{U} = \{(x, y, z, t) \in \mathbb{R}^4 \mid \cos t \neq 0\} \subset \mathbb{R}^4$ with coordinates (x, y, z, t) and metric tensor

$$g_+ = ae^{2z}(2dx \circ dt + \cos^2 t\, dy \circ dy) + bdz \circ dz + 2cdz \circ dt + qdt \circ dt,$$

where a, b, c, and q are constants with $ab > 0$ and $b \neq -q$.

(A3$^-$) The space $\mathbb{R}^4$ with coordinates (x, y, z, t) and metric tensor

$$g_- = ae^{2z}(2dx \circ dt + \cosh^2 t\, dy \circ dy) + bdz \circ dz + 2cdz \circ dt + qdt \circ dt,$$

where a, b, c, and q are constants with $ab > 0$, $b \neq q$.

Moreover, in all cases, the metric is conformally Einstein.

All cases in Theorem 3 admit a two-parameter family of conformal metrics which are Ricci-flat. Hence, they are in the conformal class of a Ricci-flat pp-wave [7].

6 Semi-Direct Extensions of the Euclidean and Poincaré Groups

A straightforward calculation shows that locally conformally flat Lorentzian left-invariant metrics on semi-direct extensions of the Euclidean or Poincaré Lie groups have diagonalizable Ricci operator (in which case they are of constant sectional curvature or locally isometric to a product $\mathbb{R} \times N(c)$) [13], or they are locally isometric to a symmetric Cahen-Wallach space.

6.1 Nontrivial Conformally Einstein Extensions

The following result describes all nontrivial conformally Einstein semi-direct Lorentzian extensions of the Euclidean and the Poincaré groups as follows. The

Lorentzian metric on $G_3 \rtimes \mathbb{R}$ restricts to a positive definite metric, to a Lorentzian metric or to a degenerate metric on the unimodular three-dimensional subgroup G_3.

Theorem 4 ([18]) *Let $(G_3 \rtimes \mathbb{R}, \langle\cdot,\cdot\rangle)$ be a semi-direct extension of the Euclidean group $G_3 = \widetilde{E}(2)$ or the Poincaré Lie group $G_3 = E(1,1)$ equipped with a left-invariant Lorentzian metric. Then, the metric is nontrivial conformally Einstein if and only if it is isomorphically homothetic to the Lie group corresponding to one of the following Lorentzian Lie algebras:*

(R) The restriction of the metric to $\mathfrak{g}_3$ is Riemannian and the metric is determined by

(i) $[e_1, e_3] = -\lambda_2 e_2, \quad [e_1, e_4] = \alpha e_1 + \lambda_2 e_2, \quad [e_2, e_3] = e_1,$
$[e_2, e_4] = -e_1 + \alpha e_2,\ \alpha > 0,\ \lambda_2 \in [-1, 1) \setminus \{0\}$, *or*

(ii) $[e_1, e_3] = e_2,\ [e_1, e_4] = [e_2, e_4] = \frac{1}{2}\sqrt{\frac{3}{2}}(e_1 + e_2),$
$[e_2, e_3] = e_1,\ [e_3, e_4] = \frac{1}{\sqrt{2}}(e_1 + e_2),$

where $\{e_i\}$ is an orthonormal basis with e_4 timelike.

(L) The restriction of the metric to $\mathfrak{g}_3$ is Lorentzian and the metric is determined by

(i) $[e_1, e_3] = -\lambda_2 e_2, \quad [e_1, e_4] = \alpha e_1 + \lambda_2 e_2, \quad [e_2, e_3] = e_1,$
$[e_2, e_4] = -e_1 + \alpha e_2,\ \alpha > 0,\ \lambda_2 \in [-1, 1) \setminus \{0\}$, *or*

(ii) $[u_1, u_2] = u_1, \quad [u_1, u_4] = \frac{1}{2}u_1,\ [u_2, u_3] = u_3,$
$[u_2, u_4] = -\frac{5}{4}u_1,\ [u_3, u_4] = u_3,$ *or*

(iii) $[u_1, u_2] = u_1, \quad [u_1, u_4] = 2\gamma u_1,\ [u_2, u_3] = u_3,$
$[u_2, u_4] = \alpha u_1 + \beta u_3,\ [u_3, u_4] = \gamma u_3,$
with $\alpha, \beta \geq 0$, $\gamma \neq 0$ and either $\beta \neq 0$ or $\alpha\gamma \neq -2$,

where $\{e_i\}$ is an orthonormal basis with e_3 timelike and $\{u_j\}$ is a pseudo-orthonormal basis with $\langle u_1, u_2\rangle = \langle u_3, u_3\rangle = \langle u_4, u_4\rangle = 1$.

(D) The restriction of the metric to $\mathfrak{g}_3$ is degenerate and the metric is determined by $[u_1, u_3] = (\gamma_4 - 1)u_2$, $[u_1, u_4] = \varepsilon u_1$, $[u_2, u_3] = -(\gamma_4 + 1)u_1$, $[u_2, u_4] = \varepsilon u_2$, where $\gamma_4 \neq 0, \pm 1$, $\varepsilon = \pm 1$, and $\{u_i\}$ is a pseudo-orthonormal basis with $\langle u_1, u_1\rangle = \langle u_2, u_2\rangle = \langle u_3, u_4\rangle = 1$.

Moreover, in all cases, the left-invariant metrics are conformally equivalent to a Ricci-flat metric.

The Weyl curvature operator $\mathcal{W} : \Lambda^2 \to \Lambda^2$ is three-step nilpotent for metrics (R.ii), and it is diagonalizable with two distinct real eigenvalues in case (L.ii). The remaining cases have two-step nilpotent Weyl curvature operator.

Remark 1 Let $\mathbb{K}$ be an associative Lie algebra. The affine Lie algebra $\mathfrak{aff}(\mathbb{K})$ is the Lie algebra given by $\mathbb{K} \oplus \mathbb{K}$ with the bracket $[(a, b), (c, d)] = (ac - ca, ad - cb)$.

Therefore, the real affine Lie algebra $\mathfrak{aff}(\mathbb{R})$ is the non-Abelian two-dimensional Lie algebra determined by $[e_1, e_2] = e_2$. The complex affine Lie algebra $\mathfrak{aff}(\mathbb{C})$ is the four-dimensional Lie algebra determined by

$$[e_1, e_3] = e_3, \quad [e_1, e_4] = e_4, \quad [e_2, e_3] = e_4, \quad [e_2, e_4] = -e_3.$$

Semi-direct extensions of $\widetilde{E}(2)$ and $E(1,1)$ are either isomorphic to the product Lie groups $\widetilde{E}(2) \times \mathbb{R}$ and $E(1,1) \times \mathbb{R}$ in the unimodular case, or otherwise, they are isomorphic to the Lie groups $Aff(\mathbb{C})$ and $Aff(\mathbb{R}) \times Aff(\mathbb{R})$ in the non-unimodular case, respectively (see, e.g., [3]). Hence, left-invariant metrics in this section are realized in some of the Lie groups above. Left-invariant metrics in Theorem 4 are realized either on the product $Aff(\mathbb{R}) \times Aff(\mathbb{R})$ or on the complex affine group $Aff(\mathbb{C})$. Moreover, while the structure operator of left-invariant Lorentzian metrics (L.i) is of Type Ia, metrics (L.ii) and (L.iii) have structure operator of Type III.

6.2 *Strictly Bach-Flat Extensions*

The strictly Bach-flat semi-direct extensions of the Euclidean and Poincaré groups are given as follows.

Theorem 5 ([18]) *Let $(G_3 \rtimes \mathbb{R}, \langle\cdot,\cdot\rangle)$ be a semi-direct extension of the Euclidean group $G_3 = \widetilde{E}(2)$ or the Poincaré group $G_3 = E(1,1)$ equipped with a left-invariant Lorentzian metric. Then, the metric is strictly Bach-flat if and only if it is isomorphically homothetic to the Lie group corresponding to one of the following Lorentzian Lie algebras:*

(i) $[e_1, e_2] = -e_2, \quad [e_1, e_3] = e_3, \qquad [e_1, e_4] = \frac{1}{2}\sqrt{\frac{5}{2}}\, e_3,$

$[e_2, e_4] = \sqrt{\frac{3}{2}}\, e_2,\ [e_3, e_4] = \frac{3}{2}\sqrt{\frac{3}{2}}\, e_3,$ *or*

(ii) $[u_1, u_2] = \frac{1}{2} u_3, \quad [u_1, u_3] = -u_2,\ [u_1, u_4] = \alpha u_2 - \frac{\alpha^2+1}{4\alpha} u_3,$

$[u_2, u_4] = -\frac{\alpha}{2} u_3,\ [u_3, u_4] = \alpha u_2,\ \alpha > 0,$

where $\{e_1, \ldots, e_4\}$ is an orthonormal basis of the Lie algebra with e_3 timelike and $\{u_1, \ldots, u_4\}$ is a pseudo-orthonormal basis of the Lie algebra with $\langle u_1, u_2\rangle = \langle u_3, u_3\rangle = \langle u_4, u_4\rangle = 1$.

The metric in Theorem 5-(i) realizes on $Aff(\mathbb{R}) \times Aff(\mathbb{R})$. Moreover, it restricts to a Lorentzian metric on $E(1,1)$ whose structure operator has complex eigenvalues. On the other hand, metrics corresponding to case (ii) in Theorem 5 are left-invariant metrics on $\widetilde{E}(2) \times \mathbb{R}$ whose restriction to $\widetilde{E}(2)$ is Lorentzian with structure operator of Type II. The Weyl curvature operator $\mathcal{W} : \Lambda^2 \to \Lambda^2$ corresponding to case (i) has two distinct real eigenvalues, one of them being a double root of the minimal polynomial, and it is three-step nilpotent in case (ii).

Remark 2 Bach-flat *pp*-waves on semi-direct extensions of the Poincaré or the Euclidean groups, which are neither Ricci-flat nor locally conformally flat, appear as the limiting cases in Theorem 4 corresponding to (R.i) and (L.i) for $\alpha = 0$

$$[e_1, e_3] = -\lambda_2 e_2, \quad [e_1, e_4] = \lambda_2 e_2, \quad [e_2, e_3] = e_1, \quad [e_2, e_4] = -e_1,$$

where $\lambda_2 \in (-1, 1) \setminus \{0\}$ and $\{e_i\}$ is an orthonormal basis with either e_3 or e_4 timelike, or they correspond to the limiting case in (L.iii) for $\gamma = 0$

$$[u_1, u_2] = u_1, \quad [u_2, u_3] = u_3, \quad [u_2, u_4] = \alpha u_1 + \beta u_3, \quad \alpha \geq 0, \; \beta \geq 0,$$

where $\{u_i\}$ is a pseudo-orthonormal basis with $\langle u_1, u_2 \rangle = 1$.

In all the situations above, the underlying structure is a plane wave on the product Lie groups $E(1, 1) \times \mathbb{R}$ or $\widetilde{E}(2) \times \mathbb{R}$.

7 Semi-Direct Extensions of the Heisenberg Group

Locally conformally flat left-invariant Lorentzian metrics on semi-direct extensions of $\mathcal{H}^3$ have diagonalizable Ricci operator (in which case are locally symmetric [13]) and are locally isometric to a locally conformally flat plane wave, or otherwise, the Ricci operator has a single eigenvalue which is a double root of the minimal polynomial. The latter case corresponds to the metrics isomorphically homothetic to those determined by

$$\begin{aligned} &[u_1, u_3] = -\varepsilon u_2, \\ &[u_1, u_4] = \gamma_1 u_1 + \gamma_2 u_2 + \gamma_3 u_3 \quad [u_2, u_4] = 3\gamma_1 u_2, \quad [u_3, u_4] = 2\gamma_1 u_3 - \gamma_3 u_2, \end{aligned}$$

where $\{u_i\}$ is a pseudo-orthonormal basis with $\langle u_1, u_2 \rangle = \langle u_3, u_3 \rangle = \langle u_4, u_4 \rangle = 1$.

7.1 Nontrivial Conformally Einstein Extensions

As well as in the case of semi-direct extensions of the Poincaré and the Euclidean groups, the Heisenberg group admits semi-direct extensions supporting Bach-flat and conformally Einstein metrics such that the restriction of the metric to the three-dimensional Heisenberg group is degenerate. The following result describes all nontrivial conformally Einstein semi-direct Lorentzian extensions of the Heisenberg group as follows.

Theorem 6 ([17]) *Let $(\mathcal{H}^3 \rtimes \mathbb{R}, \langle \cdot, \cdot \rangle)$ be a nonsymmetric semi-direct extension of the Heisenberg group equipped with a left-invariant Lorentzian metric. Then,* *the metric is nontrivial conformally Einstein if and only if it is isomorphically homothetic to one of the following:*

(D) The restriction of the metric to $\mathfrak{h}^3$ *is degenerate and the metric is determined by*

$$[u_1, u_4] = u_1, \quad [u_2, u_3] = u_1, \quad [u_2, u_4] = u_2,$$

where $\{u_i\}$ *denotes a pseudo-orthonormal basis with* $\langle u_1, u_1\rangle = \langle u_2, u_2\rangle = \langle u_3, u_4\rangle = 1$.

(R) The restriction of the metric to $\mathfrak{h}^3$ *is Riemannian and the metric is determined by*

$$[e_1, e_2] = e_3, \quad [e_1, e_4] = \alpha e_1 + e_3, \quad [e_3, e_4] = \alpha e_3, \quad \text{with} \quad \alpha > 0,$$

where $\{e_i\}$ *denotes an orthonormal basis with* e_4 *timelike.*

(L) The restriction of the metric to $\mathfrak{h}^3$ *is Lorentzian and the metric is determined by*

(i) $[e_1, e_3] = -\alpha e_2$, $[e_1, e_4] = e_1 + \alpha e_2$, $[e_2, e_4] = e_2$, *where* $\alpha > 0$ *and* $\{e_i\}$ *denotes an orthonormal basis with* e_3 *timelike.*

(ii) $[u_1, u_3] = -u_2$, $[u_1, u_4] = 4u_1 + \alpha u_2$, $[u_2, u_4] = 4u_2$, *where* $\alpha \in \mathbb{R}$ *and* $\{u_i\}$ *denotes a pseudo-orthonormal basis with* $\langle u_1, u_2\rangle = \langle u_3, u_3\rangle = \langle u_4, u_4\rangle = 1$.

(iii) $[u_1, u_3] = -u_2$, $[u_1, u_4] = -u_1$, $[u_2, u_4] = 3u_2$, $[u_3, u_4] = 4u_3$, *where* $\{u_i\}$ *denotes a pseudo-orthonormal basis with* $\langle u_1, u_2\rangle = \langle u_3, u_3\rangle = \langle u_4, u_4\rangle = 1$.

Moreover, in all cases but (L.iii), the left-invariant metrics are conformally equivalent to a Ricci-flat pp-wave.

The derived algebra $\mathfrak{g}' = [\mathfrak{g}, \mathfrak{g}]$ is one dimensional for metric (D), which induce a degenerate inner product on the Heisenberg algebra $\mathfrak{h}^3$. Metrics (L) induce a Lorentzian metric on the Heisenberg group which is of Type Ia in case (L.i) and of Type II otherwise. Moreover, the Weyl curvature operator $\mathcal{W} : \Lambda^2 \to \Lambda^2$ is two-step nilpotent in all cases but (L.iii) where it has two distinct real eigenvalues, one of them being a double root of the minimal polynomial.

7.2 Strictly Bach-Flat Extensions

In contrast with Theorem 6 and the case of *pp*-waves in Remark 3 below, strictly Bach-flat semi-direct extensions $\mathcal{H}^3 \rtimes \mathbb{R}$ are quite rare, and they are described in the following.

Theorem 7 ([17]) *Let $(\mathcal{H}^3 \rtimes \mathbb{R}, \langle\cdot,\cdot\rangle)$ be a nonsymmetric semi-direct extension of the Heisenberg group equipped with a left-invariant Lorentzian metric. Then, the metric is strictly Bach-flat if and only if it is isomorphically homothetic to one of the following Lie algebras:*

(i) $[v_1, v_2] = -3\sqrt{14}v_2 + 2\varepsilon\sqrt{11}v_3$, $[v_1, v_3] = \sqrt{14}v_3$, $[v_1, v_4] = -2\sqrt{14}v_4$, $[v_2, v_3] = 4v_4$, *where* $\varepsilon^2 = 1$ *and* $\{v_1, v_2, v_3, v_4\}$ *denotes an orthonormal basis of* $\mathfrak{h}^3 \rtimes \mathfrak{r}$ *with* v_1 *and* v_4 *spacelike and* $\langle v_2, v_2\rangle = -\langle v_3, v_3\rangle = \bar{\varepsilon} = \pm 1$.

(ii) $[u_1, u_3] = -\frac{1}{\sqrt{-(\alpha+\beta^2)}}u_2$, $[u_1, u_4] = \frac{1}{\sqrt{-(\alpha+\beta^2)}}(\beta u_1 + \gamma u_2 + u_3)$, $[u_3, u_4] = \frac{1}{\sqrt{-(\alpha+\beta^2)}}\left(\alpha u_1 - \frac{\alpha+3\beta^2}{4\alpha}u_2 - \beta u_3\right)$, *where the constants* α, β, γ *satisfy* $\gamma = \frac{1}{4\alpha^2}\left((3\alpha+\beta^2)\beta \pm 2\sqrt{-4\alpha^4 - 2(\alpha+\beta^2)^3}\right)$, *with* $-\frac{1}{2} \le \alpha < 0$ *and* $\beta^2 \le -\alpha\left(\sqrt[3]{2\alpha} + 1\right)$. *Moreover,* $\{u_1, u_2, u_3, u_4\}$ *denotes a pseudo-orthonormal basis of* $\mathfrak{h}^3 \rtimes \mathfrak{r}$ *with* $\langle u_1, u_2\rangle = \langle u_3, u_3\rangle = \langle u_4, u_4\rangle = 1$.

The induced metric on the Heisenberg group $\mathcal{H}^3$ is Lorentzian of Type Ia in case (i) and of Type II in case (ii). The Weyl curvature operator has complex eigenvalues in case (i) and it is three-step nilpotent in case (ii).

Remark 3 $(\mathcal{H}^3 \rtimes \mathbb{R}, \langle\cdot,\cdot\rangle)$ is a nonsymmetric Bach-flat *pp*-wave which is not locally conformally flat if and only if it is isomorphically homothetic to a metric determined by one of the following:

(D) $[u_1, u_2] = u_3$, $[u_1, u_4] = \kappa_1 u_1 + \kappa_2 u_2 + \kappa_4 u_3$, $[u_2, u_4] = -\kappa_2 u_1 + \kappa_3 u_2 + \kappa_5 u_3$, $[u_3, u_4] = (\kappa_1 + \kappa_3)u_3$, with $\kappa_1, \kappa_2, \kappa_3, \kappa_4, \kappa_5 \in \mathbb{R}$ satisfying $(\kappa_1 - \kappa_3)(2\kappa_2 + 1) \neq 0$, where $\{u_i\}$ is a pseudo-orthonormal basis with $\langle u_1, u_1\rangle = \langle u_2, u_2\rangle = \langle u_3, u_4\rangle = 1$.

(L) $[u_1, u_3] = -\varepsilon u_2$, $[u_1, u_4] = \kappa_1 u_2 + \kappa_2 u_3$, $[u_3, u_4] = \kappa_3 u_2$, with $\kappa_1, \kappa_2, \kappa_3 \in \mathbb{R}$ such that $\kappa_2(\kappa_2 + \kappa_3) \neq 0$, and $\varepsilon = \pm 1$, where $\{u_i\}$ is a pseudo-orthonormal basis with $\langle u_1, u_2\rangle = \langle u_3, u_3\rangle = \langle u_4, u_4\rangle = 1$.

Moreover, all the cases above are plane waves and thus conformally Einstein.

8 Semi-Direct Extensions of the Abelian Group

Since the structure operator L of the Abelian Lie algebra $\mathfrak{r}^3$ vanishes identically, any linear map $D : \mathfrak{r}^3 \to \mathfrak{r}^3$ is a derivation, and thus, it determines a semi-direct extension $\mathbb{R}^3 \rtimes_D \mathbb{R}$. In order to present a description of all left-invariant Lorentzian metrics on semi-direct extensions of the Abelian Lie group, we consider the restriction of the metric to the Abelian ideal $\mathbb{R}^3$. If the restriction of the metric is degenerate, then any Bach-flat metric corresponds to a plane wave (cf. Remark 4). Whenever the induced metric on $\mathbb{R}^3$ is nondegenerate, let $\Phi(x, y) = \langle Dx, y\rangle$ be the bilinear form associated to D, and let $\Phi_s = \frac{1}{2}(\Phi + {}^t\Phi)$ and $\Phi_a = \frac{1}{2}(\Phi - {}^t\Phi)$ be the symmetric and skew-symmetric parts of Φ, respectively. Moreover, let D_{sad} and D_{asad} defined by $\Phi_s(x, y) = \langle D_{sad}x, y\rangle$ and $\Phi_a(x, y) = \langle D_{asad}x, y\rangle$ be the corresponding self-adjoint and anti-self-adjoint endomorphisms. If the induced metric on $\mathbb{R}^3$ is Riemannian, one may rotate it to consider an orthonormal basis of eigenvectors of D_{sad}. If the restriction of the left-invariant metric to $\mathbb{R}^3$ is of Lorentzian signature, one analyzes separately the different Jordan normal forms of the self-adjoint part of the derivation, as in Sect. 3.

Locally conformally flat left-invariant Lorentzian metrics on semi-direct extensions of the Abelian group have diagonalizable Ricci operator (in which case are locally symmetric [13]), are locally isometric to a plane wave, or otherwise are isomorphically homothetic to a left-invariant metric determined by (see [20]):

(II): $[u_1, u_4] = \varepsilon u_2 - \lambda u_3 + \frac{1}{2}\kappa u_1, [u_2, u_4] = \frac{3}{2}\kappa u_2,\ [u_3, u_4] = \lambda u_2 + \kappa u_3,$

(III): $[u_1, u_4] = 2\kappa u_1,\ [u_3, u_4] = 2u_1 + \kappa u_3,$

where $\{u_i\}$ is a pseudo-orthonormal basis with $\langle u_1, u_2\rangle = \langle u_3, u_3\rangle = \langle u_4, u_4\rangle = 1$. Moreover, the Ricci operator has a single eigenvalue $-3\kappa^2$ which is a double root of the minimal polynomial for metrics II and a triple root for metrics given by III.

8.1 Nontrivial Conformally Einstein Extensions

Nontrivial conformally Einstein Lorentzian semi-direct extensions of the Abelian Lie group are described as follows.

Theorem 8 ([20]) *Let* $(\mathbb{R}^3 \rtimes \mathbb{R}, \langle\cdot,\cdot\rangle)$ *be a nonsymmetric semi-direct extension of the Abelian group equipped with a left-invariant Lorentzian metric. Then, the metric is nontrivial conformally Einstein if and only if it is isomorphically homothetic to one of the following:*

(R) The restriction of the metric to $\mathfrak{r}^3$ *is Riemannian and*

(i) $[e_1, e_4] = e_1 - \gamma_1 e_2,\quad [e_2, e_4] = \gamma_1 e_1 + e_2,\quad [e_3, e_4] = 4e_3,\quad \gamma_1 \geq 0.$
(ii) $[e_1, e_4] = e_1,\quad [e_2, e_4] = \eta_2 e_2,\quad [e_3, e_4] = (\varepsilon + \sqrt{\eta_2})^2 e_3,\quad \varepsilon^2 = 1,$
with $\eta_2 > 0$, $\eta_2 \neq 1$ *and, if* $\varepsilon = -1$, $\eta_2 \neq \frac{1}{4}, 4$.

Here, $\{e_i\}$ *is an orthonormal basis with* e_4 *timelike.*

(L) The restriction of the metric to $\mathfrak{r}^3$ *is Lorentzian and*

(Ia.i) $[e_1, e_4] = e_1 - \gamma_1 e_2,\quad [e_2, e_4] = \gamma_1 e_1 + e_2,\quad [e_3, e_4] = 4e_3,\quad \gamma_1 \geq 0.$
(Ia.ii) $[e_1, e_4] = e_1 + \gamma_2 e_3,\quad [e_2, e_4] = 4e_2,\quad [e_3, e_4] = \gamma_2 e_1 + e_3,\quad \gamma_2 \geq 0.$
(Ia.iii) $[e_1, e_4] = e_1,\quad [e_2, e_4] = \eta_2 e_2,\quad [e_3, e_4] = (\varepsilon + \sqrt{\eta_2})^2 e_3,\quad \varepsilon^2 = 1,$
with $\eta_2 > 0$, $\eta_2 \neq 1$ *and, if* $\varepsilon = -1$, $\eta_2 \neq \frac{1}{4}, 4$.
(Ib) $[e_1, e_4] = \eta e_1,\ [e_2, e_4] = \frac{\eta^2-4}{4\eta} e_2 - e_3,\ [e_3, e_4] = e_2 + \frac{\eta^2-4}{4\eta} e_3,\quad 0 < \eta \neq \frac{2}{\sqrt{3}}.$
(II.i) $[u_1, u_4] = u_1 + \varepsilon u_2,\quad [u_2, u_4] = u_2,\quad \varepsilon^2 = 1.$

(II.ii) $[u_1, u_4] = u_2 - \gamma_3 u_3$, $[u_2, u_4] = 2\eta_1 u_2$, $[u_3, u_4] = \gamma_3 u_2 + \eta_1 u_3$, *with* $\gamma_3 \in \mathbb{R}$, $\eta_1 \neq 0$.

(II.iii) $[u_1, u_4] = u_1 + \varepsilon u_2$, $[u_2, u_4] = u_2$, $[u_3, u_4] = 4u_3$, $\varepsilon^2 = 1$.

(III) $[u_1, u_4] = 2u_1$, $[u_2, u_4] = (1 - \gamma_2)u_3$, $[u_3, u_4] = (\gamma_2 + 1)u_1 + u_3$, $\gamma_2 \neq 1$.

Here, $\{e_i\}$ *is an orthonormal basis with* e_3 *timelike, while* $\{u_i\}$ *is a pseudo-orthonormal basis with* $\langle u_1, u_2 \rangle = \langle u_3, u_3 \rangle = \langle u_4, u_4 \rangle = 1$.

Moreover, in all cases, the left-invariant metrics are conformally equivalent to a Ricci-flat metric.

Left-invariant metrics inducing degenerate ones on $\mathbb{R}^3$ reduce to plane waves, and thus, they are trivially conformally Einstein (cf. Remark 4). On the other hand, Theorem 8 shows the existence of conformally Einstein Lorentzian semi-direct extensions of the Abelian group, $\mathbb{R}^3 \rtimes_D \mathbb{R}$, for all possible Jordan normal forms of the self-adjoint part of the derivation.

The Weyl curvature operator $\mathcal{W} : \Lambda^2 \to \Lambda^2$ is two-step nilpotent in cases (II.i), (II.ii), and (III). It is diagonalizable with real eigenvalues in all cases (R) and (Ia), and it is diagonalizable with complex eigenvalues in case (Ib). The Weyl curvature operator has two distinct eigenvalues in cases (R.i), (Ia.i), and (Ia.ii), while it has three distinct eigenvalues in cases (R.ii) and (Ia.iii). In case (II.iii), the Weyl curvature operator has two distinct real eigenvalues -4 and 2, the latter with multiplicity four being a double root of the minimal polynomial. Left-invariant metrics given by (Ib) are the only conformally Einstein Lorentzian left-invariant metrics whose Weyl curvature operator has complex eigenvalues. Analogously, metrics in case (II.iii) are the only conformally Einstein ones whose Weyl curvature operator is not diagonalizable with nonzero real eigenvalues.

8.2 Strictly Bach-Flat Extensions

In contrast with the previously considered cases, strictly Bach-flat extensions of the Abelian group constitute a large family.

Remark 4 The special case of Bach-flat *pp*-wave semi-direct extensions corresponds to the following situations, when the restriction of the metric to $\mathbb{R}^3$ is degenerate or Lorentzian and the self-adjoint part of the derivation has a single eigenvalue which is a triple root of the minimal polynomial (see [20]):

(III) $[u_2, u_4] = (1 - \gamma_2)u_3$, $[u_3, u_4] = (1 + \gamma_2)u_1$ where $\{u_i\}$ is a pseudo-orthonormal basis with $\langle u_1, u_2 \rangle = \langle u_3, u_3 \rangle = \langle u_4, u_4 \rangle = 1$.

(D) $[u_1, u_4] = \gamma_1 u_1 + \gamma_2 u_2 + \gamma_3 u_3$, $[u_2, u_4] = \gamma_4 u_1 + \gamma_5 u_2 + \gamma_6 u_3$, $[u_3, u_4] = \gamma_7 u_1$, where $\{u_i\}$ is a pseudo-orthonormal basis with $\langle u_1, u_1\rangle = \langle u_2, u_2\rangle = \langle u_3, u_4\rangle = 1$.

Moreover, the two cases above are conformally Einstein plane waves.

In order to describe the class of strictly Bach-flat Lorentzian extensions of $\mathbb{R}^3$, we consider the different possibilities for the metric. Left-invariant metrics on $\mathbb{R}^3 \rtimes \mathbb{R}$ inducing a degenerate one on $\mathbb{R}^3$ were considered in Remark 4, and the case when the induced metric on $\mathbb{R}^3$ is Riemannian does not provide strictly Bach-flat metrics. Hence, the discussion reduces to Lorentzian metrics on $\mathbb{R}^3 \rtimes \mathbb{R}$ which restrict to a Lorentzian metric on $\mathbb{R}^3$. As in the case of the conformally Einstein metrics, there exist strictly Bach-flat metrics for all possible Jordan normal forms of the self-adjoint part of the derivation. The strictly Bach-flat metrics on $\mathbb{R}^3 \rtimes \mathbb{R}$ are given as follows, which briefly summarizes the results in [20].

8.2.1 The Case of Diagonalizable D_{sad}

It is shown in [20] that a left-invariant metric in this case is strictly Bach-flat if and only if it is isomorphically homothetic to one of the metrics described by

(Ia.i) $[e_1, e_4] = e_1$, $[e_2, e_4] = \frac{4+\varepsilon\sqrt{30}}{4} e_2 + \frac{\sqrt{6}}{4} e_3$, $[e_3, e_4] = \frac{\sqrt{6}}{4} e_2 + \frac{4-\varepsilon\sqrt{30}}{4} e_3$.

(Ia.ii) $[e_1, e_4] = e_1$, $[e_2, e_4] = \frac{\varepsilon\sqrt{6}-2}{4} e_2 + \frac{3\sqrt{2}}{4} e_3$, $[e_3, e_4] = \frac{3\sqrt{2}}{4} e_2 - \frac{2+\varepsilon\sqrt{6}}{4} e_3$.

(Ia.iii) $[e_1, e_4] = -\frac{7+\varepsilon\sqrt{609}}{14} e_1 + \sqrt{\frac{4089-137\varepsilon\sqrt{609}}{406}}\, e_3$,

$[e_2, e_4] = -\frac{7-\varepsilon\sqrt{609}}{14} e_2 + \sqrt{\frac{4089+137\varepsilon\sqrt{609}}{406}}\, e_3$,

$[e_3, e_4] = \sqrt{\frac{4089-137\varepsilon\sqrt{609}}{406}}\, e_1 + \sqrt{\frac{4089+137\varepsilon\sqrt{609}}{406}}\, e_2 + e_3$.

(Ia.iv) $[e_1, e_4] = \left(1 - \varepsilon\sqrt{\frac{30}{7}}\right) e_1 + \sqrt{\frac{33}{14}} e_3$, $\quad [e_2, e_4] = \left(1 + \varepsilon\sqrt{\frac{30}{7}}\right) e_2 + \sqrt{\frac{33}{14}} e_3$,

$[e_3, e_4] = \sqrt{\frac{33}{14}}(e_1 + e_2) + e_3$.

(Ia.v) $[e_1, e_4] = e_1 - \gamma_1 e_2 + \gamma_2 e_3$,

$[e_2, e_4] = \gamma_1 e_1 + \eta_2 e_2 + \gamma_3 e_3$, $\quad [e_3, e_4] = \gamma_2 e_1 + \gamma_3 e_2 + \eta_3 e_3$,

where $\eta_3 = -0.707406\ldots$ or $\eta_3 = 0.376659\ldots$, which correspond to the smaller and larger real solutions of

$$8\eta_3^6 + 12\eta_3^5 - 45\eta_3^4 - 100\eta_3^3 - 45\eta_3^2 + 12\eta_3 + 8 = 0$$

in the interval $(-1, \frac{1}{2})$. For those values of η_3, the parameter η_2 is given by

$$10\eta_2 = 56\eta_3^5 + 36\eta_3^4 - 339\eta_3^3 - 414\eta_3^2 + \eta_3 + 70,$$

and γ_2 and γ_3 correspond to the positive solutions of

$$15\gamma_2^2 = -16\eta_3^5 - 24\eta_3^4 + 106\eta_3^3 + 196\eta_3^2 + 21\eta_3 - 28 \quad \text{and}$$
$$15\gamma_3^2 = -88\eta_3^5 - 60\eta_3^4 + 535\eta_3^3 + 670\eta_3^2 - 15\eta_3 - 112,$$

respectively. Moreover, γ_1 is uniquely determined by

$$40\gamma_1 = -\gamma_2\gamma_3\left(408\eta_3^5 + 452\eta_3^4 - 2535\eta_3^3 - 4072\eta_3^2 - 359\eta_3 + 856\right).$$

(Ia.vi) $[e_1, e_4] = e_1 - \gamma_1 e_2 + \gamma_2 e_3$,
$[e_2, e_4] = \gamma_1 e_1 + \eta_2 e_2 + \gamma_3 e_3$, $\quad [e_3, e_4] = \gamma_2 e_1 + \gamma_3 e_2 + \eta_3 e_3$,
with $\eta_3 = -0.425775\ldots$ or $\eta_3 = 0.169957\ldots$, which correspond to the smaller and larger solutions of

$$128\eta_3^6 - 48\eta_3^5 - 2985\eta_3^4 - 6340\eta_3^3 - 2985\eta_3^2 - 48\eta_3 + 128 = 0$$

in the interval $(-\frac{1}{2}, \frac{1}{4})$. For those values of η_3, the parameter η_2 is given by

$$760\eta_2 = 5504\eta_3^5 - 5136\eta_3^4 - 124131\eta_3^3 - 207636\eta_3^2 - 33731\eta_3 + 12040,$$

and γ_2 and γ_3 correspond to the positive solutions of

$$180\gamma_2^2 = -1664\eta_3^5 + 624\eta_3^4 + 40469\eta_3^3 + 77204\eta_3^2 + 12969\eta_3 - 4592,$$

and

$$3420\gamma_3^2 = -7040\eta_3^5 + 7248\eta_3^4 + 156623\eta_3^3 + 254948\eta_3^2 + 40023\eta_3 - 12512,$$

respectively. Finally, γ_1 is uniquely determined by

$$82080\gamma_1 = \gamma_2\gamma_3\big(477824\eta_3^5 - 328688\eta_3^4 - 11131997\eta_3^3 - 19885316\eta_3^2 - 3537085\eta_3 + 1274912\big).$$

Here, $\{e_i\}$ is an orthonormal basis with e_3 timelike, and the parameter $\varepsilon = \pm 1$.

The left-invariant metric in cases (Ia.i)–(Ia.iv) and (Ia.vi) is weakly generic with complex eigenvalues. In opposition, the Weyl curvature operator is two-step nilpotent in case (Ia.v).

8.2.2 D_{sad} Has Complex Eigenvalues

A left-invariant metric corresponding to a semi-direct extension $\mathbb{R}^3 \rtimes_D \mathbb{R}$ where the self-adjoint part D_{sad} has complex eigenvalues is strictly Bach-flat if and only if it is isomorphically homothetic to the metric described by (see [20])

$$\begin{aligned}
[e_1, e_4] &= \tfrac{2}{\sqrt{15}} e_1 + \tfrac{1}{15}\sqrt{6\left(8\sqrt{10}+25\right)}\, e_2 + \tfrac{1}{15}\sqrt{6\left(8\sqrt{10}-25\right)}\, e_3,\\
[e_2, e_4] &= -\tfrac{1}{15}\sqrt{6\left(8\sqrt{10}+25\right)}\, e_1 - \tfrac{1}{\sqrt{15}} e_2 - e_3,\\
[e_3, e_4] &= \tfrac{1}{15}\sqrt{6\left(8\sqrt{10}-25\right)}\, e_1 + e_2 - \tfrac{1}{\sqrt{15}} e_3,
\end{aligned}$$

where $\{e_i\}$ is an orthonormal basis with e_3 timelike. Moreover, the Weyl curvature operator of the left-invariant metric above is three-step nilpotent.

8.2.3 D_{sad} Has a Double Root of the Minimal Polynomial

It follows from the work in [20] that a left-invariant metric in this case is strictly Bach-flat if and only if it is isomorphically homothetic to one of the metrics described by

$$\begin{aligned}
&\text{(II.i)} && [u_1, u_4] = -\tfrac{5}{24}u_1 + u_2 - u_3, && [u_2, u_4] = \tfrac{5}{12}u_2, && [u_3, u_4] = u_2,\\
&\text{(II.ii)} && [u_1, u_4] = -2u_1 + \varepsilon u_2, && [u_2, u_4] = 4u_2, && [u_3, u_4] = 4u_3, \quad \varepsilon^2 = 1,\\
&\text{(II.iii)} && [u_1, u_4] = \tfrac{5}{18}u_1 + u_2 - u_3, && [u_2, u_4] = -\tfrac{25}{72}u_2, && [u_3, u_4] = u_2 - \tfrac{5}{36}u_3,
\end{aligned}$$

where $\{u_i\}$ is a pseudo-orthonormal basis with $\langle u_1, u_2\rangle = \langle u_3, u_3\rangle = \langle u_4, u_4\rangle = 1$.

The Weyl curvature operator of metrics determined by (II.i) is three-step nilpotent. The cases (II.ii) and (II.iii) are weakly generic with two distinct nonzero real eigenvalues, one with multiplicity four which is a double root of the minimal polynomial.

8.2.4 D_{sad} Has a Triple Root of the Minimal Polynomial

Left-invariant metrics in this situation are strictly Bach-flat if and only if they are isomorphically homothetic to a metric described by (see [20])

$$[u_1, u_4] = 3u_1, \quad [u_2, u_4] = -u_2 + u_3, \quad [u_3, u_4] = u_1 + u_3,$$

where $\{u_i\}$ is a pseudo-orthonormal basis with $\langle u_1, u_2\rangle = \langle u_3, u_3\rangle = \langle u_4, u_4\rangle = 1$. Moreover, the self-dual Weyl curvature operator is three-step nilpotent.

9 The Non-solvable Cases $\widetilde{SL}(2,\mathbb{R}) \times \mathbb{R}$ and $SU(2) \times \mathbb{R}$

In sharp contrast with the previously considered cases, the product Lie groups $\widetilde{SL}(2,\mathbb{R}) \times \mathbb{R}$ and $SU(2) \times \mathbb{R}$ do not admit nontrivial left-invariant conformally Einstein metrics [19]. However, there are plenty of strictly Bach-flat metrics which are somehow rare in the solvable cases.

In order to describe these Bach-flat metrics, we consider separately the different possibilities for the induced metric on the semisimple Lie group G_3, where $G_3 = \widetilde{SL}(2,\mathbb{R})$ or $G_3 = SU(2)$. We describe Bach-flat left-invariant metrics in terms of three polynomial equations involving five parameters. Generically, the corresponding map $F : \mathbb{R}^5 \to \mathbb{R}^3$ is of rank three, which shows that generically there is a surface of strictly Bach-flat metrics in each of the cases below. Bach-flat metrics which are not locally conformally flat have nonzero scalar curvature in all cases. Hence they cannot correspond to any pp-wave left-invariant metric. Furthermore, the Weyl curvature operator $\mathcal{W} : \Lambda^2 \to \Lambda^2$ is three-step nilpotent in all cases.

Remark 5 A locally conformally flat left-invariant Lorentzian metric on the product Lie groups $\widetilde{SL}(2,\mathbb{R}) \times \mathbb{R}$ or $SU(2) \times \mathbb{R}$ has diagonalizable Ricci operator (in which case is locally symmetric [13]), it is locally isometric to a plane wave, or otherwise it is isomorphically homothetic to a metric $g_{D.i}$ as in (10) described by

$$[v_1, v_2] = v_3,\ [v_1, v_3] = \lambda v_1,\ [v_2, v_3] = -\lambda v_2, \quad (\lambda \neq 0),$$

where $\{v_1, v_2, v_3, v_4\}$ denotes a pseudo-orthonormal basis with $\langle v_1, v_2\rangle = \kappa \neq \pm 1$, $\langle v_1, v_1\rangle = \langle v_2, v_2\rangle = \langle v_3, v_4\rangle = 1$. Moreover, the Ricci operator has eigenvalues $\{\pm\lambda, \pm i\lambda\}$.

We briefly summarize the results in [19] in the remaining of this section as follows.

9.1 Left-Invariant Metrics with Riemannian G_3

If $\langle\cdot,\cdot\rangle$ is a Lorentzian inner product on $\mathfrak{g} = \mathfrak{sl}(2,\mathbb{R}) \times \mathfrak{r}$ or $\mathfrak{g} = \mathfrak{su}(2) \times \mathfrak{r}$ whose restriction to $\mathfrak{g}_3$ is of Riemannian signature, then the structure operator of $\mathfrak{g}_3$, defined by Milnor in [35], is self-adjoint and diagonalizable. As a consequence, there exists an orthonormal basis $\{e_1, e_2, e_3, e_4\}$ of $\mathfrak{g}$, with e_4 timelike, where $\mathfrak{g}_3 = \operatorname{span}\{e_1, e_2, e_3\}$ and $\mathfrak{r} = \operatorname{span}\{e_4\}$, so that the left-invariant metric is described by

$$\mathfrak{g}_R \begin{cases} [e_1, e_2] = \lambda_3 e_3, \quad [e_1, e_3] = -\lambda_2 e_2, \quad [e_2, e_3] = \lambda_1 e_1, \\ [e_1, e_4] = \gamma_1\lambda_2 e_2 + \gamma_2\lambda_3 e_3, \quad [e_2, e_4] = -\gamma_1\lambda_1 e_1 + \gamma_3\lambda_3 e_3, \\ [e_3, e_4] = -\gamma_2\lambda_1 e_1 - \gamma_3\lambda_2 e_2, \end{cases} \tag{5}$$

for certain $\gamma_1, \gamma_2, \gamma_3 \in \mathbb{R}$, where $\lambda_1\lambda_2\lambda_3 \neq 0$ determine the Lie algebra $\mathfrak{su}(2)$ if they have the same sign, and $\mathfrak{sl}(2, \mathbb{R})$ otherwise.

A left-invariant metric $\mathfrak{g}_R$ is Bach-flat if and only if it is locally conformally flat (whenever $\lambda_1 = \lambda_2 = \lambda_3$, in which case the Ricci operator is diagonalizable), or, otherwise, it is isomorphically homothetic to a metric determined by (5) where $\lambda_1 = 1$ and the remaining parameters satisfy the equations $\{\mathfrak{P}_i = 0\}$, where

$$\begin{aligned}
\mathfrak{P}_1 &= \gamma_1^2 + \gamma_2^2 + \gamma_3^2 - 1, \\
\mathfrak{P}_2 &= (\lambda_2 - \lambda_3)(2\lambda_2 + 2\lambda_3 - 1)\gamma_2^2 + (\lambda_3 - 1)(\lambda_2 - 2\lambda_3 - 2)\gamma_3^2 \\
&\quad - \lambda_3(\lambda_2 - 2\lambda_3 + 1), \\
\mathfrak{P}_3 &= 3(\lambda_2 - \lambda_3)\gamma_2^2 - 3(\lambda_3 - 1)\left(2(\lambda_3 + 1)\lambda_2^2 + \lambda_2 - 2\lambda_3 - 2\right)\gamma_3^2 \\
&\quad + \lambda_3\left(2(3\lambda_3 - 2)\lambda_2^2 - (4\lambda_3 + 1)\lambda_2 + 3\right).
\end{aligned}$$

Moreover, in the latter case, the Bach-flat metric is always strict, and it is realized on the product Lie group $SU(2) \times \mathbb{R}$. Furthermore, the Weyl curvature operator $\mathcal{W} : \Lambda^2 \to \Lambda^2$ is three-step nilpotent.

9.2 *Left-Invariant Metrics with Lorentzian G_3*

The left-invariant metrics occurring in this situation arise from the different possibilities of the Jordan normal form of the structure operator L of the three-dimensional unimodular Lie group G_3. There are no strictly Bach-flat metrics whose structure operator has a triple root of its minimal polynomial. All the remaining possibilities admit strictly Bach-flat metrics as follows.

9.2.1 The Structure Operator L Is Diagonalizable

There exists an orthonormal basis $\{e_1, e_2, e_3, e_4\}$ of $\mathfrak{g} = \mathfrak{g}_3 \times \mathfrak{r}$, with e_3 timelike, where $\mathfrak{g}_3 = \operatorname{span}\{e_1, e_2, e_3\}$ and $\mathfrak{r} = \operatorname{span}\{e_4\}$, so that the left-invariant metric is determined by

$$\mathfrak{g}_{L.Ia} \begin{cases} [e_1, e_2] = -\lambda_3 e_3, \quad [e_1, e_3] = -\lambda_2 e_2, \quad [e_2, e_3] = \lambda_1 e_1, \\ [e_1, e_4] = \gamma_1\lambda_2 e_2 + \gamma_2\lambda_3 e_3, \quad [e_2, e_4] = -\gamma_1\lambda_1 e_1 + \gamma_3\lambda_3 e_3, \\ [e_3, e_4] = \gamma_2\lambda_1 e_1 + \gamma_3\lambda_2 e_2, \end{cases} \tag{6}$$

for certain $\gamma_1, \gamma_2, \gamma_3 \in \mathbb{R}$, where $\lambda_1\lambda_2\lambda_3 \neq 0$ determine the Lie algebra $\mathfrak{su}(2)$ if $\varepsilon_1\lambda_1, \varepsilon_2\lambda_2$, and $\varepsilon_3\lambda_3$ have the same sign, or $\mathfrak{sl}(2, \mathbb{R})$ otherwise, where $\varepsilon_k = \langle e_k, e_k \rangle$.

A Lorentzian metric $\mathfrak{g}_{L.Ia}$ is Bach-flat if and only if it is locally conformally flat (whenever $\lambda_1 = \lambda_2 = \lambda_3$, in which case the Ricci operator is diagonalizable) or, otherwise, it is isomorphically homothetic to a metric determined by (6) where

$\lambda_1 = 1$ and the remaining parameters satisfy the equations $\{\mathfrak{P}_i = 0\}$, where

$$\begin{aligned}
\mathfrak{P}_1 &= \gamma_1^2 - \gamma_2^2 - \gamma_3^2 - 1,\\
\mathfrak{P}_2 &= (\lambda_2 - \lambda_3)(2\lambda_2 + 2\lambda_3 - 1)\gamma_2^2 + (\lambda_3 - 1)(\lambda_2 - 2\lambda_3 - 2)\gamma_3^2\\
&\quad + \lambda_3(\lambda_2 - 2\lambda_3 + 1),\\
\mathfrak{P}_3 &= 3(\lambda_2 - \lambda_3)\gamma_2^2 - 3(\lambda_3 - 1)\left(2(\lambda_3 + 1)\lambda_2^2 + \lambda_2 - 2\lambda_3 - 2\right)\gamma_3^2\\
&\quad - \lambda_3\left(2(3\lambda_3 - 2)\lambda_2^2 - (4\lambda_3 + 1)\lambda_2 + 3\right).
\end{aligned} \tag{7}$$

Moreover, in the latter case, the Bach-flat metric is always strict, and it is realized on the product Lie group $SU(2) \times \mathbb{R}$ or $\widetilde{SL}(2, \mathbb{R}) \times \mathbb{R}$. Furthermore, the Weyl curvature operator is three-step nilpotent.

Explicit solutions to Eq. (7) can be given as follows. Setting $\lambda_1 = 1$, for any nonzero value of the parameter $\lambda_2 \in \mathbb{R} \setminus [\frac{4}{5}, \frac{5}{4}]$ with $\lambda_2 \neq -1$, there exist some $\lambda_3 \in (-\infty, 1)$ such that the remaining parameters $\gamma_1, \gamma_2, \gamma_3$ are determined by

$$\begin{aligned}
\gamma_1^2 &= \frac{\lambda_2\left((\lambda_2(4\lambda_2 + 11) + 4)\lambda_3 - (\lambda_2 + 1)\lambda_3^2 - 6\lambda_2(\lambda_2 + 1) - 5\lambda_3^3\right)}{3(\lambda_3 - 1)(\lambda_2 - \lambda_3)\left(2\lambda_2^2(\lambda_3 + 1) + \lambda_2(\lambda_3(2\lambda_3 + 3) + 2) + 2\lambda_3(\lambda_3 + 1)\right)},\\
\gamma_2^2 &= \frac{\lambda_3\left(\lambda_2(\lambda_3(4\lambda_3 + 11) + 4) - 5\lambda_2^3 - \lambda_2^2(\lambda_3 + 1) - 6\lambda_3(\lambda_3 + 1)\right)}{3(\lambda_2 - 1)(\lambda_2 - \lambda_3)\left(2\lambda_2^2(\lambda_3 + 1) + \lambda_2(\lambda_3(2\lambda_3 + 3) + 2) + 2\lambda_3(\lambda_3 + 1)\right)},\\
\gamma_3^2 &= -\frac{\lambda_2\lambda_3\left(\lambda_2^2(6\lambda_3 - 4) + \lambda_2\lambda_3(6\lambda_3 - 11) + \lambda_2 - 4\lambda_3^2 + \lambda_3 + 5\right)}{3(\lambda_2 - 1)(\lambda_3 - 1)\left(2\lambda_2^2(\lambda_3 + 1) + \lambda_2(\lambda_3(2\lambda_3 + 3) + 2) + 2\lambda_3(\lambda_3 + 1)\right)}.
\end{aligned}$$

9.2.2 The Structure Operator L Has Complex Eigenvalues

If the structure operator L is of Type Ib, then there exists an orthonormal basis $\{e_1, e_2, e_3, e_4\}$ of $\mathfrak{g} = \mathfrak{g}_3 \times \mathfrak{r}$, with e_3 timelike, where $\mathfrak{g}_3 = \operatorname{span}\{e_1, e_2, e_3\}$ and $\mathfrak{r} = \operatorname{span}\{e_4\}$, such that the Lie algebra structure is given by

$$\mathfrak{g}_{L.Ib}\begin{cases}
[e_1, e_2] = -\beta e_2 - \alpha e_3, \quad [e_1, e_3] = -\alpha e_2 + \beta e_3, \quad [e_2, e_3] = \lambda e_1,\\
[e_1, e_4] = (\alpha^2 + \beta^2)(\gamma_1 e_2 + \gamma_2 e_3),\\
[e_2, e_4] = -(\gamma_1\alpha - \gamma_2\beta)\lambda e_1 + \gamma_3\beta e_2 + \gamma_3\alpha e_3,\\
[e_3, e_4] = (\gamma_2\alpha + \gamma_1\beta)\lambda e_1 + \gamma_3\alpha e_2 - \gamma_3\beta e_3,
\end{cases} \tag{8}$$

for certain $\gamma_1, \gamma_2, \gamma_3 \in \mathbb{R}$, where $\alpha \in \mathbb{R}$, $\beta \neq 0$ and $\lambda \neq 0$. In this case, the underlying unimodular Lie group corresponds to $\widetilde{SL}(2, \mathbb{R}) \times \mathbb{R}$. A straightforward calculation shows that $\mathfrak{g}_{L.Ib}$ is never Einstein, nor locally conformally flat, nor locally symmetric. The characterization of the Bach-flatness is given as follows.

A left-invariant Lorentzian metric $g_{L.Ib}$ is Bach-flat if and only if it corresponds to a metric determined by (8) with parameters satisfying the equations $\{\mathfrak{P}_i = 0\}$, where

$$\begin{aligned}
\mathfrak{P}_1 &= (\alpha^2-\beta^2)(\gamma_1^2-\gamma_2^2)-\gamma_3^2-4\alpha\beta\gamma_1\gamma_2-1,\\
\mathfrak{P}_2 &= -2\beta^2(\alpha\lambda-4\beta^2)(\gamma_1^2-\gamma_2^2)+(2\lambda^2-\alpha\lambda-\alpha^2+11\beta^2)\gamma_3^2\\
&\quad -2\beta\big((\alpha^2-\beta^2)\lambda-4\alpha(\alpha^2+3\beta^2)\big)\gamma_1\gamma_2+\alpha\lambda-\alpha^2+11\beta^2,\\
\mathfrak{P}_3 &= 6\alpha\beta^2\lambda^3(\gamma_1^2-\gamma_2^2)-3\big(2\lambda^4-\alpha\lambda^3-(3\alpha^2-5\beta^2)\lambda^2+2(\alpha^2+\beta^2)^2\big)\gamma_3^2\\
&\quad +6\beta(\alpha^2-\beta^2)\lambda^3\gamma_1\gamma_2-3\alpha\lambda^3+(\alpha^2+\beta^2)\lambda^2+8\alpha(\alpha^2+\beta^2)\lambda-6(\alpha^2+\beta^2)^2.
\end{aligned}$$

Moreover, the Bach-flat metric is always strict and the Weyl curvature operator is three-step nilpotent.

9.2.3 The Structure Operator *L* Has a Double Root of Its Minimal Polynomial

If L is of Type II, there exists a pseudo-orthonormal basis $\{u_1, u_2, u_3, u_4\}$ of $\mathfrak{g} = \mathfrak{g}_3 \times \mathfrak{r}$, with $\langle u_1, u_2\rangle = \langle u_3, u_3\rangle = \langle u_4, u_4\rangle = 1$, where $\mathfrak{g}_3 = \operatorname{span}\{u_1, u_2, u_3\}$ and $\mathfrak{r} = \operatorname{span}\{u_4\}$, so that the Lie algebra is given by

$$\mathfrak{g}_{L.II}\begin{cases}[u_1,u_2]=\lambda_2u_3, \quad [u_1,u_3]=-\lambda_1u_1-\varepsilon u_2, \quad [u_2,u_3]=\lambda_1u_2,\\ [u_1,u_4]=\lambda_1\gamma_1u_1+\varepsilon\gamma_1u_2+\gamma_2\lambda_2u_3, \quad [u_2,u_4]=-\gamma_1\lambda_1u_2+\gamma_3\lambda_2u_3,\\ [u_3,u_4]=-\gamma_3\lambda_1u_1-(\gamma_2\lambda_1+\varepsilon\gamma_3)u_2,\end{cases} \tag{9}$$

with $\varepsilon^2 = 1$, for certain $\gamma_1, \gamma_2, \gamma_3 \in \mathbb{R}$, where $\lambda_1\lambda_2 \neq 0$. The underlying Lie group corresponds to $\widetilde{SL}(2,\mathbb{R})\times\mathbb{R}$. A straightforward calculation shows that $g_{L.II}$ is never Einstein, nor locally conformally flat, nor locally symmetric. We characterize the Bach-flatness as follows.

A left-invariant Lorentzian metric $g_{L.II}$ is Bach-flat if and only if it is determined by (9) with parameters satisfying the equations $\{\mathfrak{P}_i = 0\}$, where

$$\begin{aligned}
&\mathfrak{P}_1 = \gamma_1^2-2\gamma_2\gamma_3+1, \qquad \mathfrak{P}_2=\lambda_1(\lambda_1-\lambda_2)\gamma_2+\varepsilon(4\lambda_1+3\lambda_2)\gamma_3,\\
&\mathfrak{P}_3 = 16\varepsilon\gamma_3^2-(17\lambda_1+28\lambda_2)\gamma_2\gamma_3+14\lambda_2.
\end{aligned}$$

Moreover, the Bach-flat metric is always strict and the Weyl curvature operator is three-step nilpotent.

9.3 *Left-Invariant Metrics with Degenerate G_3*

Let $\langle\cdot,\cdot\rangle$ be a Lorentzian inner product on $\mathfrak{g} = \mathfrak{g}_3 \times \mathfrak{r}$ which restricts to a degenerate inner product on $\mathfrak{g}_3$. Let $\mathfrak{g}_3' = [\mathfrak{g}_3, \mathfrak{g}_3]$ be the derived subalgebra of $\mathfrak{g}_3$. Let $\{u_1, u_2, u_3, u_4\}$ be a pseudo-orthonormal basis of $\mathfrak{g} = \mathfrak{g}_3 \times \operatorname{span}\{u_4\}$ with $\langle u_1, u_1\rangle = \langle u_2, u_2\rangle = \langle u_3, u_4\rangle$ and $\operatorname{ad}_{u_3} : \mathfrak{g}_3 \to \mathfrak{g}_3$. Since the derived algebra $\mathfrak{g}_3' = \mathfrak{g}_3$, ad_{u_3} must be of rank 2 and, apart from 0, it must have either two real eigenvalues or two conjugate complex eigenvalues. We consider these two possibilities separately, omitting the case of ad_{u3} being three-step nilpotent which does not support Bach-flat metrics.

9.3.1 ad_{u_3} Has Real Eigenvalues $\{0, \lambda, -\lambda\}$, with $\lambda \neq 0$

In this case, let v_1 and v_2 be unit eigenvectors, i.e., $[v_1, u_3] = \lambda v_1$ and $[v_2, u_3] = -\lambda v_2$. The Jacobi identity implies $[v_1, v_2] \in \operatorname{span}\{u_3\}$. Thus, rescaling u_3 if necessary, we get a basis $\{v_1, v_2, v_3, v_4\}$ of $\mathfrak{g} = \mathfrak{g}_3 \times \mathfrak{r}$, with $\langle v_1, v_1\rangle = \langle v_2, v_2\rangle = \langle v_3, v_4\rangle = 1$, $\langle v_1, v_2\rangle = \kappa \neq \pm 1$, where $\mathfrak{g}_3 = \operatorname{span}\{v_1, v_2, v_3\}$ and $\mathfrak{r} = \operatorname{span}\{v_4\}$, so that the corresponding left-invariant metric is determined by

$$\mathfrak{g}_{D.i} \begin{cases} [v_1, v_2] = v_3, & [v_1, v_3] = \lambda v_1, & [v_1, v_4] = \gamma_1 v_1 + \gamma_2 v_3, \\ [v_2, v_3] = -\lambda v_2, & [v_2, v_4] = -\gamma_1 v_2 + \gamma_3 v_3, & [v_3, v_4] = \gamma_3 \lambda v_1 + \gamma_2 \lambda v_2, \end{cases} \tag{10}$$

for certain $\gamma_1, \gamma_2, \gamma_3 \in \mathbb{R}$, where $\lambda \neq 0$, which are realized on $\widetilde{SL}(2,\mathbb{R}) \times \mathbb{R}$. A straightforward calculation shows that $\mathfrak{g}_{D.i}$ is neither Einstein nor locally symmetric. Moreover, it is locally conformally flat if and only if $\kappa = \gamma_1 = \gamma_2 = \gamma_3 = 0$ (cf. Remark 5).

A metric $\mathfrak{g}_{D.i}$ is Bach-flat if and only if it is locally conformally flat or, otherwise, it corresponds to a metric determined by (10) with parameters satisfying the equations $\{\mathfrak{P}_i = 0\}$, where

$$\begin{aligned} &\mathfrak{P}_1 = 3(\kappa^2 - 1)\gamma_1 + 2\kappa, \qquad \mathfrak{P}_2 = 24\gamma_1^2 - 8\gamma_2\gamma_3\lambda + 9\kappa\gamma_1 + 6, \\ &\mathfrak{P}_3 = 24\kappa\gamma_1^2 - 4\lambda(\gamma_2^2 + \gamma_3^2) + 17\gamma_1. \end{aligned} \tag{11}$$

Moreover, in the latter case, the Bach-flat metric is always strict and the Weyl curvature operator is three-step nilpotent.

Explicit solutions to Eq. (11) can be given as follows. Setting $\gamma_2 \neq 0$, for any value of the parameter $\kappa \in \mathbb{R} \setminus (-9, 9)$, the remaining parameters $\gamma_1, \gamma_3, \lambda$ are determined by

$$\gamma_1 = \frac{2\kappa}{3(1-\kappa^2)}, \quad \gamma_3 = \frac{\gamma_2\kappa(17-\kappa^2) \pm \sqrt{\gamma_2^2(\kappa^2-81)(\kappa^2-1)^2}}{7\kappa^2+9},$$
$$\lambda = \frac{7\kappa^2+9}{12\gamma_2\gamma_3(\kappa^2-1)^2}.$$

9.3.2 ad_{u_3} Has Complex Eigenvalues $\{0, i\beta, -i\beta\}$, with $\beta \neq 0$

In this case, we consider unit vectors v_1 and v_2 so that $[v_1, u_3] = \beta v_2$ and $[v_2, u_3] = -\beta v_1$. The Jacobi identity implies $[v_1, v_2] \in \mathrm{span}\{u_3\}$. Thus, rescaling u_3 if necessary, we get a basis $\{v_1, v_2, v_3, v_4\}$ of $\mathfrak{g} = \mathfrak{g}_3 \times \mathfrak{r}$, with $\langle v_1, v_1\rangle = \langle v_2, v_2\rangle = \langle v_3, v_4\rangle = 1$, $\langle v_1, v_2\rangle = \kappa \neq \pm 1$, where $\mathfrak{g}_3 = \mathrm{span}\{v_1, v_2, v_3\}$ and $\mathfrak{r} = \mathrm{span}\{v_4\}$, such that the left-invariant metric is determined by

$$\mathfrak{g}_{D.ii} \begin{cases} [v_1, v_2] = v_3, & [v_1, v_3] = \beta v_2, \qquad\qquad [v_1, v_4] = \gamma_1 v_2 + \gamma_2 v_3, \\ [v_2, v_3] = -\beta v_1, & [v_2, v_4] = -\gamma_1 v_1 + \gamma_3 v_3, \ [v_3, v_4] = \gamma_2\beta v_1 + \gamma_3\beta v_2, \end{cases} \tag{12}$$

for certain $\gamma_1, \gamma_2, \gamma_3 \in \mathbb{R}$, where $\beta \neq 0$, which are realized on $SU(2) \times \mathbb{R}$ if $\beta < 0$, and on $\widetilde{SL}(2, \mathbb{R}) \times \mathbb{R}$ if $\beta > 0$. A straightforward calculation shows that $\mathfrak{g}_{D.ii}$ is never Einstein, nor locally conformally flat, nor locally symmetric.

A left-invariant metric $\mathfrak{g}_{D.ii}$ is Bach-flat if and only if it corresponds to a metric determined by (12) with parameters satisfying the equations $\{\mathfrak{P}_i = 0\}$, where

$$\mathfrak{P}_1 = 3(\kappa^2-1)\gamma_1 - 2, \qquad \mathfrak{P}_2 = 24\gamma_1^2 + 4\beta(\gamma_2^2+\gamma_3^2) + 9\gamma_1,$$
$$\mathfrak{P}_3 = 24\gamma_1^2 + 8\beta\kappa\gamma_2\gamma_3 + 17\gamma_1.$$

Moreover, the Bach-flat metric is always strict and the Weyl curvature operator is three-step nilpotent.

10 Conclusions

- Nontrivial four-dimensional conformally Einstein left-invariant Lorentzian metrics exist only on solvable Lie groups. Moreover, the conformally Einstein Lorentzian metric is always Ricci-flat.
- The nontrivial conformally Einstein left-invariant Lorentzian metrics are of Petrov Type I in cases (R.ii), (L.Ia.iii), and (L.Ib) of Theorem 8. The left-invariant metrics in Theorem 4-(L.ii) as well as those in cases (R.i), (L.Ia.i), (L.Ia.ii) of Theorem 8 are of Petrov Type D. The Petrov Type is II for metrics in Theorems 6-(L.iii) and 8-(L.II.iii), and it is Type III for metrics corresponding

to Theorem 4-(R.ii). The remaining cases in Theorems 4, 6, and 8 are of Petrov Type N.

Since *pp*-waves have Petrov Type N, one has that the conformal Ricci-flat metric is not necessarily a *pp*-wave. This shows uniqueness for the conformal deformation to an Einstein metric in all cases with Petrov Type different from N [7].

- There is a single family of conformally Einstein left-invariant Lorentzian metrics whose Weyl curvature operator has complex eigenvalues. These metrics, which correspond to case (Ib) in Theorem 8, are locally isometric to $\mathbb{R}^4$ with coordinates (x, y, z, t) and metric tensor

$$g = e^{2kt}\, dx \circ dx + e^{2\mu t}\, \{\cos(2t)(dy \circ dy - dz \circ dz) + 2\sin(2t) dy \circ dz\} + dt \circ dt,$$

with $\mu = \frac{k^2-4}{4k}$, $k \neq 0, \pm\frac{2}{\sqrt{3}}$. Moreover, the Ricci-flat conformal metric is given (up to a multiplicative constant) by $\widetilde{g} = e^{\sigma(t)} g$, where $\sigma(t) = \frac{4-3k^2}{2k} t$.

- Up to homotheties, there is a single conformally Einstein left-invariant Lorentzian metric of Petrov Type III, which corresponds to Theorem 4-(R.ii). This metric is locally isometric to $\mathbb{R}^4$ with coordinates (x, y, z, t) and metric

$$g = dx \circ dx + 2e^{-2x} dy \circ dy + 3e^{-2z} dt \circ dt - 2\sqrt{2} e^{-z}(dx \circ dt + dz \circ dt).$$

Moreover, the Ricci-flat conformal metric is given (up to a multiplicative constant) by $\widetilde{g} = e^{2\sigma(x,z)} g$, where $\sigma(x, z) = z + \frac{1}{2}x$.

- There are two families of conformally Einstein left-invariant Lorentzian metrics with Petrov Type II. One family corresponds, up to homotheties, to the metric in Theorem 6-(L.iii), which is locally isometric to $\mathbb{R}^4$ with coordinates (x, y, z, t) and metric

$$g_1 = e^{8t} dx \circ dx - x e^{2t} dy \circ dy + 2e^{2t} dy \circ dz + dt \circ dt.$$

The other case corresponds to metrics in Theorem 8-(II.iii), which are locally isometric to $\mathbb{R}^4$ with coordinates (x, y, z, t) and metric

$$g_2 = t\varepsilon e^{2t} dx \circ dx + 2e^{2t} dx \circ dy + e^{8t} dz \circ dz + dt \circ dt.$$

Moreover, the Ricci-flat conformal metrics are given (up to a multiplicative constant) by $\widetilde{g_1} = e^{-6t} g_1$ and $\widetilde{g_2} = e^{-6t} g_2$, respectively.

- Strictly Bach-flat Lorentzian metrics corresponding to cases (Ia.i) to (Ia.iv) as well as case (Ia.vi) of Sect. 8.2.1 and those in Theorem 7-(i) are of Petrov Type I. Metrics in cases (II.ii) and (II.iii) of Sect. 8.2.3 as well as those in Theorem 5-(i) are of Petrov Type II with real eigenvalues. Strictly Bach-flat Lorentzian metrics corresponding to case (Ia.v) in Sect. 8.2.1 are of Petrov Type N. All the remaining cases of strictly Bach-flat metrics have Petrov Type III.

Acknowledgments This work was supported by the Grant ED431C 2023/31 (Xunta de Galicia, Spain) and the Grants PID2021-127075NA-I00 and PID2022-138988NB-I00 funded by MICIU/AEI/10.13039/501100011033 and by ERDF, EU.

References

1. E. Abbena, S. Garbiero, and S. Salamon, Bach-flat Lie groups in dimension 4, *C. R. Math. Acad. Sci. Paris* **351** (2013), 303–306.
2. I. Anderson and Ch. Torre, Spacetime groups, *J. Math. Phys.* **61** (2020), 072501.
3. A. Andrada, M. L. Barberis, I. G. Dotti, and G. P. Ovando, Product structures on four dimensional solvable Lie algebras, *Homology Homotopy Appl.* **7** (2005), 9–37.
4. R. Bach, Zur Weylschen Relativitätstheorie und der Weylschen Erweiterung des Krümmungstensorbegriffs, *Math. Z.* **9** (1921), 110–135.
5. L. Bérard-Bergery, *Les spaces homogènes Riemanniens de dimension* 4, Riemannian geometry in dimension 4 (Paris 1978/1979) **3**, 40–60, 1981.
6. M. Blau and M. O'Loughlin, Homogeneous plane waves, *Nuclear Physics B* **654** (2003), 135–176.
7. H. W Brinkmann, Einstein spaces which are mapped conformally on each other, *Math. Ann.* **94** (1925), 119–145.
8. H. W. Brinkmann, Riemann spaces conformal to Einstein spaces, *Math. Ann.* **91** (1924), 269–278.
9. M. Brozos-Vázquez, E. García-Río, and X. Valle-Regueiro, Isotropic quasi-Einstein manifolds, *Classical Quantum Gravity* **36** (2019), 245005 (13pp).
10. M. Brozos-Vázquez, S. Caeiro-Oliveira, E. García-Río, and R. Vázquez-Lorenzo, Four-dimensional homogeneous critical metrics for quadratic curvature functionals, *Trans. Amer. Math. Soc.* **377** (2024), 8563–8604.
11. M. Cahen, J. Leroy, M. Parker, F. Tricerri, and L. Vanhecke, Lorentz manifolds modelled on a Lorentz symmetric space, *J. Geom. Phys.* **7** (1990), 571–581.
12. G. Calvaruso and M. Castrillón, Cyclic Lorentzian Lie groups, *Geom. Dedicata* **181** (2016), 119–136.
13. G. Calvaruso and A. Zaeim, Conformally flat homogeneous pseudo-Riemannian four-manifolds, *Tohoku Math. J.* **66** (2014), 31–54.
14. G. Calvaruso and A. Zaeim, Four-dimensional Lorentzian Lie groups, *Differential Geom. Appl.* **31** (2013), 496–509.
15. E. Calviño-Louzao, X. García-Martínez, E. García-Río, I. Gutiérrez-Rodríguez, and R. Vázquez-Lorenzo, Conformally Einstein and Bach-flat four-dimensional homogeneous manifolds, *J. Math. Pures Appl. (9)* **130** (2019), 347–374.
16. E. Calviño-Louzao, E. García-Río, I. Gutiérrez-Rodríguez, and R. Vázquez-Lorenzo, Conformal geometry of non-reductive four-dimensional homogeneous spaces, *Math. Nachr.* **290** (2017), 1470–1490.
17. E. Calviño-Louzao, E. García-Río, I. Gutiérrez-Rodríguez, and R. Vázquez-Lorenzo, Conformally Einstein Lorentzian Lie groups with Heisenberg symmetry, *Results Math.* **79** (2024), no. 8, Paper No. 264.
18. E. Calviño-Louzao, E. García-Río, I. Gutiérrez-Rodríguez, and R. Vázquez-Lorenzo, Conformally Einstein Lorentzian Lie groups: extensions of the Euclidean and Poincaré groups, *Classical Quantum Gravity* **41** (2024), no. 5, Paper No. 055006, 23 pp.
19. E. Calviño-Louzao, E. García-Río, I. Gutiérrez-Rodríguez, and R. Vázquez-Lorenzo, Conformally Einstein non-solvable Lorentzian Lie groups. *J. Nonlinear Sci.* 35 (2025), Paper No. 86, 31 pp.
20. E. Calviño-Louzao, E. García-Río, I. Gutiérrez-Rodríguez, and R. Vázquez-Lorenzo, Conformally Einstein Lorentzian Lie groups: the almost Abelian case. In preparation.

21. D. Cox, D. Little, and D. O'Shea, Ideals, varieties, and algorithms. An introduction to computational algebraic geometry and commutative algebra, Undergraduate Texts in Mathematics. Springer, Cham, 2015.
22. V. de Smedt and S. Salamon, Anti-self-dual metrics on Lie groups. *Amer. Math. Soc., Providence, RI* **308** (2002), 63–75.
23. M. E. Fels and A. G. Renner, Non-reductive homogeneous pseudo-Riemannian manifolds of dimension four, *Canadian J. Math.* **58** (2006), 282–311.
24. W. Globke and Th. Leistner, Locally homogeneous pp-waves, *J. Geom. Phys.* **108** (2016) 83–101.
25. A. R. Gover and P. A. Nagy, Four-dimensional conformal C-spaces, *Q. J. Math.* **58** (2007), 443–462.
26. G. R. Jensen, Homogeneous Einstein Spaces of Dimension Four, *J. Diff. Geom.* **3** (1969), 309–349.
27. A. W. Knapp, *Lie groups beyond an introduction*, Progr. Math., **140**, Birkhäuser Boston, Inc., Boston, MA, 2002.
28. C. N. Kozameh, E. T. Newman, and K. P. Tod, Conformal Einstein spaces, *Gen. Relativity Gravitation* **17** (1985), 343–352.
29. W. Kühnel and H.-B. Rademacher, Conformally Einstein product spaces, *Differential Geom. Appl.* **49** (2016), 65–96.
30. Th. Leistner, Conformal holonomy of C-spaces, Ricci-flat, and Lorentzian manifolds, *Differential Geom. Appl.* **24** (2006), 458–478.
31. Th. Leistner and P. Nurowski, Ambient Metrics for n-dimensional pp-waves *Comm. Math. Phys.* **296** (2010), 881–898.
32. H.-S. Liu, H. Lü, C. N. Pope, and J. F. Vázquez-Poritz, Not conformally Einstein metrics in conformal gravity, *Classical Quantum Gravity* **30** (2013) 165015 (15pp).
33. P. D. Mannheim, Making the case for conformal gravity, *Found. Phys.* **42** (2012), 388–420.
34. P. D. Mannheim, Alternatives to dark matter and dark energy, *Prog. Part. Nucl. Phys.* **56** (2006), 340–445.
35. J. Milnor, Curvatures of left invariant metrics on Lie groups, *Adv. Math.* **21** (1976), 293–329.
36. T. Otero-Casal, *Métricas de Einstein en grupos de Lie lorentzianos*, Publicaciones del Departamento de Geometría y Topología **152** (2022), Univ. Santiago de Compostela.
37. J. Patera, R. T. Sharp, P. Winternitz, and H. Zassenhaus, Invariants of real low dimension Lie algebras, *J. Math. Phys.* **17** (1976), 986–994.
38. S. Rahmani, Métriques de Lorentz sur les groupes de Lie unimodulaires, de dimension trois, *J. Geom. Phys.* **9** (1992), 295–302.
39. H. Takagi, Conformally flat Riemannian manifolds admitting a transitive group of isometries, *Tohoku Math. J. (2)* **27** (1975), 103–110.

On the Propagation of Singularities of Constant Curvature, Convex Hypersurfaces

Graham Andrew Smith

Dedicated to Professor Marcos Dajczer on his 75th birthday.

Abstract We describe the structure of the singular sets of constant curvature, convex hypersurfaces in hyperbolic space for general convex curvature functions. We apply this structure theorem to the study of the ideal Plateau problem in hyperbolic space for such curvature functions.

1 Introduction

In the study of certain nonlinear partial differential equations, we are in the fortunate position of having complete geometric descriptions of the singular sets of their solutions. In certain favorable situations, this even allows us to eliminate these singular sets entirely. This is the case, for example, for equations of real Monge-Ampère type (see, e.g., [3, 12], and [19]), where the singular sets are unions of convex hulls of closed subsets of the boundary of the domain of the function in question, as in Fig. 1. In this note, we study the singular sets of convex hypersurfaces of constant curvature for a general class of curvature functions to be specified presently. The corresponding partial differential equations are of hessian type, and we will see that the general structure of the singular sets of their solutions is similar to that just described. We will work in hyperbolic space, although analogous results can also be shown to hold in the sphere and in Euclidean space.

In order to state our main result, a few definitions will be required. Let $\mathbb{H}^{m+1}$ denote $(m+1)$-dimensional hyperbolic space, and let $\partial_\infty\mathbb{H}^{m+1}$ denote its ideal boundary. Let Conv denote the space of closed, convex subsets of $\mathbb{H}^{m+1} \cup \partial_\infty\mathbb{H}^{m+1}$ furnished with the Hausdorff topology, and let Conv^o denote the subspace consisting of those closed, convex subsets which have nontrivial interior.

G. A. Smith (✉)
Pontifícia Universidade Católica do Rio de Janeiro (PUC-Rio), Rio de Janeiro, Brazil

© The Author(s), under exclusive license to Springer Nature Switzerland AG 2025
R. Tojeiro (ed.), *Advances in Submanifold Theory and Related Topics*,
Latin American Mathematics Series – UFSCar subseries,
https://doi.org/10.1007/978-3-032-00918-0_5

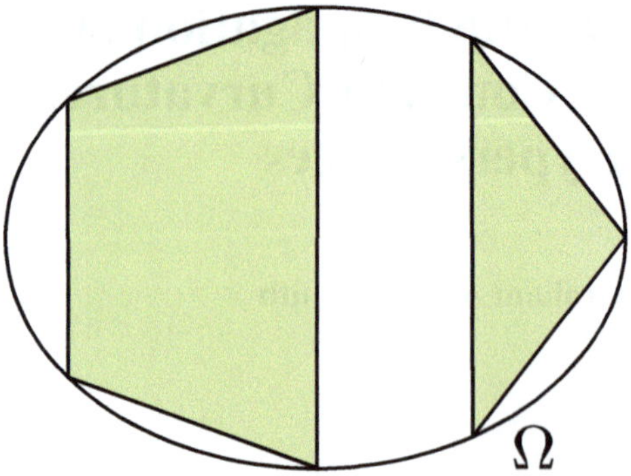

Fig. 1 The structure of the singular set. Given a solution of the real Monge-Ampère equation defined in some convex domain Ω, its singular set is a union of convex hulls of closed subsets of the boundary of this domain. Note, however, that this figure is deceptive in that the singularity set actually has high codimension (c.f. [2])

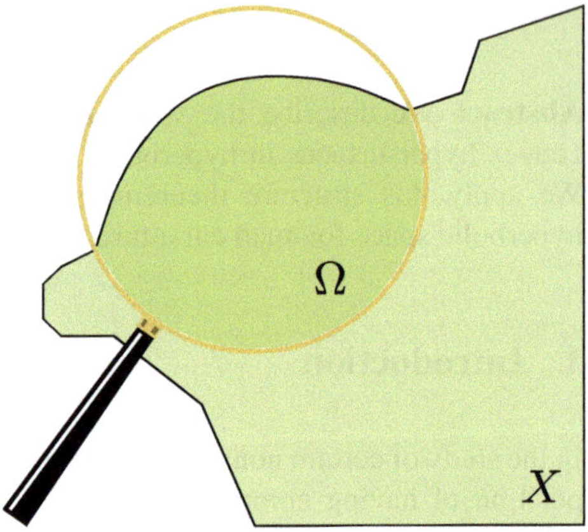

Fig. 2 Definition of $\mathrm{Conv}^o_\kappa(\Omega)$. Heuristically, we think of Ω as the aperture of a glass through which the convex set X is examined, so that Ω is the region where the structure of ∂X is better understood

Let K be a convex curvature function, as defined in Sect. 2. Given an open subset Ω of $\mathbb{H}^{m+1}$, and a positive real number κ, let $\mathrm{Conv}^o_\kappa(\Omega)$ denote the space of closed, convex subsets of $\mathbb{H}^{m+1} \cup \partial_\infty \mathbb{H}^{m+1}$, with nontrivial interior, such that $\partial X \cap \Omega$ is smooth with constant K-curvature equal to κ (see Fig. 2).

Let $\partial\mathrm{Conv}^o_\kappa(\Omega)$ denote the topological boundary of $\mathrm{Conv}^o_\kappa(\Omega)$ in Conv^o. Given $X \in \partial\mathrm{Conv}^o_\kappa(\Omega)$ and $x \in \partial X \cap \Omega$, we say that X is *regular* at x whenever $X \in \mathrm{Conv}^o_\kappa(\Omega')$ for some smaller neighborhood Ω' of x, and we say that X is *singular* at this point otherwise. Let $\mathrm{Reg}(X, \Omega)$ denote the set of points of $\partial X \cap \Omega$ at which X is regular, and let $\mathrm{Sing}(X, \Omega)$ denote the set of points at which it is singular.

Finally, given a closed subset Y of $\mathbb{H}^{m+1} \cup \partial_\infty \mathbb{H}^{m+1}$, let $\mathrm{Hull}(Y)$ denote its convex hull in $\mathbb{H}^{m+1} \cup \partial_\infty \mathbb{H}^{m+1}$. The topological boundaries of convex hulls have precise structures which we now describe (see Section 4.5 of [19]). Note first that it is natural to consider convex hulls as generalized, possibly ideal, polyhedra. Indeed, convex polyhedra in $\mathbb{H}^{m+1} \cup \partial_\infty \mathbb{H}^{m+1}$ are precisely the convex hulls of finite sets, and, just as the relative boundary of any polyhedron is a union of polyhedra of lower dimension, so too is the relative boundary of any convex hull the union of convex hulls of lower dimension. We make this precise as follows. Given a closed subset Y of $\mathbb{H}^{m+1} \cup \partial_\infty \mathbb{H}^{m+1}$, and a boundary point x of $\mathrm{Hull}(Y)$ lying away from Y, every supporting totally geodesic hyperplane H of $\mathrm{Hull}(Y)$ at x satisfies

$$F_H(Y) := H \cap \mathrm{Hull}(Y) = \mathrm{Hull}(H \cap Y)\ . \tag{1}$$

We call $F_H(Y)$ the *face* of $\mathrm{Hull}(Y)$ supported on H, and we denote by $\mathrm{Faces}(Y)$ the set of all faces of $\mathrm{Hull}(Y)$.

We are now ready to state our main result.

Theorem 1 (Propagation of Singularities) *Let K be a convex curvature function, let $\kappa > 0$ be a positive real number, let Ω be a convex, open subset of $\mathbb{H}^{m+1}$, and let X be an element of $\partial\mathrm{Conv}_\kappa^o(\Omega)$ with nontrivial interior. There exists a closed subset Y of $\partial X \cap \partial\Omega$, and a subset $\mathcal{F}$ of $\mathrm{Faces}(Y)$ such that*

$$\mathrm{Sing}(X, \Omega) = \bigcup_{F \in \mathcal{F}} F \cap \Omega\ . \tag{2}$$

In other words, the singularity set of ∂X in Ω is a union of faces of the convex hull of Y.

Theorem 1 is proven in Sect. 4.

The structure result of Theorem 1 often allows us to remove singularities through elementary geometric arguments. For example, a nice application concerns the study of the ideal Plateau problem for hypersurfaces of constant K-curvature in hyperbolic space. Indeed, we obtain a simpler, more geometric approach to the final steps of the following result (c.f. [10] and [11]).

Theorem 2 (Ideal Plateau Problem) *Let K be a convex curvature function, and let $\Omega \subseteq \partial_\infty\mathbb{H}^{m+1}$ be a proper open subset with C^2 boundary. For all $\kappa \in]0, 1[$, there exists a convex subset $X \subseteq \mathbb{H}^{m+1} \cup \partial_\infty\mathbb{H}^{m+1}$ such that:*

(1) $X \cap \partial_\infty\mathbb{H}^{m+1} = \Omega^c$.
(2) $\partial X \cap \mathbb{H}^{m+1}$ is smooth and of constant K-curvature equal to κ.

Remark 1 Denoting $\Sigma := \partial X \cap \mathbb{H}^{m+1}$, we see that $\partial_\infty\Sigma = \partial\Omega$, so that the boundary of X does indeed yield a solution of the ideal Plateau problem in the usual sense.

Remark 2 In fact, we prove a slightly stronger result, using a weaker hypothesis on the boundary of Ω, not addressed in [10] and [11], known as the exterior ball condition. This will be discussed in Sect. 5.

Remark 3 We underline that our proof of Theorem 2 relies on Theorem 1.3 of [11] or, equivalently, Theorem 1.2 of [20]. These two results concern the finite Plateau problem, that is, the Plateau problem over a horosphere, which in the present context is fairly classical, differing little from the original result [4] of Caffarelli–Nirenberg–Spruck. In [11], the authors derive Theorem 2 from these results by proving a priori estimates for solutions of the finite Plateau problem which are uniform as the horosphere tends to infinity. Our approach removes the need for such estimates by providing simple geometric criteria for the nonexistence of singularities in the limit.

Remark 4 An elementary compactness argument shows that the principal curvatures of ∂X are globally bounded above and below. Likewise, in the upper half-space model of $\mathbb{H}^{m+1}$, where $\partial_\infty \mathbb{H}^{m+1}$ identifies with $\mathbb{R}^m \cup \{\infty\}$, if Ω is taken to be a proper, open subset of $\mathbb{R}^m$, then an elementary convexity argument also shows that $\partial X \cap \mathbb{H}^{m+1}$ is a graph of some smooth function $u : \Omega \to [0, \infty[$.

Theorem 2 is proven in Sect. 5.

Theorem 1 is a straightforward consequence of two ingredients. The first is an adaptation of the remarkable estimates obtained by Sheng–Urbas–Wang in [18], which build on the works [13] and [14] of Ivochkina and [16] of Pogorelov. This will be discussed in Sect. 3 where, for the reader's convenience, we will describe their construction in some detail. The second ingredient concerns the geometric properties of convex subsets of affine flat spaces, which also hold for convex subsets of hyperbolic space via the Kleinian parametrization (see, e.g., [7]). These properties will be discussed in Sect. 4, where frequent reference will be made to our text [19]. The key result relating these two theories is Lemma 8, which transforms the analytic estimates obtained in Sect. 3 into local geometric properties of singular points.

2 Convex Curvature Functions

We now introduce the class of curvature functions that will be of interest to us (c.f. [4]). Let $\Gamma^+ \subseteq \mathbb{R}^m$ denote the open cone of m-dimensional vectors all of whose components are strictly positive, and let $\overline{\Gamma}^+$ denote its closure in $\mathbb{R}^m$. We say that a function $K \in C^0(\overline{\Gamma}^+) \cap C^\infty(\Gamma^+)$ is a *convex curvature function* whenever it possesses the following properties:

(1) $K(x)$ is invariant under permutation of the components of x.
(2) K is homogeneous of order 1.
(3) $K(1, \cdots, 1) = 1$.
(4) $K(x) > 0$ for all $x \in \Gamma^+$ and $K(x) = 0$ for all $x \in \partial\Gamma^+$.
(5) For all $x \in \Gamma^+$ and for all $1 \leqslant i \leqslant m$

$$(\partial_i K)(x) > 0\ ;\ \text{and} \tag{3}$$

(6) K is concave.

Convex curvature functions are used in the study of immersed hypersurfaces as follows. Let $Y := Y^{m+1}$ be an $(m+1)$-dimensional Riemannian manifold, let $X := X^m$ be an m-dimensional manifold, let $e : X \to Y$ be an immersion, and let $\kappa_{1,e}, \cdots, \kappa_{m,e}$ denote its principal curvatures. We say that e is *infinitesimally strictly convex* (ISC) whenever every one of its principal curvatures is strictly positive at every point. When this holds, we define its *K-curvature* $K_e : X \to \mathbb{R}$ by

$$K_e := K(\kappa_{1,e}, \cdots, \kappa_{m,e}) . \tag{4}$$

The geometric significance of Properties 1 to 4 now becomes clear. Property 1 ensures that K_e is well defined, Property 2 ensures that it scales appropriately under rescalings of the ambient metric, Property 3 is a normalization condition according to which the unit sphere in Euclidean space has unit curvature, and Property 4 ensures that the limit of a sequence of ISC hypersurfaces can only fail to be ISC at points where its K-curvature vanishes. In contrast, Properties 5 and 6 arise from analytic considerations. Indeed, Property 5 ensures that the Jacobi operator of K-curvature is elliptic, a fact that we will only use implicitly in what follows, while Property 6 is natural, and usually indispensable, with the current technology of nonlinear analysis.

In what follows, it will be more convenient to study K-curvature in terms of the shape operator of the immersion and thus to view K as a function of symmetric matrices. Abusing terminology, we therefore denote also by Γ^+ the open cone of positive-definite matrices in $\mathrm{End}(\mathbb{R}^m)$, and by $K : \overline{\Gamma}^+ \to \mathbb{R}$ the function given by

$$K(A) := K(\lambda_1(A), \cdots, \lambda_m(A)) ,$$

where, for any symmetric matrix A, $\lambda_1(A), \cdots, \lambda_m(A)$ denote its eigenvalues. Note that Property 1 ensures that this new function is well defined and invariant under the action of $\mathrm{O}(m)$ by conjugation on $\overline{\Gamma}^+$. An alternative approach consists of identifying $\mathbb{R}^m$ with the subspace of diagonal matrices in $\mathrm{End}(\mathbb{R}^m)$ and then viewing this new function as the unique $\mathrm{O}(m)$-invariant extension of the former. Any resulting ambiguity should be easily resolved by context.

Now, let A be a diagonal matrix with eigenvalues $\lambda_1 \geqslant \cdots \geqslant \lambda_m > 0$. The derivative of K at A satisfies

$$DK(A) \cdot M = \mathrm{Tr}(BM) ,$$

where B is also diagonal, with respective eigenvalues

$$\mu_i := (\partial_i K)(\lambda_1, \cdots, \lambda_m) .$$

Furthermore, by Property 6 (concavity)

$$0 < \mu_1 \leqslant \cdots \leqslant \mu_m ,$$

and by Property 2 (homogeneity)

$$\sum_k \mu_k \lambda_k = DK(A) \cdot A = K(A) . \tag{5}$$

In what follows, it will also be useful to denote

$$\mu := \mathrm{Tr}(B) = \sum_k \mu_k \text{ and } \hat{\mu} := \mathrm{Tr}(BA^2) = \sum_k \mu_k \lambda_k^2 \; . \tag{6}$$

In particular, by (5) and Property 6 again

$$\mu = K(A) + \sum_k \mu_k(1 - \lambda_k) \geqslant K(\mathrm{Id}) = 1 \; . \tag{7}$$

Finally (see, e.g., Lemma 1.1 of [9] and Lemma 2.3 of [18]), the second derivative of K at A satisfies for every symmetric matrix M

$$D^2K(A)^{mn,pq}M_{mn}M_{pq} \leqslant \sum_{i,j}(\partial_i\partial_j K)M_{ii}M_{jj} + 2\sum_{\lambda_j<\lambda_i}\frac{(\mu_i - \mu_j)}{(\lambda_i - \lambda_j)}M_{ij}^2 \; , \tag{8}$$

from which it follows, in particular, that this function is also concave.

3 Controlling the Principal Curvatures

It will henceforth be convenient to view $\mathbb{H}^{m+1}$ as a quadric in $\mathbb{R}^{m+1,1}$, that is,

$$\mathbb{H}^{m+1} := \left\{x \in \mathbb{R}^{m+1,1} \mid \langle x, x\rangle_{m+1,1} = -1 \text{ and } x_{m+2} > 0\right\} , \tag{9}$$

where $\langle\cdot,\cdot\rangle_{m+1,1}$ here denotes the Lorentzian inner product with signature $(m+1,1)$. In particular, every unit tangent vector of $\mathbb{H}^{m+1}$ is an element of $(m,1)$-dimensional de Sitter space, that is,

$$\mathrm{d}\mathbb{S}^{m,1} := \left\{x \in \mathbb{R}^{m+1,1} \mid \langle x, x\rangle_{m+1,1} = 1\right\} . \tag{10}$$

We will make considerable use of these identifications in what follows.

Let K be a convex curvature function, let X be an m-dimensional manifold, let $e : X \to \mathbb{H}^{m+1}$ be an ISC immersion of constant K-curvature equal to $\kappa > 0$, and let $\hat{e} : X \to \mathrm{d}\mathbb{S}^{m,1}$ denote its unit normal vector field, which we view as an immersion in its own right. Recall (see, e.g., [8]) the natural duality of $\mathbb{H}^{m+1}$ and $\mathrm{d}\mathbb{S}^{m,1}$ whereby the three fundamental forms of $\hat{e}$ are related to those of e by

$$\mathrm{I}_{\hat{e}} = \mathrm{III}_e \; , \; \mathrm{II}_{\hat{e}} = \mathrm{II}_e \; , \; \mathrm{III}_{\hat{e}} = \mathrm{I}_e \; . \tag{11}$$

In particular, the principal curvatures of $\hat{e}$ are related to those of e by

$$\kappa_{i,\hat{e}} = \frac{1}{\kappa_{i,e}} \; \forall 1 \leqslant i \leqslant m \; , \tag{12}$$

so that $\hat{e}$ is also an ISC immersion with constant $\hat{K}$-curvature equal to $1/\kappa$, where the curvature function $\hat{K}$ is given by

$$\hat{K}(x_1, \cdots, x_m) := K(1/x_1, \cdots, 1/x_m)^{-1} . \tag{13}$$

We will only make implicit use of this fact in what follows.

We aim to control the principal curvatures of e in terms of the geometries of e and $\hat{e}$, using a modification of the estimate [18] of Sheng–Urbas–Wang. In what follows, we aim to emphasize the symmetries of the roles played by the auxiliary functions f and $\hat{f}$ that will be introduced presently.

Let $A := A_e$ denote the shape operator of e, and let $\lambda_{1,e} \geqslant \cdots \geqslant \lambda_{m,e} > 0$ denote its eigenvalues, that is, the principal curvatures of e. In what follows, we will use a semicolon ";" to denote covariant differentiation with respect to the Levi-Civita covariant derivative of the metric induced by e (this is explained, e.g., in [15]). In particular, with this notation, the Codazzi–Mainardi equations read for all (i, j, k)

$$A_{ij;k} = A_{kj;i} . \tag{14}$$

That is, $A_{ij;k}$ is symmetric under every permutation of its components. On account of this symmetry, we will suppress the semicolon in the first covariant derivative of A in what follows.

Now, let x_0 be a point of X. Let f be a positive, convex function defined in a neighborhood V of $e(x_0)$ in $\mathbb{H}^{m+1}$, and let $\hat{f}$ be a positive, concave function defined in a neighborhood $\hat{V}$ of $\hat{e}(x_0)$ in $\mathrm{d}\mathbb{S}^{m,1}$. For $\alpha > 0$ a positive constant to be determined presently, consider the function:

$$\Phi_\alpha := \log(\lambda_1) + \alpha \log(f \circ e) - \log(\hat{f} \circ \hat{e}) . \tag{15}$$

We aim to estimate a certain Laplacian of this function at x_0 whenever x_0 is a critical point. Let $0 < \mu_{1,e} \leqslant \cdots \leqslant \mu_{m,e}$ denote the eigenvalues of $B_e := DK(A_e)$, as in Sect. 2. We define the *K-laplacian* of any smooth function $\phi : X \to \mathbb{R}$ by

$$\Delta^K \phi := \sum_k \mu_k \mathrm{Hess}^X(\phi)_{kk} , \tag{16}$$

where Hess^X denotes the Hessian operator of X with respect to the metric induced by e. We require the following elementary formula for the K-Laplacian of the logarithm of a positive function ϕ:

$$\Delta^K \log(\phi) = \frac{1}{\phi} \Delta^K \phi - \frac{1}{\phi} \sum_{k=1}^{m} \mu_k \phi_k^2 . \tag{17}$$

We will also require the following *generalized Simons identity* (c.f. [17]).

Lemma 1 (Generalized Simons identity) *Let Y be a Riemannian space-form of constant curvature equal to c, let $e : X \to Y$ be a smooth codimension 1 immersion, and let A denote its shape operator. Let x be a point of X, let $e_1, \cdots, e_m \in T_xX$ be an orthonormal basis of eigenvectors of $A(x)$, and let $\lambda_1, \cdots, \lambda_m$ denote its corresponding eigenvalues. With respect to this basis, for all (i, j)*

$$A_{ii;jj} = A_{jj;ii} + (\lambda_i\lambda_j + c)(\lambda_i - \lambda_j) . \tag{18}$$

Proof Indeed, let $\overline{R}$ denote the Riemann curvature tensor of Y and let R denote the Riemann curvature tensor of the metric that e induces over X. Using the Codazzi-Mainardi equations (14), we find that, for all (i, j)

$$A_{ii;jj} = A_{ji;ij} = A_{ji;ji} + R^p_{ijj}A_{pi} + R^p_{iji}A_{jp} = A_{jj;ii} + R^p_{ijj}A_{pi} + R^p_{iji}A_{jp} ,$$

where Einstein's summation convention is used in the index p but not in the indices i and j. By Gauss' equation

$$R^l_{ijk} = \overline{R}^l_{ijk} + A^l_i A_{jk} - A^l_j A_{ik} ,$$

which, upon substitution into the preceding relation, yields, for all (i, j)

$$\begin{aligned} A_{ii;jj} &= A_{jj;ii} + \overline{R}^p_{ijj}A_{pi} + \overline{R}^p_{iji}A_{jp} \\ &\qquad + A^p_i A_{jj}A_{pi} - A^p_j A_{ij}A_{pi} + A^p_i A_{ji}A_{jp} - A^p_j A_{ii}A_{jp} \\ &= A_{jj;ii} + c\lambda_i - c\lambda_j + \lambda_i^2\lambda_j - \delta_{ij}\lambda_i^3 + \delta_{ij}\lambda_i^3 - \lambda_j^2\lambda_i \\ &= A_{jj;ii} + (\lambda_i\lambda_j + c)(\lambda_i - \lambda_j) , \end{aligned}$$

as desired. □

Finally, we recall (see [5]) that a continuous function ϕ satisfies the differential inequality:

$$\Delta^K\phi \geqslant a$$

in the *weak sense* at some point x whenever there exists a smooth function ψ defined in a neighborhood of this point such that

$$\phi \geqslant \psi , \ \phi(x) = \psi(x) , \ \text{and } (\Delta^K\psi)(x) \geqslant a .$$

Note that this is stronger than satisfying the above differential inequality in the *viscosity sense* (see [6]), since the latter does not require the function ψ to exist.

Lemma 2 *When e has constant K-curvature equal to κ, the function λ_1 satisfies*

$$\Delta^K \log(\lambda_1) \geqslant \frac{2}{\lambda_1}\sum_{\lambda_k<\lambda_1}\frac{(\mu_k-\mu_1)}{(\lambda_1-\lambda_k)}A_{11k}^2-\frac{1}{\lambda_1^2}\sum_k \mu_k A_{11k}^2-(\hat{\mu}+\mu)+\kappa\left(\lambda_1+\frac{1}{\lambda_1}\right), \tag{19}$$

in the weak sense.

Remark 5 Since we aim to apply a maximum principle, we will be interested in establishing mechanisms to ensure that the right-hand side of (19) is positive. We already note that, when $\mu_k \geqslant 2\mu_1$, the k-th summand of the second term on the right-hand side is dominated by the k-th summand of the first term. The case where $\mu_k < 2\mu_1$, as well as the third term on the right-hand side, will be addressed via the auxiliary functions $\log(f \circ e)$ and $\log(\hat{f} \circ \hat{e})$.

Proof Let x_0 be a point of X, let $e_1, \cdots, e_m \in T_{x_0}X$ be an orthonormal basis of eigenvectors of $A(x_0)$, and denote also by $e_1, \cdots, e_m$ the orthonormal frame obtained by parallel transport of this basis along geodesics passing through x_0. Note that, with the frame defined in this manner, at x_0, for all (i, j)

$$\nabla_{e_i} e_j = \nabla_{e_i}\nabla_{e_i} e_j = 0\,. \tag{20}$$

Consider now the function:

$$a_{11} := \langle Ae_1, e_1\rangle\,.$$

Trivially, $\lambda_1 \geqslant a_{11}$ near x_0, and $\lambda_1(x_0) = a_{11}(x_0)$, so that, by the above definition

$$(\Delta^K \lambda_1)(x_0) \geqslant (\Delta^K a_{11})(x_0)$$

in the weak sense.

In order to estimate $(\Delta^K a_{11})(x_0)$, we consider, more generally, the matrix-valued function:

$$a_{ij} := \langle Ae_i, e_j\rangle\,.$$

By (20), at x_0, for all (i, j, k),

$$\begin{aligned} D_{e_k} a_{ij} &= A_{ijk}\,,\quad D_{e_k} D_{e_k} a_{ij} = A_{ij;kk}\,,\ \text{and}\\ \mathrm{Hess}^X(a_{11})(e_k, e_k) &= D_{e_k} D_{e_k} a_{11} = A_{11;kk}\,. \end{aligned} \tag{21}$$

Twice differentiating the identity $K(a) = \kappa$ yields, for all k

$$\sum_{p,q} DK(a)^{pq} D_{e_k} D_{e_k} a_{pq} + \sum_{p,q,m,n} D^2K(a)^{pq,mn}\big(D_{e_k} a_{pq}\big)\big(D_{e_k} a_{mn}\big) = 0\,,$$

so that, at x_0, for all k

$$\sum_{p,q} DK(a)^{pq} A_{pq;kk} = -\sum_{p,q,m,n} D^2K(a)^{pq,mn} A_{pqk} A_{mnk} \,. \tag{22}$$

Using (5), (6), (21), (22), and the generalized Simons identity (18), we thus obtain

$$\begin{aligned}
\Delta^K a_{11} &= \sum_k \mu_k A_{11;kk} \\
&= \sum_k \mu_k \big(A_{kk;11} + (\lambda_1\lambda_k - 1)(\lambda_1 - \lambda_k)\big) \,, \\
&= -\sum_{p,q,m,n,k} D^2K(\alpha)^{pq,mn} A_{pqk} A_{mnk} + \lambda_1^2\kappa - \lambda_1\mu - \lambda_1\hat{\mu} + \kappa \,,
\end{aligned}$$

so that, by (8)

$$\Delta^K a_{11} \geqslant 2 \sum_{\lambda_k<\lambda_1} \frac{(\mu_k - \mu_1)}{(\lambda_1 - \lambda_k)} A_{11k}^2 - \lambda_1(\hat{\mu} + \mu) + \kappa\lambda_1\left(\lambda_1 + \frac{1}{\lambda_1}\right),$$

and the result now follows by (17). □

The calculations of the K-Laplacians of the remaining two components of Φ_α are more straightforward. We will use the following elementary but useful formula for hessians of restrictions of functions.

Lemma 3 *Let Y be a semi-Riemannian manifold, let X be a nondegenerate hypersurface in Y, let ν denote its unit normal vector field, and let ϵ denote the norm-squared of ν. Given a point x of X, and a C^2 function ϕ defined over some neighborhood of this point in Y*

$$\mathrm{Hess}^X(\phi) = \mathrm{Hess}^Y(\phi)|_{TX} - \epsilon d\phi(\nu)\mathrm{II}^X \,, \tag{23}$$

where II^X here denotes the second fundamental form of X with respect to the normal ν.

Proof Indeed, let ∇ and $\overline{\nabla}$ denote, respectively, the Levi-Civita covariant derivatives of X and Y. For all tangent vector fields u and v over X

$$\begin{aligned}
\mathrm{Hess}^X(\phi)(u,v) &= D_v D_u\phi - \phi(\nabla_u v) \\
&= D_v D_u\phi - d\phi(\overline{\nabla}_u v - \epsilon\langle\overline{\nabla}_u v, \nu\rangle\nu) \\
&= \mathrm{Hess}^Y(\phi)(u,v) - \epsilon d\phi(\nu)\langle\mu, \overline{\nabla}_u v\rangle \\
&= \mathrm{Hess}^Y(\phi)(u,v) - \epsilon d\phi(\nu)\mathrm{II}^X(u,v) \,,
\end{aligned}$$

as desired. □

Lemma 4 *Let $C > 0$ be such that*

$$\mathrm{Hess}^{\mathbb{H}^{m+1}}(f) \geqslant \frac{1}{C} f \mathrm{Id} . \tag{24}$$

Then

$$\Delta^K \log(f \circ e) \geqslant \frac{1}{C}\mu - \frac{1}{f^2}\sum_k \mu_k f_k^2 - \frac{1}{f} df(\nu)\kappa . \tag{25}$$

Proof Indeed, by (5), (6), (23), and (24)

$$\begin{aligned}\Delta^K(f \circ e) &= \sum_k \mu_k \mathrm{Hess}^X(f \circ e)_{kk} \geqslant \frac{1}{C} f \sum_k \mu_k - df(\nu)\sum_k \mu_k \lambda_k \\ &= \frac{1}{C} f\mu - df(\nu)\kappa ,\end{aligned}$$

and the result now follows by (17). □

Lemma 5 *Let $C > 0$ be such that*

$$\mathrm{Hess}^{\mathrm{d}\mathbb{S}^{m,1}}(\hat{f}) \leqslant -\frac{1}{C}\hat{f}\mathrm{Id} . \tag{26}$$

Then

$$\Delta^K \log(\hat{f} \circ \hat{e}) \leqslant -\frac{1}{C}\hat{\mu} - \frac{1}{\hat{f}^2}\sum_k \mu_k \lambda_k^2 \hat{f}_k^2 + \frac{1}{\hat{f}}\kappa d\hat{f}(\hat{\nu}) . \tag{27}$$

Remark 6 We understand (26) in the sense that

$$\mathrm{Hess}^{\mathrm{d}\mathbb{S}^{m,1}}(\hat{f})(\xi, \xi) \leqslant -\frac{1}{B}\hat{f}\langle \xi, \xi\rangle_{m+1,1}$$

for every spacelike vector ξ.

Remark 7 In fact, the K-Laplacian of e coincides with the $\hat{K}$-Laplacian of $\hat{e}$, and the proof of Lemma 5 then reduces to an elementary calculation involving (23) almost identical to that used to prove Lemma 4. This is the main idea used below.

Proof We will make use here of the natural duality between $\mathbb{H}^{m+1}$ and $\mathrm{d}\mathbb{S}^{m,1}$ mentioned above. Since the shape operator A_e of e satisfies the Codazzi–Mainardi equations, we readily verify that the covariant derivative $\hat{\nabla}$ of $\mathrm{I}_{\hat{e}} = \mathrm{III}_e$ is related to that of I_e by

$$\nabla_u v = \hat{\nabla}_u v + A^{-1}(\nabla_u A)v .$$

Consequently, for any function ϕ defined over X, and for all (u, v)

$$\mathrm{Hess}^X(\phi)(u,v) = \widehat{\mathrm{Hess}}^X(\phi)(u,v) + d\phi(A^{-1}(\nabla_u A)v)\ ,$$

where here $\widehat{\mathrm{Hess}}^X$ denotes the hessian operator of $\hat{\nabla}$. However, differentiating the identity $K(A) = \kappa$ yields, for all i

$$\sum_k \mu_k A_{kki} = \mathrm{Tr}(B\nabla_{e_i}A) = D_{e_i}K = 0\ ,$$

so that, for any such ϕ

$$\Delta^K\phi = \sum_k \mu_k \mathrm{Hess}^X(\phi)_{kk} = \sum_k \mu_k \widehat{\mathrm{Hess}}^X(\phi)_{kk}\ .$$

Note now that the second fundamental form of $\hat{e}$ is equal to that of e. Since the unit normal $\hat{\nu}$ of $\hat{e}$ has negative norm-squared, using (5), (6), and (23), we therefore obtain

$$\begin{aligned}\Delta^K(\hat{f}\circ\hat{e}) &= \sum_k \mu_k \widehat{\mathrm{Hess}}^X(\hat{f}\circ\hat{e})_{kk}\\ &\leqslant -\frac{1}{B}\hat{f}\sum_k \mu_k\lambda_k^2 + d\hat{f}(\hat{\nu})\sum_k \mu_k\lambda_k\\ &= -\frac{1}{B}\hat{f}\hat{\mu} + d\hat{f}(\hat{\nu})\kappa\ ,\end{aligned}$$

and the result now follows by (17). □

We are now ready to provide our estimate for the K-Laplacian of Φ_α.

Lemma 6 *Let $K \subseteq \mathbb{H}^{m+1}$ and $\hat{K} \subseteq \mathrm{d}\mathbb{S}^{m,1}$ be, respectively, compact subsets of $\mathbb{H}^{m+1}$ and $\mathrm{d}\mathbb{S}^{m,1}$. Suppose that $e(X) \subseteq K$ and $\hat{e}(X) \subseteq \hat{K}$ and that there exists $C > 0$ such that over K and $\hat{K}$*

$$\mathrm{Hess}(f) \geqslant \frac{1}{C}f\ ,\ \hat{f} \geqslant \frac{1}{C}\ ,\ \textit{and}\ \mathrm{Hess}(\hat{f}) \leqslant -\left(1+\frac{1}{C}\right)\hat{f}\ . \tag{28}$$

For all $\alpha > \mathrm{Max}(2, C)$, there exists $D > 0$, which only depends on α, K, $\hat{K}$, f, and $\hat{f}$, with the following property. At any critical point x_0 of Φ_α with $\Phi_\alpha(x_0) > D$

$$\Delta^K\Phi_\alpha > 0 \tag{29}$$

in the weak sense.

Proof Note first that $\nu_e := \hat{e}$ and $\nu_{\hat{e}} := e$ are, respectively, the unit normal vector fields over e and $\hat{e}$. Since the images of these functions are relatively compact, upon increasing C is necessary, we may suppose that

$$|f|, \left|\hat{f}\right| < C ,$$

that

$$|df(\nu_e)|, |d\hat{f}(\nu_{\hat{e}})| < C ,$$

and that, for all k

$$|f_k|, |\hat{f}_k| < C .$$

It now follows by (19), (25), and (27) that

$$\begin{aligned}\Delta^K \Phi_\alpha \geqslant \kappa\lambda_1 + \frac{2}{\lambda_1}\sum_{\lambda_k<\lambda_1}\frac{(\mu_k-\mu_1)}{(\lambda_1-\lambda_k)}A_{11k}^2 - \frac{1}{\lambda_1^2}\sum_k \mu_k A_{11k}^2\\ -\mu + \frac{\alpha}{C}\mu - \frac{\alpha}{f^2}\sum_k \mu_k f_k^2\\ -\hat{\mu} + \left(1+\frac{1}{C}\right)\hat{\mu} + \frac{1}{\hat{f}^2}\sum_k \mu_k\lambda_k^2\hat{f}_k^2 - \mathrm{O}\left(1+\frac{\alpha}{f}\right),\end{aligned}$$

so that, for $\alpha \geqslant C$

$$\Delta^K\Phi \geqslant \kappa\lambda_1 - \mathrm{O}\left(1+\frac{\alpha}{f}\right) + \sum_k \left(P_{1,k}+P_{2,k}-N_{1,k}-N_{2,k}\right),$$

where the summands

$$\begin{aligned}P_{1,k} &:= \frac{2}{\lambda_1}\frac{(\mu_k-\mu_1)}{(\lambda_1-\lambda_k)}A_{11k}^2 , \text{ and}\\ P_{2,k} &:= \frac{1}{\hat{f}^2}\left(1+\frac{1}{C^5}\right)\mu_k\lambda_k^2\hat{f}_k^2\end{aligned}$$

yield nonnegative contributions, while the summands

$$\begin{aligned}N_{1,k} &:= \frac{1}{\lambda_1^2}\mu_k A_{11k}^2 , \text{ and}\\ N_{2,k} &:= \frac{\alpha}{f^2}\mu_k f_k^2\end{aligned}$$

yield nonpositive contributions (by convention, we have defined $P_{1,k} := 0$ whenever $\lambda_k = \lambda_1$).

We will show that, for all k, the terms $N_{1,k}$ and $N_{2,k}$ are dominated by the terms $P_{1,k}$ and $P_{2,k}$ provided that α and Φ_α are sufficiently large. To this end, we observe first that, at any critical point x_0 of Φ_α, for all k

$$\frac{1}{\lambda_1}A_{11k} + \frac{\alpha f_k}{f} - \frac{\lambda_k \hat{f}_k}{\hat{f}} = 0\,. \tag{30}$$

We observe also that, since all principal curvatures are positive, by (5)

$$\mu_1\lambda_1 \leqslant \sum_k \mu_k\lambda_k = \kappa\,,$$

so that

$$\mu_1 \leqslant \frac{\kappa}{\lambda_1}\,. \tag{31}$$

Now, choose $0 < \epsilon < 1$. The desired dominations of $N_{1,k}$ and $N_{2,k}$ will be obtained upon using (30) in two different ways, depending on the size of μ_k. In the first case, when $\mu_k < 2\kappa/(1-\epsilon)\lambda_1$, we use (30) to obtain

$$\frac{1}{\lambda_1^2}A_{11k}^2 \leqslant \frac{(1+C^5)\alpha^2 f_k^2}{f^2} + \frac{1}{\hat{f}^2}\left(1+\frac{1}{C^5}\right)\lambda_k^2\hat{f}_k^2\,,$$

which yields

$$\begin{aligned} P_{2,k} - N_{1,k} &= \frac{1}{\hat{f}^2}\left(1+\frac{1}{C^5}\right)\mu_k\lambda_k^2\hat{f}_k^2 - \frac{1}{\lambda_1^2}\mu_k A_{11k}^2 \geqslant -\frac{\mu_k(1+C^5)\alpha^2 f_k^2}{f^2} \\ &= -\mathrm{O}\left(\frac{\alpha^2}{\lambda_1 f^2}\right), \end{aligned}$$

so that, for $\alpha > 1$

$$P_{1,k} + P_{2,k} - N_{1,k} - N_{2,k} \geqslant -\mathrm{O}\left(\frac{\alpha^2}{\lambda_1 f^2}\right).$$

In the second case, when $\mu_k \geqslant 2\kappa/(1-\epsilon)\lambda_1$, we use (30) to obtain

$$\frac{\alpha f_k^2}{f^2} \leqslant \frac{2}{\alpha\lambda_1^2}A_{11k}^2 + \frac{2}{\alpha\hat{f}^2}\lambda_k^2\hat{f}_k^2\,,$$

which yields, for $\alpha > 2$

$$P_{2,k} - N_{1,k} - N_{2,k} = \frac{1}{\hat{f}^2}\left(1+\frac{1}{C^5}\right)\mu_k\lambda_k^2\hat{f}_k^2 - \frac{1}{\lambda_1^2}\mu_k A_{11k}^2 - \frac{\alpha\mu_k f_k^2}{f^2}$$

$$\geqslant -\frac{1}{\lambda_1^2}\mu_k\left(1+\frac{2}{\alpha}\right)A_{11k}^2 .$$

Bearing in mind (31), this yields

$$P_{2,k} - N_{1,k} - N_{2,k} \geqslant -\frac{2(\mu_k - \mu_1)}{\lambda_1^2}\frac{(1+2/\alpha)}{(1+\epsilon)}A_{11k}^2 ,$$

and, since $\lambda_k < \lambda_1$,

$$P_{1,k} + P_{2,k} - N_{1,k} - N_{2,k} \geqslant \frac{2(\mu_k - \mu_1)}{\lambda_1(\lambda_1 - \lambda_k)}\frac{(\epsilon - 2/\alpha)}{(1+\epsilon)}A_{11k}^2 ,$$

which is non-negative for $\epsilon > 2/\alpha$.

Combining these relations, we find that, for $\epsilon > 2/\alpha$,

$$\Delta^K\phi \geq \lambda_1\kappa - \lambda_1 \mathrm{O}\left(\frac{1}{\lambda_1} + \frac{1}{\lambda_1 \mathrm{f}} + \frac{1}{\lambda_1^2 \mathrm{f}^2}\right).$$

Note now that

$$\lambda_1 = \frac{e^{\Phi}\hat{f}}{f^{\alpha}} \geq \frac{e^{\Phi}}{C^{1+\alpha}} ,$$

so that λ_1 is large when Φ is large. There therefore exists $D_1 > 0$ such that, for $\Phi > D_1$,

$$\Delta^K\Phi \geq \lambda_1\left(\frac{\kappa}{2} - \frac{D_1}{\lambda_1^2 f^2}\right) \geq \lambda_1\left(\frac{\kappa}{2} - \frac{D_1 C^{2(\alpha-1)}}{\lambda_1^2 f^{2\alpha}}\right).$$

Hence

$$\Delta^K\Phi \geq \lambda_1(\kappa/2 - D_1 C^{2\alpha}e^{-2\Phi}) ,$$

and the result follows.

□

4 Propagation of Singularities

We now use Lemma 6 to show how the singularities of ISC hypersurfaces of constant K-curvature propagate along totally geodesic submanifolds. We limit ourselves to hypersurfaces which bound convex sets, and we recall the formalism given in the introduction. Let Conv denote the space of closed, convex subsets of

$\mathbb{H}^{m+1} \cup \partial_\infty \mathbb{H}^{m+1}$, furnished with the Hausdorff topology. Recall (see, e.g., [7]) that the Kleinian parametrization maps $\mathbb{H}^{m+1} \cup \partial_\infty \mathbb{H}^{m+1}$ to the closed unit ball $\overline{\mathbb{B}}^{m+1}$ in $\mathbb{R}^{m+1}$, sending geodesics to straight lines, so that Conv identifies with the space of closed, convex subsets of $\overline{\mathbb{B}}^{m+1}$. In particular, Conv is compact (see, e.g., Lemma 4.3 of [19]). Let Conv^o denote the subspace of Conv consisting of those closed, convex subsets which have nontrivial interior. Given an open subset Ω and a positive real number κ, let $\mathrm{Conv}^o_\kappa(\Omega)$ denote the subspace consisting of those convex subsets X, with nontrivial interior, such that $\partial X \cap \Omega$ is smooth and of constant K-curvature equal to κ.

We will be interested in the topological boundary of $\mathrm{Conv}^o_k(\Omega)$ in Conv^o. Given an element $X \in \partial\mathrm{Conv}^o_k(\Omega)$ with nontrivial interior, and a point $x \in \partial X \cap \Omega$, we will say that X is *regular* at x whenever $X \in \mathrm{Conv}^o_k(\Omega')$, for some smaller neighborhood Ω' of x, and we will say that it is *singular* at this point otherwise. Let $\mathrm{Reg}(X, \Omega)$ denote the set of points of $\partial X \cap \Omega$ at which X is regular, and let $\mathrm{Sing}(X, \Omega)$ denote the set of points at which it is singular. Note that, by definition, $\mathrm{Reg}(X, \Omega)$ is an open subset of $\partial X \cap \Omega$, while $\mathrm{Sing}(X, \Omega)$ is closed. Finally, given a closed subset Y of $\mathbb{H}^{m+1} \cup \partial_\infty \mathbb{H}^{m+1}$, let $\mathrm{Hull}(Y)$ denote its convex hull in $\mathbb{H}^{m+1} \cup \partial_\infty \mathbb{H}^{m+1}$, and let $\mathrm{Faces}(Y)$ denote the set of all its faces.

The remainder of this section will be devoted to proving Theorem 1.

Theorem 1 *Let K be a convex curvature function, let $\kappa > 0$ be a positive real number, let Ω be a convex, open subset of $\mathbb{H}^{m+1}$, and let X be an element of $\partial\mathrm{Conv}^o_\kappa(\Omega)$ with nontrivial interior. There exists a closed subset Y of $\partial X \cap \partial\Omega$ and a subset $\mathcal{F}$ of* $\mathrm{Faces}(Y)$ *such that*

$$\mathrm{Sing}(X, \Omega) = \bigcup_{F \in \mathcal{F}} F \cap \Omega\ . \tag{32}$$

This theorem will follow as a straightforward consequence of Lemma 6 together with the following characterization of convex hulls.

Definition 1 Let $X \in \mathrm{Conv}^o$ be a convex subset of $\mathbb{H}^{m+1}$ with nontrivial interior. We say that X satisfies the *local geodesic property* (LGP) at a point $x \in \partial X$ whenever there exists an *open* geodesic segment Γ such that

$$x \in \Gamma \subseteq \partial X\ . \tag{33}$$

Remark 8 The set X is trivially non-strictly convex at any point where it satisfies the LGP, but the converse is not true. Indeed, consider the union of a square and a semi-disk, as in Fig. 3. This set is non-strictly convex at x since its unique supporting tangent at this point also contains one side of the square, but it trivially does not satisfy the LGP at this point.

The LGP characterizes convex hulls, as the following result shows (see Theorem 4.8 of [19]).

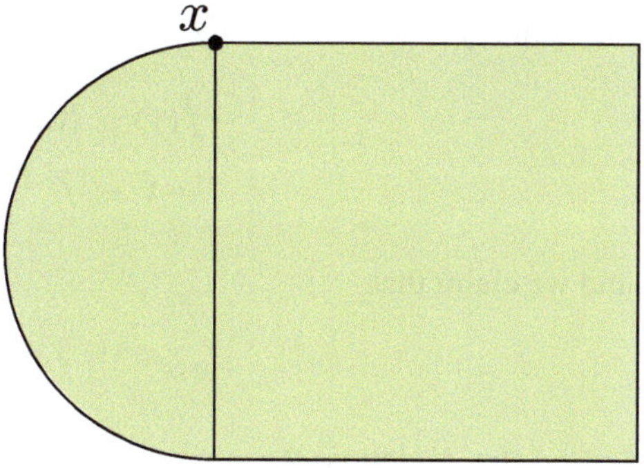

Fig. 3 The local geodesic property. Although the local geodesic property is not satisfied at the point x, this set is non-strictly convex at this point

Lemma 7 *Let X be a compact, convex subset of $\mathbb{R}^{m+1}$ with nontrivial interior. Let $Y \subseteq \partial X$ denote the set of points of ∂X at which X satisfies the LGP, and let ∂Y denote its topological boundary. Then*

$$Y = \partial X \cap \mathrm{Hull}(\partial Y) .$$

Remark 9 Since the Kleinian parametrization sends $\mathbb{H}^{m+1} \cup \partial_\infty \mathbb{H}^{m+1}$ to the closed unit ball $\overline{\mathbb{B}}^{m+1}$ in $\mathbb{R}^{m+1}$, sending geodesic segments to geodesic segments, this result also applies to the closed, convex subsets studied here.

By Lemma 7, Theorem 1 will follow as soon as we show that the singular points of X in Ω are precisely those points at which it satisfies the LGP.

Lemma 8 *Let Ω by an open subset of $\mathbb{H}^{m+1}$, let X be an element of $\mathrm{Conv}^o_k(\Omega)$ with nontrivial interior, and let x_0 be a point of $\partial X \cap \Omega$. X is singular at x_0 if and only if it satisfies the LGP at this point.*

Proof Note first that if X satisfies the LGP at x_0, then it is singular at this point, for otherwise the least principal curvature of $\partial X \cap \Omega$ at this point would vanish, and therefore so too would its K-curvature, which is absurd. Suppose now that X does not satisfy the LGP at x_0. We will show that X is regular at this point.

First note that, since X has nontrivial interior, its set of supporting normals at x_0 is contained in an open hemisphere of $\mathrm{T}^1_{x_0}\mathbb{H}$ and that there therefore exists a supporting normal $\nu_0 \in \mathrm{d}\mathbb{S}^{m,1}$ at x_0 and $\epsilon > 0$ such that, for every other supporting normal ν of X at this point

$$\langle \nu_0, \nu \rangle > 4\epsilon .$$

By continuity of supporting normal sets (see Lemma 4.4 of [19]), there exists a neighborhood U of x_0 in $\mathbb{H}^{m+1}$ such that, for all $x \in \partial X \cap V$, and for every supporting normal $\nu \in \mathrm{d}\mathbb{S}^{m,1}$ of X at x

$$\langle \nu_0, \nu \rangle > 3\epsilon .$$

We define $\hat{f} : \mathrm{d}\mathbb{S}^{m,1} \to \mathbb{R}$ and $\hat{V} \subseteq \mathrm{d}\mathbb{S}^{m,1}$ by

$$\begin{aligned}\hat{f}(y) &= \langle \nu_0, y\rangle - \epsilon \text{ and}\\ \hat{V} &= \hat{f}^{-1}(]2\epsilon, +\infty[)\ ,\end{aligned}$$

and we claim that

$$\mathrm{Hess}^{\mathrm{d}\mathbb{S}^{m,1}}(\hat{f}) = -(\hat{f}+\epsilon)\mathrm{Id}\ .$$

Indeed, let N denote the outward-pointing unit normal vector field over $\mathrm{d}\mathbb{S}^{m,1}$ in $\mathbb{R}^{m+1,1}$. Since N is spacelike, it follows by (23) that

$$\mathrm{Hess}^{\mathrm{d}\mathbb{S}^{m,1}}(\hat{f}) = \mathrm{Hess}^{\mathbb{R}^{m+1,1}}(\hat{f})|_{\mathrm{d}\mathbb{S}^{m,1}} - d\hat{f}(N)\mathrm{II}^{\mathrm{d}\mathbb{S}^{m,1}} = -(\hat{f}+\epsilon)\mathrm{Id}\ ,$$

as asserted.

Now, let $V_0 \subseteq \mathbb{H}^{m+1}$ denote the unique open half-space such that ∂V_0 passes through x_0 with inward-pointing normal ν_0 at this point. For all $r > 0$, let $\overline{B}_r(x_0)$ denote the closed ball of radius r about x_0 in $\mathbb{H}^{m+1}$. Let $r > 0$ be such that $\overline{B}_r(x_0) \subseteq U$. Since X does not satisfy the LGP at x_0, $\partial X \cap \partial V_0 \cap \partial B_r(x_0)$ is contained in an open hemisphere of $\partial B_r(x_0)$, lying to one side of some totally geodesic hypersurface $H_0 \subseteq \partial V_0$, say (see Theorem 4.19 and Lemma 4.21 of [19]). Rotating V_0 slightly about H_0 and then translating slightly away from ν_0 thus yields an open half-space V, containing x_0, such that

$$\partial X \cap \overline{V} \subseteq U\ ,$$

and that this intersection is compact (see Fig. 4).

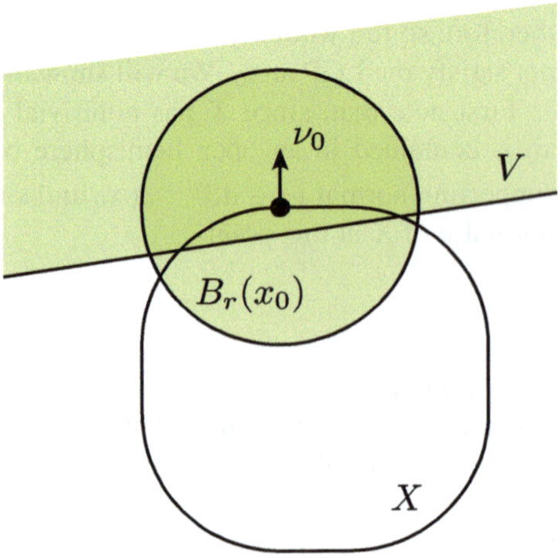

Fig. 4 Constructing a suitable half-space. Upon slightly rotating, and then slightly translating a suitable supporting open half-space of X at x_0, we obtain an open half-space V, containing x_0, whose intersection with X is a relatively compact subset of U.

Recall that ∂V is the intersection of the quadric $\mathbb{H}^{m+1}$ with a nondegenerate, linear hyperplane $P \subseteq \mathbb{R}^{m+1,1}$ of mixed signature. Let $\nu_1 \in \mathbb{R}^{m+1,1}$ denote the unit vector normal to this hyperplane which points into V, and define $f : \mathbb{H}^{m+1} \to \mathbb{R}$ by

$$f(x) := \langle x, \nu_1 \rangle \ .$$

Since the unit normal vector over $\mathbb{H}^{m+1}$ in $\mathbb{R}^{m+1,1}$ is timelike, using (23), we obtain, as before

$$\mathrm{Hess}^{\mathbb{H}^{m+1}}(f) = f\mathrm{Id} \ .$$

Now, let $(X_m)_{m\in\mathbb{N}}$ be a sequence of elements of $\mathrm{Conv}^o_\kappa(\Omega)$ converging in the Hausdorff sense to X. We claim that $(\partial X_m \cap V)_{m\in\mathbb{N}}$ converges in the C^∞_{loc} sense toward $\partial X \cap V$, from which the regularity of X at x_0 will immediately follow. Indeed, for all m, let ν_m denote the outward-pointing, unit, normal vector field over $\partial X_m \cap V$, let A_m denote its shape operator, and, for all $\alpha > 0$, denote

$$\Phi_{\alpha,m} := \log(\|A_m\|) + \alpha \log(f) - \log(\hat{f} \circ \nu_m) \ .$$

Since $(X_m)_{m\in\mathbb{N}}$ converges in the Hausdorff sense to X, we may suppose that, for all m

$$\partial X_m \cap \overline{V} \subseteq U \ ,$$

and that this intersection is compact. Furthermore, by Lemma 4.4 of [19], we may also suppose that for all m

$$\nu_m(\partial X_m \cap \overline{V}) \subseteq \hat{V} \ .$$

Note now that, for all m, $\Phi_{\alpha,m}$ converges to $-\infty$ near the boundary of $\partial X_m \cap \overline{V}$ and therefore attains its maximum at some interior point of this set. With α as in Lemma 6, it then follows by the maximum principle that there exists $C > 0$ such that for all m

$$\|A_m\| f^\alpha (\hat{f} \circ \nu_m)^{-1} = \mathrm{Exp}(\Phi_{\alpha,m}) \leqslant C \ ,$$

so that $(\|A_m\|)_{m\in\mathbb{N}}$ is locally uniformly bounded over $\partial X_m \cap V$. It now follows by elliptic regularity that $(\partial X_m \cap V)_{m\in\mathbb{N}}$ converges to $\partial X \cap V$ in the C^∞_{loc} sense, as asserted. Regularity follows, and this completes the proof. □

We now complete the proof of Theorem 1.

Proof of Theorem 1 Indeed, by Lemma 8, $\mathrm{Sing}(X, \Omega)$ is precisely the set of points of $\partial X \cap \Omega$ at which X satisfies the LGP. It follows by Lemma 7 that

$$\mathrm{Sing}(X, \Omega) = \partial X \cap \mathrm{Hull}(Y) \cap \Omega \ ,$$

where

$$Y := \overline{\mathrm{Sing}}(X, \Omega) \cap \partial\Omega .$$

Now, let x be a point of $\mathrm{Sing}(X, \Omega)$, and let H be a supporting totally geodesic hypersurface of X at x. Since H is also a supporting totally geodesic hypersurface of $\mathrm{Hull}(Y)$ at x

$$x \in H \cap \mathrm{Hull}(Y) \cap \Omega = F_H(Y) \cap \Omega ,$$

so that x lies on a face of Y. Since$\mathrm{Hull}(Y) \cap \Omega$ is contained in X, so too is $F_H(Y) \cap \Omega$, and since the latter lies on a supporting hyperplane of X

$$F_H(Y) \cap \Omega \subseteq \partial X \cap \Omega .$$

Since every point of $F_H(Y) \cap \Omega$ satisfies the LGP, this subset is contained in $\mathrm{Sing}(X, \Omega)$. That is, the entire face on which x lies is contained in $\mathrm{Sing}(X, \Omega)$. We conclude that $\mathrm{Sing}(X, \Omega)$ is indeed the union of intersections with Ω of faces of $\mathrm{Hull}(Y)$, and this completes the proof. □

5 The Asymptotic Plateau Problem

We conclude this note by applying Theorem 1 to the study of the asymptotic Plateau problem for convex curvature functions in $\mathbb{H}^{m+1}$. We require the following definition.

Definition 2 Let Ω be a proper open subset of $\partial_\infty \mathbb{H}^{m+1}$, and let x be a boundary point of this set. We say that Ω satisfies the *exterior ball condition* (EBC) at this point whenever there exists a closed round ball $B \subseteq \partial_\infty \mathbb{H}^{m+1}$, lying in the complement of Ω, which contains x in its boundary (see Fig. 5).

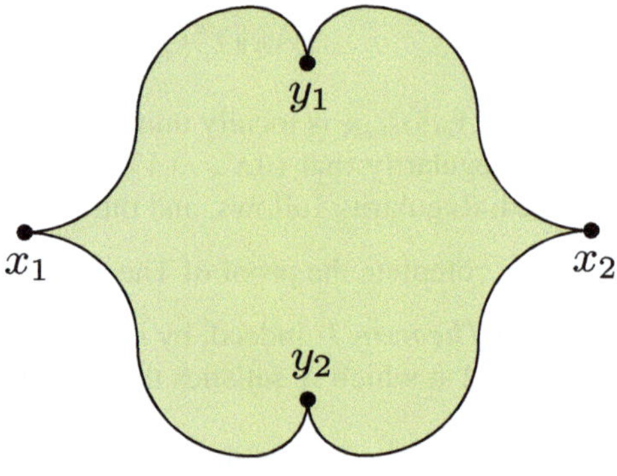

Fig. 5 The exterior ball condition. This set satisfies the exterior ball condition at every point of its boundary where it is smooth. It also satisfies the exterior ball condition at the non-smooth points x_1 and x_2, but not at y_1 and y_2

We now obtain the desired simplification of the following result of Guan–Spruck and Guan–Spruck–Szapiel (see [10] and [11]).

Theorem 2 *Let K be a convex curvature function, and let $\Omega \subseteq \partial_\infty \mathbb{H}^{m+1}$ be a proper open subset satisfying the EBC at every point of its boundary. For all $\kappa \in$ $]0, 1[$, there exists a convex subset $X \subseteq \mathbb{H}^{m+1} \cup \partial_\infty \mathbb{H}^{m+1}$ such that:*

(1) $X \cap \partial_\infty \mathbb{H}^{m+1} = \Omega^c$.
(2) $\partial X \cap \mathbb{H}^{m+1}$ *is smooth and of constant K-curvature equal to κ.*

Remark 10 Theorem 2 is actually slightly stronger than the results of [10] and [11], since the latter only address the case where the boundary is C^2.

The proof of Theorem 2 makes use of the geometric maximum principle, which for our present purposes can be expressed as follows.

Lemma 9 (Geometric Maximum Principle) *Let $X, X' \subseteq \mathbb{H}^{m+1} \cup \partial_\infty \mathbb{H}^{m+1}$ be convex subsets such that*

$$X \subseteq X' . \tag{34}$$

If $x \in \partial X \cap \partial X'$, if ∂X and $\partial X'$ are smooth near x, and if K and K' denote the respective K-curvatures of ∂X and $\partial X'$ near this point, then

$$K(x) \geqslant K'(x) . \tag{35}$$

Proof We recall the elementary proof for the reader's convenience. Let U be a neighborhood of x in $\mathbb{H}^{m+1}$ such that $U \cap \partial X$ and $U \cap \partial X'$ are smooth. Let d and d' denote the respective signed distances in $\mathbb{H}^3$ to ∂X and $\partial X'$. Upon reducing U if necessary, we may suppose that these functions are smooth over this set. Since $X \subseteq X'$

$$d \geqslant d' .$$

Let II and II$'$ denote the respective second fundamental forms of ∂X and $\partial X'$ in U. By (23), and bearing in mind that $T_x \partial X = T_x \partial X'$,

$$\mathrm{II}(x) = \mathrm{Hess}(d)(x)|_{T_x X} \geqslant \mathrm{Hess}(d')(x)|_{T_x X'} = \mathrm{II}'(x)$$

and the result follows by (3). □

Proof of Theorem 2 We use the upper half-space model of $\mathbb{H}^{m+1}$, namely,

$$\mathbb{H}^{m+1} := \{(x_1, \cdots, x_{m+1}) \mid x_{m+1} > 0\} ,$$

and we identify $\partial_\infty \mathbb{H}^{m+1}$ with $\mathbb{R}^m \cup \{\infty\}$. In particular, since Ω is a proper open subset of $\partial_\infty \mathbb{H}^{m+1}$ which satisfies the EBC at every point of its boundary, we may view it as a relatively compact open subset of $\mathbb{R}^m$. Let $(\Omega_i)_{i\in\mathbb{N}}$ be a sequence of open subsets of $\mathbb{R}^m$ with smooth boundary such that

$$\Omega := \bigcup_{i\in\mathbb{N}} \Omega_i ,$$

and such that for all i

$$\overline{\Omega}_i \subseteq \Omega_{i+1} .$$

For all $i \in \mathbb{N}$, let B_i denote the open horoball

$$B_i := \{(x_1, \cdots, x_{m+1}) \mid x_{m+1} > 1/i\} ,$$

and let $\hat{\Omega}_i$ denote the set of points of ∂B_i lying directly above Ω_i. For all i, solving the finite Plateau problem (see, e.g., Theorem 1.3 of [11] or Theorem 1.2 of [20]) yields a closed, convex subset $X_i \in \text{Conv}^o_\kappa(B_i)$ such that

$$X_i \subseteq \overline{B}_i \text{ and } X_i \cap \partial B_i = \hat{\Omega}_i^c .$$

By compactness of Conv, there then exists a closed, convex subset $X \in$ Conv toward which $(X_i)_{i\in\mathbb{N}}$ subconverges in the Hausdorff sense. It remains to show that X has the desired properties.

Before proceeding, it is worth recalling the structure of totally umbilic submanifolds of $\mathbb{H}^{m+1}$ in the upper half-space model (c.f. [1]). This will yield a clear picture of how smoothness almost immediately follows from the EBC. In the upper half-space model, the intersections with $\mathbb{H}^{m+1}$ of hyperplanes and hyperspheres of $\mathbb{R}^{m+1}$ yield totally umbilic hypersurfaces of $\mathbb{H}^{m+1}$, and, furthermore, all totally umbilic hypersurfaces of $\mathbb{H}^{m+1}$ are accounted for in this manner. In addition, if H is a hyperplane or a hypersphere which intersects $\mathbb{R}^m$ nontrivially, making an angle of $\theta \in [0, \pi/2[$ with the vertical axis along this intersection, then the hyperbolic shape operator of $H\cap\mathbb{H}^{m+1}$ is equal to $\sin(\theta)\text{Id}$ and the K-curvature of this submanifold is equal to $\sin(\theta)$. Finally, for any such H, $H\cap\mathbb{H}^{m+1}$ bounds a convex subset of $\mathbb{H}^{m+1}$. In the case where H is a hyperplane, its outward-pointing normal points downward, in the case where it is a hypersphere with center lying below $\mathbb{R}^m$, its outward-pointing normal also points downward, and in the case where it is a hypersphere with center lying above $\mathbb{R}^m$, its outward-pointing normal points upward.

We now show that $X \cap \partial_\infty\mathbb{H}^{m+1} = \Omega^c$. Indeed, trivially

$$\Omega^c \subseteq X \cap \partial_\infty\mathbb{H}^{m+1} .$$

Conversely, choose $x \in \Omega$, and let B be an open ball in $\mathbb{R}^m$ such that

$$x \in B \subseteq \overline{B} \subseteq \Omega ,$$

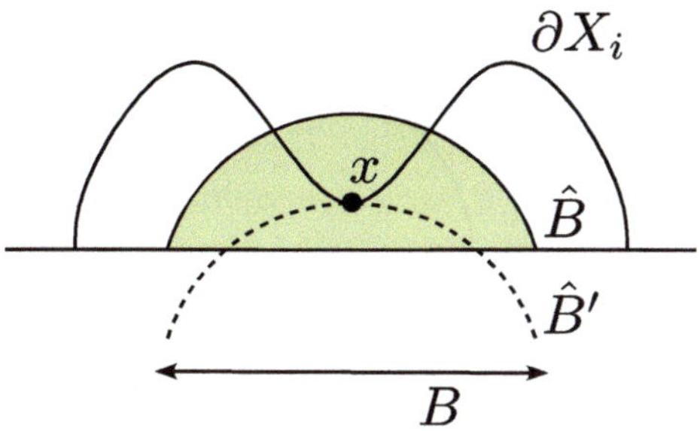

Fig. 6 The geometric maximum principle. Since $\hat{B}'$ lies below $\hat{B}$, the curvature of its boundary with respect to the hyperbolic metric is greater than that of the boundary of $\hat{B}$ and therefore also greater than that of ∂X_i

and note that for sufficiently large i

$$\overline{B} \subseteq \Omega_i \ .$$

Now, let $\hat{B}$ denote the open ball in $\mathbb{R}^{m+1}$ with center lying below $\mathbb{R}^m$, whose intersection with $\mathbb{R}^m$ is equal to B and whose boundary makes an angle of $\arcsin(\kappa)$ with the vertical axis along ∂B. We claim that for all i

$$X_i \cap \hat{B} = \emptyset \ .$$

Indeed, otherwise, for some i, there exists an open ball $\hat{B}'$, of radius equal to that of $\hat{B}$, and with center lying directly below that of $\hat{B}$, such that $X_i \cap \hat{B}' = \emptyset$ and $\partial \hat{B}'$ meets ∂X_i at some point (see Fig. 6). Since $\partial \hat{B}'$ makes an angle strictly greater than $\arcsin(\kappa)$ with the vertical axis along $\partial \hat{B}' \cap \mathbb{R}^m$, its K-curvature is strictly greater than κ. By the geometric maximum principle, this is absurd, and it follows that $X_i \cap \hat{B} = \emptyset$, as asserted. Taking limits now yields

$$X \cap \hat{B} = \emptyset \ ,$$

so that $x \notin X$. Since $x \in \Omega$ is arbitrary, it follows that the intersection of $X \cap \partial_\infty \mathbb{H}^{m+1}$ with Ω is trivial, so that $X \cap \partial_\infty \mathbb{H}^{m+1} = \Omega^c$, as asserted.

We now show that $X \in \mathrm{Conv}_\kappa^0(\mathbb{H}^{m+1})$. To this end, we first show that $\partial X \cap \mathbb{H}^{m+1}$ makes an angle of at least $\theta := \arcsin(\kappa)$ with the vertical axis along $\partial\Omega$. Indeed, let y be a point of $\partial\Omega$. By the EBC, there exists an open ball B in $\mathbb{R}^m$ contained in the complement of Ω which contains y in its closure. For all $r \in]0, 1]$, let B_r denote the open ball in $\mathbb{R}^m$ obtained upon contracting B by a factor r about its center, and let $\hat{B}_r$ denote the open ball in $\mathbb{R}^{m+1}$ with center lying above $\mathbb{R}^m$, whose intersection with $\mathbb{R}^m$ is equal to B_r, and which makes an angle of θ with the vertical axis along ∂B_r. Using the geometric maximum principle again, we show that for all r and for all sufficiently large i

$$\hat{B}_r \cap B_i \subseteq X_i \ .$$

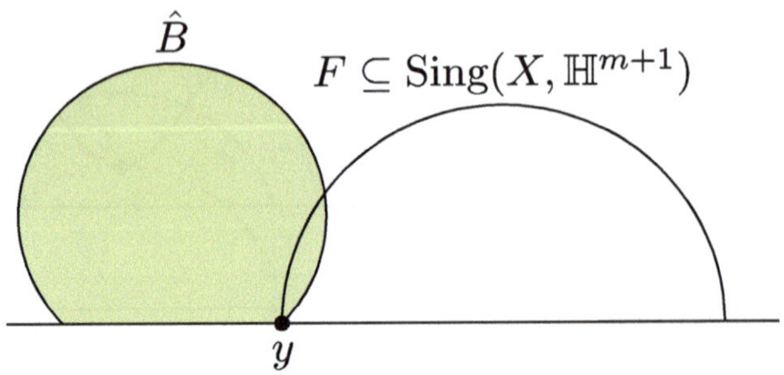

Fig. 7 Toward a contradiction The set $\mathrm{Sing}(X, \mathbb{H}^{m+1})$ meets $\mathbb{H}^{m+1}$ orthogonally at y, contradicting the fact that $\partial X \cap \mathbb{H}^{m+1}$ makes an angle of at least $\arcsin(\kappa)$ with $\mathbb{R}^m$ at every point

Upon taking limits, first in i, and then in r, it follows that

$$\hat{B} \cap \mathbb{H}^{m+1} \subseteq X ,$$

so that, indeed, $\partial X \cap \mathbb{H}^{m+1}$ makes an angle of at least θ with $\mathbb{R}^m$ at y.

It remains only to show that $\mathrm{Sing}(X, \mathbb{H}^{m+1})$ is empty. However, by Theorem 1, there exists a closed subset Y of $\partial\Omega$ and a subset $\mathcal{F} \subseteq \mathrm{Faces}(Y)$ such that

$$\mathrm{Sing}(X, \mathbb{H}^{m+1}) = \bigcup_{F \in \mathcal{F}} F \cap \mathbb{H}^{m+1} .$$

Choose $x \in \mathrm{Sing}(X, \mathbb{H}^{m+1})$, let $F \in \mathcal{F}$ be a face of $\mathrm{Hull}(Y)$ containing this point, let H denote the totally geodesic hypersurface on which it lies, and let $y \in (\partial_\infty H) \cap Y$ be an extremity of F. With $\hat{B}$ as above, by the preceding discussion

$$F \cap \mathbb{H}^{m+1} \cap \hat{B} \subseteq \partial X \cap \mathbb{H}^{m+1} \cap \hat{B} = \emptyset ,$$

which is absurd, since H, and therefore also F, meets $\mathbb{R}^m$ orthogonally at y (see Fig. 7). It follows that $\mathrm{Sing}(X, \mathbb{H}^{m+1})$ is indeed trivial, so that $X \in \mathrm{Conv}^0_\kappa(\mathbb{H}^{m+1})$, and this completes the proof. □

Acknowledgments The author is grateful to Thierry Barbot for enlightening conversations and to Sébastien Alvarez and François Fillastre for helpful comments. This note was written while the author was benefitting from a CNRS Poste Rouge hosted by the Fédération de Recherche des Unités de Mathématiques de Marseille (FRUMAM) and the Laboratoire de Mathématiques of the Université d'Avignon.

References

1. Benedetti R., Petronio C., *Lectures on Hyperbolic Geometry*, Universitext, Springer-Verlag, (2012)
2. Caffarelli L., A note on the degeneracy of convex solutions to Monge–Ampére equation, *Comm. PDE.*, **18**, (1993), 1213–1217
3. Caffarelli L. A., Interior a priori estimates for solutions of fully nonlinear elliptic equations, *Comm. Pure. Appl. Math.*, **50**, no. 7, (1997), 623–665
4. Caffarelli L., Nirenberg L., Spruck J., The Dirichlet problem for nonlinear second order elliptic equations, III: Functions of the eigenvalues of the Hessian, *Acta Math.*, **155**, (1985), 261–301
5. Calabi E., An extension of E. Hopf's maximum principle with an application to Riemannian geometry, *Duke Math. J.*, **25**, no. 1, 45–56
6. Crandall M. G., Ishii H., Lions P.-L., User's guide to viscosity solutions of second order partial differential equations, *Bull. Amer. Math. Soc.*, **27**, (1992), 1–67
7. Fillastre F., Barbot T., Quasi-Fuchsian co-Minkowski manifolds, in *In the Tradition of Thurston*, Ohshika K., Papadopoulos A., (eds.), Springer-Verlag, (2020)
8. Fillastre F., Seppi A., Spherical, hyperbolic, and other projective geometries: convexity, duality, transitions, in *Eighteen essays in non-euclidean geometry*, Alberge V., Papadopoulos A., (eds.), EMS Press, (2019), 321–409
9. Gerhardt C., Closed Weingarten hypersurfaces in Riemannian manifolds, *J. Diff. Geom.*, **43**, (1996), 612–641
10. Guan B., Spruck J., Hypersurfaces of constant curvature in hyperbolic space, *Adv. Lect. Math. (ALM)*, **20** International Press, Somerville, MA, (2011)
11. Guan B., Spruck J., Szapiel M., Hypersurfaces of Constant Curvature in Hyperbolic Space I, *J. Geom. Anal.*, **19**, (2009), 772–795
12. Gutiérrez C. E., *The Monge-Ampère equation*, Progress in Nonlinear Differential Equations and Their Applications (PNLDE), **89**, Birkhäuser-Verlag, (2016)
13. Ivochkina N. M., Construction of a priori bounds for convex solutions of the Monge-Ampère equation by integral methods, *Ukranian Math. J.*, **30**, (1978), 32–38
14. Ivochkina N. M., Solution of the Dirichlet problem for equations of m'th order curvature (Russian), *Mat. Sb.*, **180**, (1989), 867–887; English translation in *Sb. Math.*, **67**, (1990), 317–339
15. Kobayashi S., Nomizu K., Foundations of differential geometry, Vol. I., Wiley classics library, Wiley, (1996)
16. Pogorelov A. V., *The Minkowski multidimensional problem*, Wiley, New York, (1978)
17. Rosenberg H., Smith G., Degree theory of immersed hypersurfaces, *Mem. Amer. Math. Soc.*, **265**, (2020)
18. Sheng W., Urbas J., Wang X. J., Interior curvature bounds for a class of curvature equations, *Duke Math. J.*, **123**, no. 2, (2004), 235–264
19. Smith G., Global singularity theory for the Gauss curvature equation, *Ensaios Matemáticos*, **28**, (2015), 1–114
20. Smith G., The Plateau problem for convex curvature functions, *Ann. Inst. Fourier*, **70**, no. 1, 1–66, (2020)

Ribaucour Partial Tubes and Hypersurfaces of Enneper Type

Sergio Chion and **Ruy Tojeiro**

Dedicated to Professor Marcos Dajczer on his 75th birthday.

Abstract In this article, we introduce the notion of a Ribaucour partial tube and use it to derive several applications. These are based on a characterization of Ribaucour partial tubes as the immersions of a product of two manifolds into a space form such that the distributions given by the tangent spaces of the factors are orthogonal to each other with respect to the induced metric and are invariant under all shape operators, and one of them is spherical. Our first application is a classification of all hypersurfaces with dimension at least three of a space form that carry a spherical foliation of codimension one, extending previous results by Dajczer, Rovenski, and the second author for the totally geodesic case. Another main application concerns the class of hypersurfaces of $\mathbb{R}^{n+1}$ that are of Enneper type, that is, hypersurfaces that carry a family of lines of curvature, corresponding to a simple principal curvature, whose orthogonal $(n-1)$-dimensional distribution is integrable and has the property that its leaves are contained in hyperspheres or affine hyperplanes of $\mathbb{R}^{n+1}$. We show how Ribaucour partial tubes in the sphere can be used to describe all n-dimensional hypersurfaces of Enneper type for which the leaves of the $(n-1)$-dimensional distribution are contained in affine hyperplanes of $\mathbb{R}^{n+1}$ and then show how a general hypersurface of Enneper type can be constructed in terms of a hypersurface in the latter class. We give an explicit description of some special hypersurfaces of Enneper type, among which are natural generalizations of Joachimsthal surfaces.

S. Chion
CENTRUM Católica Graduate Business School, Pontificia Universidad Católica del Perú, Lima, Peru
e-mail: sjchiona@pucp.edu.pe

R. Tojeiro (✉)
Instituto de Ciências Matemáticas e de Computação, Universidade de São Paulo, São Carlos, Brazil
e-mail: tojeiro@icmc.usp.br

© The Author(s), under exclusive license to Springer Nature Switzerland AG 2025
R. Tojeiro (ed.), *Advances in Submanifold Theory and Related Topics*,
Latin American Mathematics Series – UFSCar subseries,
https://doi.org/10.1007/978-3-032-00918-0_6

1 Introduction

In [9] and [10], the authors studied the problem of determining the hypersurfaces of dimension at least three of a space form that carry a totally geodesic foliation of codimension one. The initial motivation for this work was to investigate the similar problem that one can pose by assuming the foliation to be spherical instead of totally geodesic. A foliation being spherical means that each leaf is an umbilical submanifold whose mean curvature vector field is parallel with respect to its normal connection.

In the solution to the problem addressed in [9] and [10], one main example of a hypersurface of a space form that carries a totally geodesic foliation of codimension one is a partial tube over a smooth regular curve. Partial tubes are submanifolds that are generated by starting with any submanifold whose normal bundle has a parallel and flat subbundle, taking a submanifold in the fiber of that subbundle at a given point, and then parallel transporting it along the former submanifold with respect to its normal connection. They are characterized by the property that the tangent spaces to the submanifold that is parallel transported along the starting submanifold give rise to a totally geodesic distribution that is invariant under all of its shape operators.

In this paper, we introduce a class of submanifolds that play the role of partial tubes in our context and which are the basis for some of our main results. They are immersions of product manifolds with two factors that contain partial tubes as special cases and whose construction is based on the extension of the Ribaucour transformation for submanifolds developed in [7] and [8], so we call them *Ribaucour partial tubes* (see Sect. 3 for the precise definition). They extend the notion of an $\mathcal{N}$-Ribaucour transform defined in [11], which was utilized to characterize submanifolds with a Dupin principal normal with integrable conullity. Dupin hypersurfaces have been a focal point of research due to their connection to Lie sphere geometry (see [4] or [5] for a recent survey).

It is a basic result of the present work that Ribaucour partial tubes are precisely the immersions of product manifolds with two factors such that the distributions given by the tangent spaces of the factors are orthogonal to each other with respect to the induced metric and are invariant under all shape operators, and one of them is spherical (Theorem 3). This immediately implies that any submanifold of a space form that carries a spherical distribution that is invariant under all shape operators and whose orthogonal distribution is integrable is locally a Ribaucour partial tube (Corollary 2). In particular, this yields an explicit description of all surfaces with a flat normal bundle of a space form such that the lines of curvature of one of the two families have constant geodesic curvature (Corollary 3). Another immediate application is a description of all foliations of a space form by spherical submanifolds whose orthogonal distributions are integrable, in particular of all foliations of a space form whose leaves are spherical hypersurfaces (Corollary 5).

In the hypersurface case, we show that the condition of being invariant by the shape operator is automatically satisfied if the rank of the distribution is greater than half of the dimension of the hypersurface. In particular, it follows that any hypersurface with dimension ≥ 3 of a space form that carries a spherical foliation of codimension one is locally a Ribaucour partial tube over a curve. This gives a complete answer to the problem that was one of the initial motivations of this paper and leads to a rather explicit description of such hypersurfaces (Corollary 6).

Hypersurfaces of $\mathbb{R}^{n+1}$ that are Ribaucour partial tubes over curves can also be characterized by the fact that they carry a family of lines of curvature, corresponding to a simple principal curvature, whose orthogonal $(n-1)$-dimensional distribution is integrable and has the property that its leaves are contained in hyperspheres of $\mathbb{R}^{n+1}$ that intersect the hypersurface orthogonally. This has led us to investigate the more general class of hypersurfaces for which such leaves are contained in hyperspheres that do not necessarily intersect the hypersurface orthogonally. For $n = 2$, this reduces to studying the class of surfaces with spherical lines of curvature corresponding to one of the principal curvatures, which was widely investigated by many geometers since the second half of the nineteenth century. The interest in such surfaces has been renewed in connection with the construction of immersed constant mean curvature tori in Euclidean three-space by Wente [18], solving a long-standing conjecture by H. Hopf. The examples were represented by a doubly periodic conformal immersion from $\mathbb{R}^2$ to $\mathbb{R}^3$, with its fundamental domain being a rectangle, thus possessing nice symmetry properties. Later, in [20], new examples were constructed, known as *twisted tori*, which no longer had a rectangular fundamental domain. This was followed by spectacular developments (see, e.g., [13]). More recently, isothermic surfaces with planar or spherical lines of curvature also appeared in connection with the construction of isometric tori with the same curvatures [2, 3], which provided examples of compact Bonnet pairs and settled another old and important problem.

In his subsequent memoir [19], Wente called any surface with spherical lines of curvature corresponding to one of the principal curvatures a *surface of Enneper type*. Accordingly, we say that a hypersurface of $\mathbb{R}^{n+1}$ is of Enneper type if it carries a family of lines of curvature, corresponding to a simple principal curvature λ, such that the $(n-1)$-dimensional distribution given by the orthogonal complement $E_\lambda^\perp$ of the eigenbundle E_λ associated with λ is integrable and such that the leaves of $E_\lambda^\perp$ are contained in hyperspheres or affine hyperplanes of $\mathbb{R}^{n+1}$.

Our approach to studying hypersurfaces of Enneper type was inspired by that presented in Bianchi's book [1] for the case $n = 2$. First, we show how Ribaucour partial tubes in $\mathbb{S}^n$ can be used to describe all hypersurfaces of Enneper type for which the leaves of the $(n-1)$-dimensional distribution are contained in affine hyperplanes of $\mathbb{R}^{n+1}$ (Theorem 6). For $n = 2$, these correspond to surfaces with planar lines of curvature associated with one of the principal curvatures. Then, we prove that any hypersurface of Enneper type can be constructed in terms of a hypersurface in the latter class (Theorem 4). For that, we first show how to parametrize any hypersurface of Enneper type in $\mathbb{R}^{n+1}$ in terms of its Gauss map and a triple (γ, α, β), where $\gamma: I \to \mathbb{R}^{n+1}$ is a smooth curve and $\alpha, \beta \in C^\infty(I)$ (Theorem 5). The next step is to determine all the triples $(\bar{\gamma}, \bar{\alpha}, \bar{\beta})$ that give rise to hypersurfaces of Enneper type in $\mathbb{R}^{n+1}$ with the same Gauss map as a given one (Proposition 4). It turns out that, among them, there always exists a hypersurface for which the hyperspheres containing the leaves of the $(n-1)$-dimensional distribution all pass through a common point. An inversion with respect to a hypersphere centered at that point then maps such hyperspheres into affine hyperplanes and, hence, maps the hypersurface into a hypersurface of Enneper type whose leaves are contained in affine hyperplanes of $\mathbb{R}^{n+1}$.

We also give an explicit description of hypersurfaces of Enneper type in $\mathbb{R}^{n+1}$ for which the leaves of the $(n-1)$-dimensional distribution are contained either in concentric hyperspheres, parallel affine hyperplanes or affine hyperplanes that intersect along a common affine $(n-1)$-dimensional subspace (Theorem 7). Surfaces in $\mathbb{R}^3$ with the last of these properties are classically known as Joachimsthal surfaces, and our result yields a new description of them (Corollary 8). It was recently shown in [14] that this last property is also shared by the so-called cyclic conformally flat hypersurfaces of $\mathbb{R}^4$ with three distinct principal curvatures.

2 The Ribaucour Transformation

This section is devoted to reviewing some basic facts on the Ribaucour transformation. For further details, we refer to [7] and [8].

Let M^n be an n-dimensional Riemannian manifold. We denote by $\mathcal{S}(M)$ the module of symmetric sections of the vector bundle of endomorphisms of TM, that is, those elements of $\Gamma(\mathrm{End}(TM))$ such that $\langle \Phi X, Y\rangle = \langle X, \Phi Y\rangle$ for all $X, Y \in \mathfrak{X}(M)$.

Let $f\colon M^n \to \mathbb{R}^m$ be an isometric immersion of a Riemannian manifold M^n. A map $\mathcal{F}\colon M^n \to \mathbb{R}^m$ is called a *Combescure transform* of f determined by $\Phi \in \mathcal{S}(M)$ when $\mathcal{F}_* = f_* \circ \Phi$. This condition forces Φ to satisfy the Codazzi equation:

$$(\nabla_X \Phi)Y = (\nabla_Y \Phi)X$$

and to commute with the second fundamental form of f, in the sense that

$$\alpha(X, \Phi Y) = \alpha(\Phi X, Y)$$

for all $X, Y \in \mathfrak{X}(M)$. Conversely, if M^n is simply connected, then any $\Phi \in \mathcal{S}(M)$ satisfying these two conditions determines a Combescure transform $\mathcal{F}\colon M^n \to \mathbb{R}^m$ of $f\colon M^n \to \mathbb{R}^m$ such that $\mathcal{F}_* = f_* \circ \Phi$.

For any Combescure transform $\mathcal{F}\colon M^n \to \mathbb{R}^m$ of an isometric immersion $f\colon M^n \to \mathbb{R}^m$ of a simply connected Riemannian manifold, there exist $\varphi \in C^\infty(M)$ and $\beta \in \Gamma(N_f M)$ satisfying

$$\alpha(\operatorname{grad}\varphi, X) + \nabla^\perp_X \beta = 0 \tag{1}$$

such that

$$\mathcal{F} = f_* \operatorname{grad} \varphi + \beta \quad \text{and} \quad \Phi = \operatorname{Hess} \varphi - A_\beta. \tag{2}$$

Conversely, any solution (φ, β) of (1) determines a Combescure transform $\mathcal{F}$ of f defined by the first equality in (2). We denote by $\mathcal{D}(f)$ the space of solutions (φ, β) of (1).

Given an isometric immersion $f : M^n \to \mathbb{R}^m$, an immersion $\tilde{f} : M^n \to \mathbb{R}^m$ is said to be a *Ribaucour transform* of f when $\|f - \tilde{f}\| \neq 0$ everywhere and there exists a triple $(\mathcal{P}, D, \delta)$, with $\mathcal{P} : f^*T\mathbb{R}^m \to \tilde{f}^*T\mathbb{R}^m$ a vector bundle isometry, $D \in \mathcal{S}(M)$ and $\delta \in \Gamma(f^*T\mathbb{R}^m)$ nowhere vanishing, such that:

(i) $\mathcal{P}Z - Z = \langle Z, \delta \rangle (f - \tilde{f})$ for all $Z \in \Gamma(f^*T\mathbb{R}^m)$.
(ii) $\tilde{f}_* = \mathcal{P} f_* D$.

Geometrically, f and $\tilde{f}$ envelop a common congruence of n-dimensional spheres, with $\mathcal{P}(x)$, $x \in M^n$, being the reflection with respect to the hyperplane orthogonal to $\tilde{f}(x) - f(x)$, $x \in M^n$. The requirement that the tensor D be symmetric implies that the shape operators A^f_ξ and $A^{\tilde{f}}_{\mathcal{P}\xi}$ with respect to corresponding normal directions commute for every $\xi \in \Gamma(N_f M)$. For surfaces in $\mathbb{R}^3$, this is equivalent to requiring f and $\tilde{f}$ to share the same lines of curvature.

A nice feature of the Ribaucour transformation is that all Ribaucour transforms of a given isometric immersion $f : M^n \to \mathbb{R}^m$ can be explicitly parametrized as follows in terms of f and the pairs $(\varphi, \beta) \in \mathcal{D}(f)$.

Theorem 1 ([8]) *Let $f : M^n \to \mathbb{R}^m$ be an isometric immersion of a simply connected Riemannian manifold and let $\tilde{f} : M^n \to \mathbb{R}^m$ be a Ribaucour transform of f with data $(\mathcal{P}, D, \delta)$. Then, there exists $(\varphi, \beta) \in \mathcal{D}(f)$ such that*

$$\tilde{f} = f - 2\nu\varphi\mathcal{F}, \tag{3}$$

where $\mathcal{F} = f_* \operatorname{grad} \varphi + \beta$ is the Combescure transform determined by (φ, β) and $\nu = \langle \mathcal{F}, \mathcal{F} \rangle^{-1}$. Moreover

$$\mathcal{P} = I - 2\nu\mathcal{F}^*\mathcal{F}, \quad D = I - 2\nu\varphi\Phi \quad \text{and} \quad \delta = -\varphi^{-1}\mathcal{F},$$

where $\Phi = \operatorname{Hess} \varphi - A_\beta$ and $\mathcal{F}^* \in \Gamma((f^*T\mathbb{R}^m)^*)$ is defined by $\mathcal{F}^* Z = \langle \mathcal{F}, Z \rangle$.

Conversely, given $(\varphi, \beta) \in \mathcal{D}(f)$ and an open subset $U \subset M^n$ where φ and $\mathcal{F} = f_* \operatorname{grad} \varphi + \beta$ are nowhere vanishing and D is invertible, the map $\tilde{f} : U \to \mathbb{R}^m$ given by (3) *is a Ribaucour transform of $f|_U$.*

The Ribaucour transform determined by (φ, β) is denoted by $\mathcal{R}_{(\varphi,\beta)} f$.

Example 1

(*i*) Given a point $P_0 \in \mathbb{R}^m$ and $r > 0$, set $2\varphi_1 = \|f - P_0\|^2 - r^2$ and $\beta_1 = (f - P_0)_{N_f M}$. Then, $\mathcal{F} = f - P_0$, $\Phi = I$, and

$$\tilde{f} = \mathcal{R}_{(\varphi_1, \beta_1)}(f) = P_0 + r^2 \|f - P_0\|^{-2}(f - P_0)$$

is the composition of f with an inversion with respect to the sphere of radius r centered at P_0.

(*ii*) Given a parallel normal vector field ξ, define (φ_2, β_2) by $2\varphi_2 = \|\xi\|^2$ and $\beta_2 = -\xi$. Then, $\mathcal{F} = -\xi$ and

$$\tilde{f} = \mathcal{R}_{(\varphi_2, \beta_2)}(f) = f + \xi$$

is the parallel translation of f by ξ.

The Ribaucour transformation can be easily extended for submanifolds of Minkowski space $\mathbb{R}^m_1$. It also admits an extension for submanifolds of any space form $\mathbb{Q}^m_c$ of constant curvature $c \neq 0$ as follows. Let $\mathbb{R}^{m+1}_\epsilon$ denote either Euclidean or Minkowski space of dimension $m + 1$, according to whether $\epsilon = 0$ or $\epsilon = 1$, respectively, and let $i : \mathbb{Q}^m_c \to \mathbb{R}^{m+1}_\epsilon$ be the umbilical inclusion of $\mathbb{Q}^m_c$ into $\mathbb{R}^{m+1}_\epsilon$, with $\epsilon = 0$ if $c > 0$ and $\epsilon = 1$ if $c < 0$. Then, an immersion $\tilde{f} \colon M^n \to \mathbb{Q}^m_c$ is a Ribaucour transform of an isometric immersion $f \colon M^n \to \mathbb{Q}^m_c$ with data $(\mathcal{P}, D, \delta)$ if $\tilde{F} := i \circ \tilde{f} \colon M^n \to \mathbb{R}^{m+1}_\epsilon$ is a Ribaucour transform of $F = i \circ f$ with data $(\hat{\mathcal{P}}, D, \hat{\delta})$, where $\hat{\delta} = \delta - cF$ and $\hat{\mathcal{P}} \colon F^*\mathbb{R}^{m+1}_\epsilon \to \tilde{F}^*\mathbb{R}^{m+1}_\epsilon$ is the extension of $\mathcal{P}$ defined by setting $\hat{\mathcal{P}}(F) = \tilde{F}$. In this setting, Theorem 1 reads as follows.

Theorem 2 ([8]) *Let $f \colon M^n \to \mathbb{Q}^m_c$ be an isometric immersion of a simply connected Riemannian manifold and let $\tilde{f} \colon M^n \to \mathbb{Q}^m_c$ be a Ribaucour transform of f with data $(\mathcal{P}, D, \delta)$. Then, there exists $(\varphi, \beta) \in \mathcal{D}(f)$ such that*

$$\tilde{F} = F - 2\nu\varphi\mathcal{G}, \tag{4}$$

where $\mathcal{G} = F_* grad\, \varphi + \beta + c\varphi F$ and $\nu = \langle \mathcal{G}, \mathcal{G} \rangle^{-1}$. Moreover

$$\hat{\mathcal{P}} = I - 2\nu\mathcal{G}\mathcal{G}^*, \quad D = I - 2\nu\varphi(Hess\, \varphi + c\varphi I - A_\beta) \quad and \quad \hat{\delta} = -\varphi^{-1}\mathcal{G}. \tag{5}$$

Conversely, given $(\varphi, \beta) \in \mathcal{D}(f)$ and an open subset $U \subset M^n$ where $\varphi\nu \neq 0$ and the tensor D given by (5) is invertible, let $\tilde{F} \colon U \to \mathbb{R}^{m+1}_\epsilon$ be defined by (4). Then, $\tilde{F} = i \circ \tilde{f}$, where $\tilde{f}$ is a Ribaucour transform of f.

3 Ribaucour Partial Tubes

In this section, we introduce the concept of a Ribaucour partial tube, on which some of the main results of this article are based.

Let $f_1 \colon M_1 \to \mathbb{R}^m$ be an isometric immersion along which there is an orthonormal set $\{\xi_1, \ldots, \xi_k\}$ of normal vector fields that are parallel in the normal connection. The subbundle $\mathcal{L} = \operatorname{span}\{\xi_1, \ldots, \xi_k\}$ of $N_{f_1}M_1$ is thus parallel and flat. Hence, the map $\Psi \colon M_1 \times \mathbb{R}^k \to \mathcal{L}$, defined by

$$\Psi_{x_1}(y) = \Psi(x_1, y) = \sum_{i=1}^{k} y_i \xi_i(x_1)$$

for all $x_1 \in M_1$ and $y = (y_1, \ldots, y_k) \in \mathbb{R}^k$, is a parallel vector bundle isometry.

For a fixed $y \in \mathbb{R}^k$, we denote by $\Psi(y)$ the parallel section of $\mathcal{L}$ given by $\Psi(y)(x_1) = \Psi(x_1, y)$ for all $x_1 \in M_1$. Given an isometric immersion $f_0 \colon M_0 \to \mathbb{R}^k$, first recall that the partial tube over f_1 with f_0 as fiber is the map $g \colon M_0 \times M_1 \to \mathbb{R}^m$ given by

$$g(x_0, x_1) = f_1(x_1) + \Psi_{x_1}(f_0(x_0)).$$

Geometrically, $g(M_0 \times M_1)$ is generated by taking the image of $f_0(M_0)$ under a fixed Ψ_{x_1}, $x_1 \in M_1$, and parallel translating it along f_1 with respect to its normal connection (see, e.g., Chapter 10 of [6] for details).

Now, take $(\varphi, \beta) \in \mathcal{D}(f_1)$ and define $f \colon M_0 \times M_1 \to \mathbb{R}^m$ by

$$f(x_0, x_1) = \big(\mathcal{R}_{(\varphi, \beta + \Psi(f_0(x_0)))} f_1\big)(x_1) = (f_1 - 2\nu_{x_0}\varphi \mathcal{F}_{x_0})(x_1), \tag{6}$$

where $\mathcal{F}_{x_0} = f_{1*}\operatorname{grad}\varphi + \beta + \Psi(f_0(x_0))$ and $\nu_{x_0} = ||\mathcal{F}_{x_0}||^{-2}$. For each $x_0 \in M_0$, we denote

$$\mathcal{P}_{x_0} = I - 2\nu_{x_0}\mathcal{F}^*_{x_0}\mathcal{F}_{x_0} \quad \text{and} \quad D_{x_0} = I - 2\nu_{x_0}\varphi\Phi_{x_0},$$

with $\Phi_{x_0} = \operatorname{Hess}\varphi - A^{f_1}_{\beta + \Psi(f_0(x_0))}$, the vector bundle isometry and the symmetric endomorphism associated with the Ribaucour transform $\mathcal{R}_{(\varphi, \beta + \Psi(f_0(x_0)))} f_1$ of f_1.

If, in particular, $(\varphi, \beta) \in \mathcal{D}(f_1)$ is given by $2\varphi = -1$ and $\beta = 0$, then

$$\mathcal{R}_{(\varphi, \beta + \Psi(f_0(x_0)))} f_1 = f_1 - 2\nu_{x_0}\varphi \mathcal{F}_{x_0} = f_1 + \Psi\left(\frac{f_0(x_0)}{\| f_0(x_0)\|^2}\right).$$

Thus $\mathcal{R}_{(\varphi, \beta + \Psi(f_0(x_0)))} f_1$ reduces to the partial tube over f_1 whose fiber is the composition of f_0 with an inversion with respect a hypersphere of unit radius centered at the origin.

Given a product manifold $M = \prod_{i=0}^{r} M_i$ and $x = (x_0, \ldots, x_r) \in M$, we denote by $\tau_i^x : M_i \to M$ the inclusion given by $\tau_i^x(y_i) = (x_0, \ldots, x_{i-1}, y_i, x_{i+1}, \ldots, x_r)$.

Proposition 1 *The differential of the map f in (6) at $x = (x_0, x_1)$ is given by*

$$f_* \tau_{0\,*}^x X_0 = -2\nu_{x_0} \varphi \mathcal{P}_{x_0}(\Psi_{x_1}(f_{0*} X_0)) \tag{7}$$

and

$$f_* \tau_{1\,*}^x X_1 = \mathcal{P}_{x_0} f_{1*} D_{x_0} X_1, \tag{8}$$

for all $X_i \in T_{x_i} M_i$, $0 \le i \le 1$.

Proof Differentiating (6), we obtain

$$\begin{aligned} f_* \tau_{0\,*}^x X_0 &= 4v_{x_0}^2 \varphi \left\langle \Psi_{x_1}(f_{0*} X_0), \mathcal{F}_{x_0} \right\rangle \mathcal{F}_{x_0} - 2\nu_{x_0} \varphi \Psi_{x_1}(f_{0*} X_0) \\ &= -2\nu_{x_0} \varphi \mathcal{P}_{x_0}(\Psi_{x_1}(f_{0*} X_0)). \end{aligned}$$

Since f a Ribaucour transformation of f_1, Eq. (8) follows. □

If the map f given by (6) is an immersion at any point, then it is called the *Ribaucour partial tube* over f_1 with fiber f_0 associated with Ψ and $(\varphi, \beta) \in \mathcal{D}(f_1)$ or the Ribaucour partial tube determined by $(f_0, f_1, \Psi, \varphi, \beta)$. It follows from Proposition 1 that f is an immersion at $(x_0, x_1) \in M$ if and only if neither φ nor $\mathcal{F}_{x_0}$ vanish at x_1 and the endomorphism D_{x_0} of $T_{x_1} M_1$ is invertible. We always assume that $f_0 : M_0 \to \mathbb{R}^k$ is a substantial immersion, for if this is not the case, say, $f_0(M_0) \subset v + \mathbb{R}^l$ for some $v \in \mathbb{R}^k$ and $l < k$, we may replace (φ, β) by $(\varphi, \beta + \Psi_{x_1}(v))$ and restrict Ψ to $M_1 \times \mathbb{R}^l$.

Proposition 2 *The following assertions on the Ribaucour partial tube f determined by $(f_0, f_1, \Psi, \varphi, \beta)$ hold:*

(i) The induced metric is given by

$$\left\langle \tau_{0\,*}^x X_0 + \tau_{1\,*}^x X_1, \tau_{0\,*}^x Y_0 + \tau_{1\,*}^x Y_1 \right\rangle_f = 4v_{x_0}^2 \varphi^2 \langle X_0, Y_0 \rangle_{f_0} + \left\langle D_{x_0}^2 X_1, Y_1 \right\rangle_{f_1}.$$

(ii) The normal space of f at (x_0, x_1) is

$$N_f M(x_0, x_1) = \mathcal{P}_{x_0}\big(\mathcal{L}^\perp(x_1) \oplus \Psi_{x_1}(N_{f_0} M_0(x_0))\big).$$

(iii) Given $\delta \in \Gamma(\mathcal{L}^\perp)$ and $\zeta \in \Gamma(N_{f_0} M_0)$, the shape operators of f with respect to $\hat{\delta}, \hat{\zeta} \in N_f M$, defined by

$$\hat{\delta}(x_0, x_1) = \mathcal{P}_{x_0} \delta(x_1) \quad \text{and} \quad \hat{\zeta}(x_0, x_1) = \mathcal{P}_{x_0}(\Psi_{x_1}(\zeta(x_0))),$$

are

$$A^f_{\hat{\delta}}\tau^x_{0*} = -\varphi^{-1}\langle\delta,\beta\rangle\,\tau^x_{0*}, \tag{9}$$

$$A^f_{\hat{\delta}}\tau^x_{1*} = \tau^x_{1*}D^{-1}_{x_0}(A^{f_1}_{\delta} + 2\nu_{x_0}\langle\beta,\delta\rangle\,\Phi_{x_0}), \tag{10}$$

$$A^f_{\hat{\zeta}}\tau^x_{0*} = -\frac{1}{\varphi}\left(\frac{1}{2\nu_{x_0}}\tau^x_{0*}A^{f_0}_{\zeta} + \left\langle\Psi_{x_1}(\zeta),\mathcal{F}_{x_0}\right\rangle\tau^x_{0*}\right) \tag{11}$$

and

$$A^f_{\hat{\zeta}}\tau^x_{1*} = \tau^x_{1*}D^{-1}_{x_0}(A^{f_1}_{\Psi_{x_1}(\zeta)} + 2\nu_{x_0}\left\langle\Psi_{x_1}(\zeta),\mathcal{F}_{x_0}\right\rangle\Phi_{x_0}). \tag{12}$$

(iv) The normal connection of f is given by

$$^f\nabla^\perp_{\tau^x_{0*}X_0}\hat{\delta} = 0, \tag{13}$$

$$^f\nabla^\perp_{\tau^x_{1*}X_1}\hat{\delta} = \mathcal{P}_{x_0}{}^{f_1}\nabla^\perp_{x_1}\delta, \tag{14}$$

$$^f\nabla^\perp_{\tau^x_{0*}X_0}\hat{\zeta} = \mathcal{P}_{x_0}(\Psi_{x_1}({}^{f_0}\nabla^\perp_{X_0}\zeta)) \tag{15}$$

and

$$^f\nabla^\perp_{\tau^x_{1*}X_1}\hat{\zeta} = 0. \tag{16}$$

Proof Items (i) and (ii) are immediate consequences of (7) and (8). For (iii) and (iv), on one hand, we have

$$\bar{\nabla}_{\tau^x_{0*}X_0}\hat{\delta} = -f_*A^f_{\hat{\delta}}\tau^x_{0*}X_0 + {}^f\nabla^\perp_{\tau^x_{0*}X_0}\hat{\delta} \quad\text{and}\quad \bar{\nabla}_{\tau^x_{1*}X_1}\hat{\delta} = -f_*A^f_{\hat{\delta}}\tau^x_{1*}X_1 + {}^f\nabla^\perp_{\tau^x_{1*}X_1}\hat{\delta}.$$

On the other hand, using (7) and (8), we obtain

$$\begin{aligned}\bar{\nabla}_{\tau^x_{0*}X_0}\hat{\delta} &= \bar{\nabla}_{\tau^x_{0*}X_0}(\delta - 2\nu_{x_0}\left\langle\mathcal{F}_{x_0},\delta\right\rangle\mathcal{F}_{x_0})\\ &= -2\nu_{x_0}\langle\beta,\delta\rangle\,\mathcal{P}_{x_0}\Psi_{x_1}(f_{0*}X_0)\\ &= \varphi^{-1}\langle\beta,\delta\rangle\,f_*\tau^x_{0*}X_0\end{aligned}$$

and

$$\begin{aligned}\bar{\nabla}_{\tau_1^x{}_* X_1}\hat{\delta} &= \bar{\nabla}_{\tau_1^x{}_* X_1}(\delta - 2\nu_{x_0}\left\langle \mathcal{F}_{x_0}, \delta\right\rangle \mathcal{F}_{x_0})\\ &= -\mathcal{P}_{x_0} f_{1*}(A_\delta^{f_1} X_1 + 2\nu_{x_0}\langle \beta, \delta\rangle \Phi_{x_0} X_1) + \mathcal{P}_{x_0}{}^{f_1}\nabla_{X_1}^{\perp}\delta\\ &= -f_*\tau_1^x{}_* D_{x_0}^{-1}(A_\delta^{f_1} X_1 + 2\nu_{x_0}\langle \beta, \delta\rangle \Phi_{x_0} X_1) + \mathcal{P}_{x_0}{}^{f_1}\nabla_{X_1}^{\perp}\delta,\end{aligned}$$

which yield (9), (10), (13), and (14). Repeating the argument for $\hat{\zeta} \in \Gamma(N_f M)$ gives

$$\bar{\nabla}_{\tau_0^x{}_* X_0}\hat{\zeta} = -f_* A_{\hat{\zeta}}^f \tau_0^x{}_* X_0 + {}^f\nabla^{\perp}_{\tau_0^x{}_* X_0}\hat{\zeta} \text{ and } \bar{\nabla}_{\tau_1^x{}_* X_1}\hat{\zeta} = -f_* A_{\hat{\zeta}}^f \tau_1^x{}_* X_1 + {}^f\nabla^{\perp}_{\tau_1^x{}_* X_1}\hat{\zeta}.$$

Using (7), (8) and the fact that Ψ is a parallel vector bundle isometry, we obtain

$$\begin{aligned}\bar{\nabla}_{\tau_0^x{}_* X_0}\hat{\zeta} &= \bar{\nabla}_{\tau_0^x{}_* X_0}(\Psi(\zeta) - 2\nu_{x_0}\left\langle \Psi(\zeta), \mathcal{F}_{x_0}\right\rangle \mathcal{F}_{x_0})\\ &= -\mathcal{P}_{x_0}\Psi_{x_1} f_{0*}(A_\zeta^{f_0} X_0 + 2\nu_{x_0}\left\langle \Psi_{x_1}(\zeta), \mathcal{F}_{x_0}\right\rangle X_0) + \mathcal{P}_{x_0}(\Psi_{x_1}({}^{f_0}\nabla_{X_0}^{\perp}\zeta))\\ &= (2\nu_{x_0}\varphi)^{-1} f_*\tau_0^x{}_* A_\zeta^{f_0} X_0 + \varphi^{-1}\left\langle \Psi_{x_1}(\zeta), \mathcal{F}_{x_0}\right\rangle f_*\tau_0^x{}_* X_0\\ &\quad + \mathcal{P}_{x_0}(\Psi_{x_1}({}^{f_0}\nabla_{X_0}^{\perp}\zeta))\end{aligned}$$

and

$$\begin{aligned}\bar{\nabla}_{\tau_1^x{}_* X_1}\hat{\zeta} &= \bar{\nabla}_{\tau_1^x{}_* X_1}(\Psi(\zeta) - 2\nu_{x_0}\left\langle \Psi(\zeta), \mathcal{F}_{x_0}\right\rangle \mathcal{F}_{x_0})\\ &= -\mathcal{P}_{x_0} f_{1*}(A^{f_1}_{\Psi_{x_1}(\zeta)} X_1 + 2\nu_{x_0}\left\langle \Psi(\zeta), \mathcal{F}_{x_0}\right\rangle \Phi_{x_0} X_1)\\ &= -f_*\tau_1^x{}_* D_{x_0}^{-1}(A^{f_1}_{\Psi_{x_1}(\zeta)} X_1 + 2\nu_{x_0}\left\langle \Psi(\zeta), \mathcal{F}_{x_0}\right\rangle \Phi_{x_0} X_1).\end{aligned}$$

Thus, (11), (12), (15), and (16) follow. □

Before we state some consequences of the preceding proposition, we first recall some terminology.

A *net* $\mathcal{E} = (E_i)_{i=0,\ldots,r}$ on a differentiable manifold M is a decomposition of its tangent bundle $TM = \oplus_{i=0}^r E_i$ as a Whitney sum of integrable distributions. If M is a Riemannian manifold, the net $\mathcal{E} = (E_i)_{i=0,\ldots,r}$ is said to be an *orthogonal net* if the distributions E_i are mutually orthogonal. Given an isometric immersion $f: M \to \mathbb{R}^m$ of a Riemannian manifold M equipped with a net $\mathcal{E} = (E_i)_{i=0,\ldots,r}$, then the second fundamental form α_f of f is said to be *adapted to the net* $\mathcal{E}$ if $\alpha_f(X_i, X_j) = 0$ whenever $X_i \in \Gamma(E_i)$ and $X_j \in \Gamma(E_j)$ with $0 \leq i \neq j \leq r$.

In a product manifold $M = \prod_{i=0}^{r} M_i$, we have a natural net $\mathcal{E} = (E_i)_{i=0,\ldots,r}$, called its *product net*, given by the tangent bundles of its factors, that is, $E_i(x) = \tau_i^x{}_* T_{x_i} M_i$ for all $x = (x_0, \ldots, x_r) \in M$. For each $0 \le i \le r$, we denote

$$M_{\perp_i} = \prod_{\substack{j=0 \\ j \neq i}}^{r} M_j$$

and define the projection $\pi_{\perp_i} \colon M \to M_{\perp_i}$ by

$$\pi_{\perp_i}(x_0, \ldots, x_r) = (x_0, \ldots, x_{i-1}, x_{i+1}, \ldots, x_r).$$

Given $x_i \in M_i$, the map $\mu_{x_i} \colon M_{\perp_i} \to M$ stands for the inclusion:

$$\mu_{x_i}(y_0, \ldots, y_{i-1}, y_{i+1}, \ldots, y_r) = (y_0, \ldots, y_{i-1}, x_i, y_{i+1}, \ldots, y_r).$$

A tangent subbundle $E \subset TM$ is *umbilical* if there exists $Z \in \Gamma(E^\perp)$ such that

$$(\nabla_T S)_{E^\perp} = \langle T, S \rangle \, Z \tag{17}$$

for all $T, S \in \Gamma(E)$, where the subscript $E^\perp$ denotes the orthogonal projection onto $E^\perp$. Umbilical distributions are integrable and the leaves are totally umbilical submanifolds with mean curvature vector field Z. If, in addition, Z satisfies

$$(\nabla_T Z)_{E^\perp} = 0 \tag{18}$$

for any $T \in \Gamma(E)$, then E is said to be *spherical* and its leaves are extrinsic spheres.

Corollary 1 *Let $f \colon M_0 \times M_1 \to \mathbb{R}^m$ be the Ribaucour partial tube determined by $(f_0, f_1, \Psi, \varphi, \beta)$ and let $\mathcal{E} = (E_0, E_1)$ be the product net of M. Then, the following assertions hold:*

(i) The net $\mathcal{E}$ is orthogonal with respect to the metric induced by f.
(ii) The second fundamental form of f is adapted to $\mathcal{E}$.
(iii) E_0 is a spherical distribution.

Proof The assertions in items (i) and (ii) are immediate consequences of parts (i) and (iii) of Proposition 2, respectively. To prove item (iii), we must show that there exists $Z \in \Gamma(E_1)$ such that

$$\langle \nabla_S T, X \rangle = \langle S, T \rangle \, \langle X, Z \rangle \tag{19}$$

and

$$\langle \nabla_S Z, X \rangle = 0 \tag{20}$$

for all $S, T \in \Gamma(E_0)$ and $X \in \Gamma(E_1)$. We have

$$\begin{aligned}
\bar{\nabla}_{\tau_{0*}^x X_0} f_* \tau_{0*}^x Y_0 &= -\bar{\nabla}_{\tau_{0*}^x X_0} 2\nu_{x_0} \varphi \mathcal{P}_{x_0}(\Psi(f_{0*}Y_0)) \\
&= \bar{\nabla}_{\tau_{0*}^x X_0}(-2\nu_{x_0}\varphi\Psi(f_{0*}Y_0) + 4\nu_{x_0}^2\varphi\left\langle\Psi(f_{0*}Y_0), \mathcal{F}_{x_0}\right\rangle\mathcal{F}_{x_0}) \\
&= 4\nu_{x_0}^2\varphi\left\langle\Psi_{x_1}(f_{0*}X_0), \mathcal{F}_{x_0}\right\rangle\mathcal{P}_{x_0}(\Psi_{x_1}(f_{0*}Y_0)) \\
&\quad +4\nu_{x_0}^2\varphi\left\langle\Psi_{x_1}(f_{0*}Y_0), \mathcal{F}_{x_0}\right\rangle\mathcal{P}_{x_0}(\Psi_{x_1}(f_{0*}X_0)) \\
&\quad -2\nu_{x_0}\varphi\mathcal{P}_{x_0}(\Psi_{x_1}(f_{0*}\nabla_{X_0}Y_0 + \alpha^{f_0}(X_0, Y_0)) \\
&\quad -\varphi^{-1}\left\langle\tau_{0*}^x X_0, \tau_{0*}^x Y_0\right\rangle_f \mathcal{P}_{x_0}(f_{1*}\mathrm{grad}_1\,\varphi + \beta + \Psi_{x_1}(f_0(x_0)),
\end{aligned}$$

where grad_1 stands for the gradient with respect to the metric induced by f_1. Using (7) and (8), we obtain

$$\left\langle\nabla_{\tau_{0*}^x X_0}\tau_{0*}^x Y_0, \tau_{1*}^x Y_1\right\rangle = -\varphi^{-1}\left\langle\tau_{0*}^x X_0, \tau_{0*}^x Y_0\right\rangle_f\left\langle\tau_{1*}^x D_{x_0}^{-1}\mathrm{grad}_1\,\varphi, \tau_{1*}^x Y_1\right\rangle_f.$$

Hence, (19) holds with $Z(x_0, x_1) = -\tau_{1*}^x D_{x_0}^{-1}\mathrm{grad}_1 \log\varphi(x_1)$. Now, (20) follows from

$$\begin{aligned}
\bar{\nabla}_{\tau_{0*}^x X_0} f_* Z &= -\bar{\nabla}_{\tau_{0*}^x X_0}\mathcal{P}_{x_0} f_{1*}\mathrm{grad}_1 \log\varphi \\
&= -4\nu_{x_0}^2\left\langle\Psi_{x_1}(f_{0*}X_0), \mathcal{F}_{x_0}\right\rangle\left\langle f_{1*}\mathrm{grad}_1 \log\varphi, \mathcal{F}_{x_0}\right\rangle\mathcal{F}_{x_0} \\
&\quad +2\nu_{x_0}\left\langle f_{1*}\mathrm{grad}_1 \log\varphi, \mathcal{F}_{x_0}\right\rangle\Psi_{x_1}(f_{0*}X_0) \\
&= -\varphi^{-1}\left\langle f_{1*}\mathrm{grad}_1 \log\varphi, \mathcal{F}_{x_0}\right\rangle f_*\tau_{0*}^x X_0 \\
&= -\left\langle f_* Z, f_* Z\right\rangle f_*\tau_{0*}^x X_0.
\end{aligned}$$

□

4 A Characterization of Ribaucour Partial Tubes

This section aims to prove that conditions (i) to (iii) in Corollary (1) characterize Ribaucour partial tubes among immersions $f\colon M_0 \times M_1 \to \mathbb{R}^m$ of product manifolds. We make use of the following lemma (see Proposition 9 of [15]).

Lemma 1 *Let $f\colon M^n \to \mathbb{R}^m$ and $g\colon L^k \to M^n$ be isometric immersions. Then, the following assertions are equivalent:*

(i) g is an extrinsic sphere whose mean curvature vector has length $1/r$, $r > 0$, and $\alpha_f(g_*X, Z) = 0$ for all $X \in \mathfrak{X}(L)$ and $Z \in \Gamma(N_gL)$.

(ii) There exists $\zeta \in \Gamma(N_gL)$ of length $1/r$, $r > 0$, such that the subbundle of f_*N_gL orthogonal to $f_*\zeta$ is constant in $\mathbb{R}^m$ and the map $f \circ g + r^2 f_*\zeta$ is constant on L^k.

(iii) $f(g(L))$ is contained in a sphere of radius r and dimension $(m-n+k)$ in $\mathbb{R}^m$ whose normal space along $f(g(L))$ is f_*N_gL.

The characterization of Ribaucour partial tubes is as follows.

Theorem 3 *Let $f: M = M_0 \times M_1 \to \mathbb{R}^m$ be an immersion satisfying conditions (i) to (iii) in Corollary 1. Then, for any fixed $\bar{x}_0 \in M_0$, the map $f_1: M_1 \to \mathbb{R}^m$ given by $f_1 = f \circ \mu_{\bar{x}_0}$ is an immersion whose normal bundle $N_{f_1}M_1$ carries a parallel flat vector subbundle $\mathcal{L}$, and there exist a parallel vector bundle isometry $\Psi: M_1 \times \mathbb{R}^k \to \mathcal{L}$, an immersion $f_0: M_0 \to \mathbb{R}^k$, and $(\varphi, \beta) \in \mathcal{D}(f_1)$ such that f is the Ribaucour partial tube determined by $(f_0, f_1, \Psi, \varphi, \beta)$.*

Proof The normal space of f_1 at $x_1 \in M_1$ splits orthogonally as

$$N_{f_1}M_1(x_1) = f_*E_0(\bar{x}_0, x_1) \oplus N_f M(\bar{x}_0, x_1).$$

Let $Z \in \Gamma(E_1)$ be the mean curvature vector field of E_0 and let $\mu: M_1 \to \mathbb{R}^m$ be defined by $\mu(x_1) = f_*Z(\bar{x}_0, x_1) = f_{1*}Z_1(x_1)$, where $\mu_{\bar{x}_0 *}Z_1 = Z \circ \mu_{\bar{x}_0}$. By Lemma 1, for each $x_1 \in M_1$, the image by f of the leaf $\sigma(x_1) = M_0 \times \{x_1\} = \mu_{x_1}(M_0)$ of E_0 is contained in an $(m - m_1)$-dimensional sphere $S(x_1)$ through $f_1(x_1)$ having f_*E_1 as its normal bundle along $\sigma(x_1)$ and whose center is the constant value

$$f_1(x_1) + \frac{\mu(x_1)}{||\mu(x_1)||^2}$$

along $\sigma(x_1)$ of the map $f + \dfrac{f_*Z}{||Z||^2}$. Thus, we can parametrize $S(x_1)$ by

$$\mu(x_1) + t \in (\mu(x_1) + N_{f_1}M_1(x_1)) \mapsto f_1(x_1) + \frac{2}{||\mu(x_1) + t||^2}(\mu(x_1) + t),$$

which can be thought of as the restriction to the affine subspace $\mu(x_1) \oplus N_{f_1}M_1(x_1)$ of the composition of an inversion with respect to a sphere of radius $\sqrt{2}$ centered at the origin and a translation by $f_1(x_1)$. Notice that the image of this parametrization misses the point $f_1(x_1) = f(\bar{x}_0, x_1) \in S(x_1)$ itself, which is achieved by letting $\|t\|$ go to infinity.

For each $x_1 \in M_1$, since $f(\sigma(x_1)) \subset S(x_1)$, there exists an immersion $h^{x_1}: M_0 \to N_{f_1}M_1(x_1)$ such that

$$f(x_0, x_1) = f_1(x_1) + \frac{2}{||\mu(x_1) + h^{x_1}(x_0)||^2}(\mu(x_1) + h^{x_1}(x_0)).$$

Denoting

$$\rho(x_0, x_1) = \mu(x_1) + h^{x_1}(x_0),$$

we may write

$$f = f_1 \circ \pi_1 + \frac{2}{||\rho||^2}\rho, \tag{21}$$

where $\pi_1 \colon M \to M_1$ is the projection.

Given $\theta \in N_{f_1}M_1(x_1) = f_*E_0(\bar{x}_0, x_1) \oplus N_f M(\bar{x}_0, x_1)$, for each $x_0 \in M_0$, let

$$\bar{\theta} = \theta - \frac{2\langle\theta, \rho(x_0, x_1)\rangle}{||\rho(x_0, x_1)||^2}\rho(x_0, x_1)$$

be the reflection of θ with respect to the hyperplane orthogonal to the vector $\rho(x_0, x_1)$. We claim that $\bar{\theta} \in T_{f(x_0,x_1)}S(x_1) = (f_*E_1(x_0, x_1))^\perp$, that is, $\langle\bar{\theta}, \gamma\rangle = 0$ for all $\gamma \in f_*E_1(x_0, x_1)$. To prove this, consider the decomposition:

$$f_*E_1(x_0, x_1) = (f_*E_1(x_0, x_1) \cap \{f_*Z(x_0, x_1)\}^\perp) \oplus \operatorname{span}\{f_*Z(x_0, x_1)\}.$$

Since the subbundle $f_*E_1 \cap \{f_*Z\}^\perp$ is parallel along the leaves of E_0 by Lemma 1

$$\begin{aligned} f_*E_1(x_0, x_1) \cap \{f_*Z(x_0, x_1)\}^\perp &= f_*E_1(\bar{x}_0, x_1) \cap \{f_*Z(\bar{x}_0, x_1)\}^\perp \\ &= f_{1*}T_{x_1}M_1 \cap \{\mu(x_1)\}^\perp. \end{aligned}$$

Thus, $\langle\bar{\theta}, \gamma\rangle = 0$ for all $\gamma \in f_*E_1(x_0, x_1) \cap \{f_*Z(x_0, x_1)\}^\perp$. On the other hand, Lemma 1 also says that

$$f(x_0, x_1) + \frac{f_*Z(x_0, x_1)}{\|Z(x_0, x_1)\|^2} = f_1(x_1) + \frac{\mu(x_1)}{\|\mu(x_1)\|^2}$$

for all $x_0 \in M_0$. Moreover, $\|Z(x_0, x_1)\|$ also does not depend on x_0 and hence coincides with $\|\mu(x_1)\|$. Hence, substituting (21) in the preceding equation gives

$$f_*Z(x_0, x_1) = \mu(x_1) - \frac{2\|\mu(x_1)\|^2}{\|\rho(x_0, x_1)\|^2}\rho(x_0, x_1).$$

It follows that $\langle\bar{\theta}, f_*Z(x_0, x_1)\rangle = 0$, and the claim follows.

Now, differentiating (21) at $x = (x_0, x_1)$ gives

$$f_*\tau_0^x{}_*X_0 = \frac{2}{||\rho||^2}(h^{x_1}{}_*X_0 - \frac{2\langle h^{x_1}{}_*X_0, \rho\rangle}{||\rho||^2}\rho)$$

and

$$f_*\tau_1^x{}_*X_1 = f_{1*}X_1 + \frac{2}{||\rho||^2}(\rho_*\tau_1^x{}_*X_1 - \frac{2\langle\rho_*\tau_1^x{}_*X_1, \rho\rangle}{||\rho||^2}\rho).$$

Given $\theta \in N_{f_1}M_1(x_1)$, $x_0 \in M_0$ and $X_1 \in T_{x_1}M_1$, endowing M_1 with the metric induced by f_1, we obtain

$$\left\langle \rho_* \tau_1^x{}_* X_1, \theta \right\rangle = \langle Z_1, X_1 \rangle \left\langle h^{x_1}(x_0), \theta \right\rangle,$$

bearing in mind that $\left\langle \bar{\theta}, f_* \tau_1^x{}_* X_1 \right\rangle = 0$. Thus

$$\alpha^{f_1}(X_1, Z_1) + \nabla^{\perp}_{X_1} h(x_0) = \langle Z_1, X_1 \rangle \, h^{x_1}(x_0) \tag{22}$$

for all $x_0 \in M_0$ and $X_1 \in T_{x_1}M_1$.

For a fixed $z_0 \in M_0$, define $\xi^{x_0} = h(x_0) - h(z_0)$. Then

$$\nabla^{\perp}_{X_1} \xi^{x_0} = \langle X_1, Z_1 \rangle \, \xi^{x_0}$$

for all $X_1 \in T_{x_1}M_1$. In particular, this implies that

$$\langle X_1, Z_1 \rangle = X_1(\log \|\xi^{x_0}\|)$$

for all $X_1 \in T_{x_1}M_1$, that is, $Z_1 = \operatorname{grad} \tau$, where $\tau = \log \|\xi^{x_0}\|$. Hence, $e^{-\tau}\xi^{x_0}$ is a parallel section of $N_{f_1}M_1$ for any $x_0 \in M_0$. It follows that the subspaces

$$\mathcal{L}(x_1) = \operatorname{span}\{\xi^{x_0}(x_1) : x_0 \in M_0\} \subset N_{f_1}M_1(x_1)$$

define a parallel and flat subbundle of $N_{f_1}M_1$.

Let $\Psi\colon M_1 \times \mathbb{R}^k \to \mathcal{L}$ be a parallel vector bundle isometry. Since $e^{-\tau}\xi^{x_0}$ is a parallel section of $\mathcal{L}$, there exists $f_0\colon M_0 \to \mathbb{R}^k$ such that $\Psi(f_0(x_0)) = e^{-\tau}\xi^{x_0}$.

Let $\varphi \in C^\infty(M_1)$ be given by $\varphi = -e^{-\tau}$ and let $\beta = e^{-\tau}h(z_0) \in \Gamma(N_{f_1}M_1)$. Then, (22) becomes

$$\alpha^{f_1}(X_1, \operatorname{grad} \varphi) + \nabla^{\perp}_{X_1} \beta = 0,$$

that is, $(\varphi, \beta) \in \mathcal{D}(f_1)$. Finally

$$\begin{aligned} \mathcal{R}_{(\varphi, \beta + \Psi(f_0(x_0)))} f_1 &= f_1 - \frac{2(-e^{-\tau})}{e^{-2\tau}\,||\mu + h(x_0)||^2}(e^{-\tau}(\mu + h(x_0))) \\ &= f_1 + \frac{2}{||\mu + h(x_0)||^2}(\mu + h(x_0)), \\ &= f, \end{aligned}$$

thus f is the Ribaucour partial tube determined by $(f_0, f_1, \Psi, \varphi, \beta)$. □

Remark 1 Theorem 3 can be regarded as a conformal counterpart of Theorem 3.5 in [16], which characterizes partial tubes as the immersions $g\colon M_0 \times M_1 \to \mathbb{R}^m$ of product manifolds whose induced metrics have properties (i) and (ii) of

Corollary 1, and, in addition, the distribution E_0 in the product net $\mathcal{E} = (E_0, E_1)$ of $M_0 \times M_1$ is totally geodesic. A preliminary step in the proof of that result is the version of Lemma 1 which states that, if $g\colon M^n \to \mathbb{R}^m$ and $h\colon L^k \to M^n$ are isometric immersions, then h is totally geodesic and $\alpha_g(h_*X, Z) = 0$ for all $X \in \mathfrak{X}(L)$ and $Z \in \Gamma(N_hL)$ if and only if $g(h(L))$ is contained in an affine subspace of $\mathbb{R}^m$ whose normal space along $g(h(L))$ is g_*N_hL. In particular, if $g\colon M_0 \times M_1 \to \mathbb{R}^m$ is a partial tube, then the image $g(M_0 \times \{x_1\})$ by g of each leaf of E_0 is contained in an affine subspace of $\mathbb{R}^m$ whose normal space along $g(M_0 \times \{x_1\})$ is g_*E_1. Similarly, it follows from Corollary 1, and Lemma 1 that if $f\colon M_0 \times M_1 \to \mathbb{R}^m$ is a Ribaucour partial tube, then the image $f(M_0 \times \{x_1\})$ by f of each leaf of E_0 is contained in a sphere of $\mathbb{R}^m$ whose normal space along $f(M_0 \times \{x_1\})$ is f_*E_1. As a consequence, the composition of a partial tube $g\colon M_0 \times M_1 \to \mathbb{R}^m$ with an inversion I with respect to a hypersphere of $\mathbb{R}^m$ is a Ribaucour partial tube, and a Ribaucour partial tube $f\colon M_0 \times M_1 \to \mathbb{R}^m$ is given in this way if and only if the spheres containing the images $f(M_0 \times \{x_1\})$ by f of the leaves of E_0 all pass through a common point.

Remark 2 The definition of a Ribaucour partial tube, as well as Proposition 2, Corollary 1 and Theorem 3, can be easily extended for the case in which the ambient space is any space form $\mathbb{Q}_c^m$, by making use of the extension of the Ribaucour transformation to this setting discussed at the end of Sect. 2. The details are left to the reader.

Corollary 2 *Let M be a Riemannian manifold carrying a spherical distribution D whose orthogonal distribution $D^\perp$ is integrable. Then, any isometric immersion $f\colon M \to \mathbb{R}^m$ whose second fundamental form is adapted to the net $\mathcal{E} = (D, D^\perp)$ is locally a Ribaucour partial tube over the restriction of f to a leaf of $D^\perp$.*

In particular, Corollary 1 and Theorem 3 yield the following explicit parametrization of any umbilic-free surface with a flat normal bundle of a space form such that the lines of curvature of one of the two families have constant geodesic curvature.

Corollary 3 *Let $\gamma\colon I \to \mathbb{R}^{n+1}$ be a unit-speed curve. Let $\xi_1, \ldots, \xi_n$ be a parallel orthonormal frame of $N_\gamma I$ and let $\varphi, \beta_i \in C^\infty(I)$, $1 \le i \le n$, satisfy $\beta_i' + \varphi' k_i = 0$, where $\gamma'' = \sum_{i=1}^n k_i \xi_i$. Let $\alpha\colon J \to \mathbb{R}^n$ be a unit-speed curve. Then, the map $f\colon I \times J \to \mathbb{R}^{n+1}$ given by*

$$f(s,t) = \gamma(s) - 2\varphi(s)\frac{\varphi'(s)\gamma'(s) + \sum_{i=1}^n(\beta_i(s) + \alpha_i(t))\xi_i(s)}{(\varphi'(s))^2 + \sum_{i=1}^n(\beta_i(s) + \alpha_i(t))^2} \tag{23}$$

parametrizes, at regular points, a surface with a flat normal bundle whose coordinate curves are lines of curvature and such that the t-coordinate curves have constant geodesic curvature.

Conversely, any umbilic-free surface with a flat normal bundle whose lines of curvature of one family have constant geodesic curvature can be locally parametrized in this way.

Proof That ξ_i is parallel along γ in the normal connection means that there exists $k_i \in C^\infty(I)$ such that $\xi_i' = -k_i\gamma'$. This implies that $\gamma'' = \sum_{i=1}^n k_i\xi_i$. Therefore, for $\varphi \in C^\infty(I)$ and $\beta = \sum_{i=1}^n \beta_i\xi_i$, Eq. (1) reduces to the set of ODEs $\beta_i' + \varphi' k_i = 0$, $1 \le i \le n$. Since the map $f : I \times J \to \mathbb{R}^{n+1}$ given by (23) is the Ribaucour partial tube over γ with $\alpha : J \to \mathbb{R}^n$ as fiber, the assertion in the direct statement is a consequence of Corollary 1, while the converse follows from Theorem 3. □

Given an isometric immersion $f : M^n \to \tilde{M}^m$, a vector $\eta \in N_f M(x)$ at $x \in M^n$ is called a *principal normal* of f at x if the subspace

$$E_\eta(x) = \{T \in T_x M : \alpha(T, X) = \langle T, X\rangle\, \eta \ \text{ for all } \ X \in T_x M\}$$

is nontrivial. A normal vector field $\eta \in \Gamma(N_f M)$ is called a *principal normal vector field of f with multiplicity* $q > 0$ if $E_\eta(x)$ has dimension q at any point $x \in M^n$. A principal normal vector field $\eta \in \Gamma(N_f M)$ is said to be a *Dupin principal normal vector field* if η is parallel in the normal connection along E_η.

As a particular case of Corollary 2, we recover one of the main results in [11].

Corollary 4 *Let* $f : M^n \to \mathbb{R}^m$ *be an isometric immersion that carries a Dupin principal normal vector field* η *with multiplicity* $q > 0$*. If the subbundle* $E_\eta^\perp$ *is integrable, then the restriction* $\mathcal{N} = E_\eta|_{M_1}$ *of* E_η *to any leaf* M_1^{n-q} *of* $E_\eta^\perp$ *is a flat parallel subbundle of* $N_{f_1}M_1$*, where* $f_1 = f|_{M_1}$*, and there exist* $(\varphi, \beta) \in \mathcal{D}(f_1)$ *and a vector bundle isometry* $\Psi : M_1 \times \mathbb{R}^q \to \mathcal{N}$ *such that* f *is locally a Ribaucour partial tube determined by* $(f_0, f_1, \Psi, \varphi, \beta)$*, where* $f_0 : \mathbb{R}^q \to \mathbb{R}^q$ *is the identity map.*

Ribaucour partial tubes also provide a parametrization of all foliations of a space form by spherical submanifolds whose orthogonal distributions are integrable, in particular of all foliations of a space form whose leaves are spherical hypersurfaces.

Corollary 5 *Let* $f : M^m \to \mathbb{Q}_c^m$ *be a local diffeomorphism of a product manifold* $M^m = M_0^{m_0} \times M_1^{m_1}$ *such that the product net* $\mathcal{E} = (E_0, E_1)$ *of* M^m *is orthogonal and* E_0 *is spherical with respect to the metric induced by* f*. Then, for any fixed* $\bar{x}_0 \in M_0$*, the map* $f_1 : M_1 \to \mathbb{R}^m$ *given by* $f_1 = f \circ \mu_{\bar{x}_0}$ *is an immersion with a flat normal bundle, and there exist a parallel vector bundle isometry* $\Psi : M_1^{m_1} \times \mathbb{R}^{m_0} \to N_{f_1}M_1$*, a local isometry* $f_0 : M_0 \to \mathbb{R}^{m_0}$*, and* $(\varphi, \beta) \in \mathcal{D}(f_1)$ *such that* f *is the Ribaucour partial tube determined by* $(f_0, f_1, \Psi, \varphi, \beta)$*.*

5 Hypersurfaces of Space Forms Carrying a Spherical Foliation

In this section, we derive some consequences of Theorem 3 for hypersurfaces of space forms; in particular, we provide a complete solution to the problem that was the initial motivation for this work.

A smooth distribution D on a Riemannian manifold M^n is said to be *curvature invariant* if $R(X, Y)Z \in \Gamma(D)$ for all $X, Y, Z \in \Gamma(D)$, where R denotes the curvature tensor of M^n. A basic observation for us is the following fact.

Lemma 2 *Any spherical distribution on a Riemannian manifold is curvature invariant.*

Proof Let D be a spherical distribution on a Riemannian manifold M^n with mean curvature vector field Z. Given $T, S, U \in \Gamma(D)$ and $X \in \Gamma(D^\perp)$ with $\langle X, Z\rangle = 0$, since $\langle \nabla_S U, X\rangle = 0$ and $\nabla_T X = 0$ by (17) and (18), respectively, it follows that $\langle R(T, S)U, X\rangle = 0$. On the other hand, again from (17) and (18), we obtain

$$\begin{aligned}\langle \nabla_T \nabla_S U, Z\rangle &= T(\langle S, U\rangle)\|Z\|^2 - \langle \nabla_S U, \nabla_T Z\rangle \\ &= (T(\langle S, U\rangle) + \langle \nabla_S U, T\rangle)\|Z\|^2.\end{aligned}$$

Similarly

$$\langle \nabla_S \nabla_T U, Z\rangle = S(\langle T, U\rangle) + \langle \nabla_T U, S\rangle)\|Z\|^2$$

and

$$\left\langle \nabla_{[T,S]} U, Z\right\rangle = \langle [T, S], U\rangle \, \|Z\|^2.$$

Subtracting the last two equations from the first one gives $\langle R(T, S)U, Z\rangle = 0$. □

For an oriented hypersurface $f : M^n \to \mathbb{Q}_c^{n+1}$, we denote by A its shape operator with respect to the Gauss map N and by $\Delta = \ker A$ its relative nullity distribution. The next algebraic lemma was proved in [10].

Lemma 3 *Let $f : M^n \to \mathbb{Q}_c^{n+1}$ be an oriented hypersurface carrying a curvature invariant distribution D of rank k. Then, one of the following holds pointwise:*

(*i*) $A(D) \subset D^\perp$.
(*ii*) $A(D) \subset D$.
(*iii*) *rank* $D \cap \Delta = k - 1$.

The two preceding lemmas have the following consequence for hypersurfaces of dimension $n \geq 3$ that carry a spherical foliation of high rank.

Proposition 3 *If $f : M^n \to \mathbb{Q}_c^{n+1}$, $n \geq 3$, is an oriented hypersurface and D is a spherical distribution on M^n of rank $k > n/2$ that is not totally geodesic on any open subset, then D is invariant by the shape operator of f.*

Proof Since the subset of M^n where D is invariant by the shape operator of f is closed, there is no loss of generality in assuming that its mean curvature vector field never vanishes. Then, $D \cap \Delta$ must be trivial, for Δ is totally geodesic. On the other hand, the distribution D being curvature invariant by Lemma 2, it follows from Lemma 2 that one of conditions (*i*) to (*iii*) must hold at any point of M^n. Since

$k > n/2$, then rank $D(x) \cap \Delta(x) > 0$ if $A(D(x)) \subset D(x)^{\perp}$ at some $x \in M^n$, and because $k-1 > 0$, then rank $D(x) \cap \Delta(x) > 0$ also if (iii) holds at x. Thus, neither condition (i) nor condition (iii) can hold at any point of M^n. □

It follows from Corollary 2 and Proposition 3 that any hypersurface $f: M^n \to \mathbb{Q}_c^{n+1}$, $n \geq 3$, that carries a spherical distribution D of rank $k > n/2$ whose mean curvature vector field never vanishes and such that $D^{\perp}$ is integrable is locally a Ribaucour partial tube over a leaf of $D^{\perp}$. In particular, this leads to the following explicit description of such hypersurfaces when $k = n-1$ (in which case integrability of $D^{\perp}$ is automatic).

Corollary 6 *Let $\gamma: I \to \mathbb{R}^{n+1}$ be a unit-speed curve. Let $\xi_1, \ldots, \xi_n$ be a parallel orthonormal frame of $N_\gamma I$ and let $\varphi, \beta_i \in C^\infty(I)$, $1 \leq i \leq n$, satisfy $\beta_i' + \varphi' k_i = 0$, where $\gamma'' = \sum_{i=1}^{n-1} k_i \xi_i$. Let $g = (g_1, \ldots, g_n): M_1^{n-1} \to \mathbb{R}^n$ be a hypersurface. Then, the map $f: M^n = I \times M_1^{n-1} \to \mathbb{R}^{n+1}$, given by*

$$f(s,t) = \gamma(s) - 2\varphi(s)\frac{\varphi'(s)\gamma'(s) + \sum_{i=1}^{n}(\beta_i(s) + g_i(t))\xi_i(s)}{(\varphi'(s))^2 + \sum_{i=1}^{n}(\beta_i(s) + g_i(t))^2},$$

parametrizes, at regular points, a hypersurface for which the tangent spaces to M_1 give rise to a spherical distribution of codimension one.

Conversely, any hypersurface of $\mathbb{R}^{n+1}$, $n \geq 3$, carrying a spherical distribution of codimension one can be locally parametrized in this way.

6 Hypersurfaces of Enneper Type

According to Theorem 3, hypersurfaces $f: M^n = M_0^{n-1} \times I \to \mathbb{R}^{n+1}$ that are Ribaucour partial tubes over curves $\gamma: I \to \mathbb{R}^{n+1}$ are characterized by the fact that their product nets $\mathcal{E} = (E_0, E_1)$ satisfy conditions (i) to (iii) in Corollary 1, with E_1 of rank one. A special class of such hypersurfaces consists of channel hypersurfaces.

Notice that, by Lemma 1, conditions (ii) and (iii) in Corollary 1 can be replaced by the requirement that the image by f of each leaf σ of E_0 be contained in a hypersphere of $\mathbb{R}^{n+1}$ that intersects $f(M)$ orthogonally along $f(\sigma)$. It is a natural problem to investigate the more general class of hypersurfaces $f: M^n = M_0^{n-1} \times I \to \mathbb{R}^{n+1}$ for which condition (iii) is replaced by the following:
(iii') The image by f of each leaf of E_0 is contained in a hypersphere of $\mathbb{R}^{n+1}$ (which does not necessarily intersect $f(M)$ orthogonally along $f(\sigma)$).

The next lemma shows that, under condition (i), conditions (ii) and (iii') together are equivalent to requiring the image by f of each leaf σ of E_0 to be contained in a hypersphere of $\mathbb{R}^{n+1}$ that intersects $f(M)$ *at a constant angle* along $f(\sigma)$.

Lemma 4 *Let $f: M^n \to \mathbb{R}^{n+1}$ be a hypersurface and let $g: L^{n-1} \to M^n$ be a hypersurface of M^n such that $f(g(L))$ is contained in a hypersphere $\mathcal{S}$ of $\mathbb{R}^{n+1}$. Then, $\mathcal{S}$ intersects $f(M)$ at a constant angle along $f(g(L))$ if and only if the shape operator A of f leaves g_*TL invariant.*

Proof Let $P_0 \in \mathbb{R}^{n+1}$ and $R > 0$ be the center and the radius of $\mathcal{S}$, respectively, and let θ be the angle between its unit normal vector field $(f \circ g - P_0)/R$ and a unit normal vector field N of f along $f(g(L))$. Then

$$f \circ g - P_0 = R\cos\theta(N \circ g) + R\sin\theta f_*\delta,$$

where δ is a unit normal vector field to g. Hence, for all $T \in \mathfrak{X}(L)$, we have

$$\begin{aligned} f_*T = {} & RT(\theta)(-\sin\theta N \circ g + \cos\theta f_*\delta) - R\cos\theta f_* A g_* T \\ & + R\sin\theta(f_*\nabla_T\delta + \langle Ag_*T, \delta\rangle N). \end{aligned}$$

Thus, $T(\theta) = 0$ for all $T \in \mathfrak{X}(L)$ if and only if $\langle Ag_*T, \delta\rangle = 0$ for all $T \in \mathfrak{X}(L)$. □

For $n = 2$, conditions (i) and (ii) in Corollary 1 mean that the leaves of E_0 and E_1 are lines of curvature of f. Condition (iii') says that those corresponding to E_0 are contained in spheres. Surfaces in $\mathbb{R}^3$ with these properties were called *surfaces of Enneper type* in [19]. Accordingly, we call a hypersurface $f: M^n = M_0^{n-1} \times I \to \mathbb{R}^{n+1}$, with product net $\mathcal{E} = (E_0, E_1)$, that satisfies conditions (i), (ii), and (iii') above, a *hypersurface of Enneper type*, or, more precisely, a hypersurface of Enneper type with respect to $\mathcal{E} = (E_0, E_1)$.

In the next subsection, we first consider the case in which condition (iii) is replaced by the following:

(iii”) The image by f of each leaf of E_0 is contained in an affine hyperplane of $\mathbb{R}^{n+1}$.

For $n = 2$, surfaces that satisfy conditions (i), (ii), and (iii'') are surfaces with planar lines of curvature corresponding to one of their principal curvatures.

6.1 Hypersurfaces of Enneper Type with Extrinsically Planar Leaves

The next result shows how all hypersurfaces $f: M^n = M_0^{n-1} \times I \to \mathbb{R}^{n+1}$ that satisfy conditions (i), (ii), and (iii'') can be constructed in terms of Ribaucour partial tubes $N: M^n \to \mathbb{S}^n$ over curves in $\mathbb{S}^n$.

Notice that, for a Ribaucour partial tube $N: M^n \to \mathbb{S}^n$ over a curve in $\mathbb{S}^n$, the product net $\mathcal{E} = (E_0, E_1)$ of M^n is a twisted product net with respect to the metric $d\sigma^2$ induced on $M^n = M_0^{n-1} \times I$ by N, that is, both E_0 and E_1 are umbilical distributions (with E_0 being, in fact, spherical), and

$$d\sigma^2 = v_0^2 d\sigma_0^2 + \nu^2 ds^2,$$

where $d\sigma_0^2$ is a metric of constant curvature 1 on M_0 and ds^2 is the standard metric on I. The mean curvature vector field of E_0 is

$$H_0 = -(\text{grad } \log v_0)_{E_1} = -\frac{1}{\nu^2}\frac{\partial(\log v_0)}{\partial s}\partial_s,$$

where ∂_s is a unit vector field (with respect to the metric ds^2) along I. Writing

$$\varphi = -\frac{1}{\nu}\frac{\partial(\log v_0)}{\partial s},$$

that E_0 is spherical is equivalent to φ depending only on s.

Theorem 4 *Let $N\colon M^n = M_0^{n-1} \times I \to \mathbb{S}^n$ be a Ribaucour partial tube over a unit-speed curve $\beta\colon I \to \mathbb{S}^n$. Given $V \in C^\infty(I)$ and $U \in C^\infty(M_0)$, define $\gamma \in C^\infty(M)$ by*

$$\gamma(x, s) = v_0(x, s)\left(U(x) + \int_0^s \frac{V(\tau)\nu(x, \tau)}{v_0(x, \tau)}d\tau\right). \tag{24}$$

Then, the map $f\colon M^n \to \mathbb{R}^{n+1}$ given by

$$f = \gamma(i \circ N) + i_* N_* grad\, \gamma, \tag{25}$$

where $i\colon \mathbb{S}^n \to \mathbb{R}^{n+1}$ is the inclusion and $grad\, \gamma$ is computed with respect to the metric $d\sigma^2$ on M^n induced by N, defines, on the subset of its regular points, a hypersurface satisfying conditions (i), (ii), and (iii'').

Conversely, any hypersurface $f\colon M_0^{n-1} \times I \to \mathbb{R}^{n+1}$ satisfying conditions (i), (ii), and (iii'') whose shape operator has rank n everywhere is given locally in this way.

Proof Let $f\colon M^n \to \mathbb{R}^{n+1}$ be given by (25) for some $\gamma \in C^\infty(M)$. Differentiating (25) gives

$$f_* = i_* N_* P, \tag{26}$$

where $P = \text{Hess}\, \gamma + \gamma I$, the Hessian being computed with respect to $d\sigma^2$.

Moreover, on the open subset where P is invertible, that is, on the open subset of regular points of f, it follows from (26) that the map N is the Gauss map of f and that the shape operator of f with respect to N is

$$A = -P^{-1}. \tag{27}$$

We claim that Hess γ is adapted to $\mathcal{E} = (E_0, E_1)$ if and only if γ is given by (24) for some $V \in C^\infty(I)$ and $U \in C^\infty(M_0)$. Using that E_0 is umbilical with mean curvature vector field $H_0 = \varphi\nu^{-1}\partial_s$, we obtain

$$\begin{aligned}\nabla_X \partial_s &= \nabla_X \nu(\nu^{-1}\partial_s)\\ &= X(\nu)\nu^{-1}\partial_s - \nu\|X\|^{-2}\left\langle \nabla_X X, \nu^{-1}\partial_s\right\rangle X\\ &= X(\log\nu)\partial_s - \nu\varphi X\end{aligned}$$

for all $X \in \Gamma(E_0)$. Hence

$$\begin{aligned}\operatorname{Hess}\gamma(X, \partial_s) &= X\left(\frac{\partial\gamma}{\partial s}\right) - (\nabla_X\partial_s)(\gamma)\\ &= X\left(\frac{\partial\gamma}{\partial s}\right) - X(\log\nu)\frac{\partial\gamma}{\partial s} + \nu\varphi X(\gamma).\end{aligned}$$

Therefore, $\operatorname{Hess}\gamma(X, \partial_s) = 0$ if and only if

$$X\left(\frac{\partial\gamma}{\partial s}\right) + \nu\varphi X(\gamma) = X(\log\nu)\frac{\partial\gamma}{\partial s}.$$

Since φ depends only on s, this can be also written as

$$X\left(\frac{\partial\gamma}{\partial s} + \varphi\nu\gamma\right) = X(\log\nu)\left(\frac{\partial\gamma}{\partial s} + \varphi\nu\gamma\right).$$

Thus, $\dfrac{\partial\gamma}{\partial s} + \varphi\nu\gamma = \nu V$ for some $V \in C^\infty(I)$, which can be written as

$$\frac{\partial(\gamma v_0^{-1})}{\partial s} = \nu V v_0^{-1},$$

taking into account that $\varphi\nu = -\dfrac{\partial(\log v_0)}{\partial s}$. This proves our claim.

Notice that Hess γ being adapted to $\mathcal{E} = (E_0, E_1)$ implies both P and P^{-1} to be also adapted to $\mathcal{E}$. This, together with (26), (27) and the fact that the product net $\mathcal{E} = (E_0, E_1)$ of M^n is orthogonal with respect to the metric induced by N, implies that $\mathcal{E}$ is also orthogonal with respect to the metric induced by f and that the second fundamental form of f is adapted to $\mathcal{E}$. Finally, the image by N of each leaf σ of E_0 is a small hypersphere of $\mathbb{S}^n$. Hence, if $\mathcal{H}$ is the hyperplane of $\mathbb{R}^{n+1}$ that is parallel to the affine hyperplane that contains $N(\sigma)$, then $N_*T \in \mathcal{H}$ for all $x \in \sigma$ and $T \in T_x\sigma$. It follows from (26) and the fact that P leaves E_0 invariant that $f_*T \in \mathcal{H}$ for all $x \in \sigma$ and $T \in T_x\sigma$. Therefore, $f(\sigma)$ is also contained in an affine hyperplane parallel to $\mathcal{H}$.

Conversely, let $f: M^n = M_0^{n-1} \times I \to \mathbb{R}^{n+1}$ be a hypersurface satisfying conditions (i), (ii), and (iii'') above and let $N: M^n \to \mathbb{S}^n$ be its Gauss map. If $\mathcal{H}$ is the hyperplane that is parallel to the affine hyperplane containing the image by f of a leaf σ of E_0, by condition (ii), we have

$$N_* T = \tilde{\nabla}_T N = -f_* AT \in f_* T_x \sigma \in \mathcal{H}$$

for all $T \in T_x\sigma$. Hence, $N(\sigma)$ is also contained in an affine hyperplane parallel to $\mathcal{H}$, and consequently, it is an open subset of the small hypersphere of $\mathbb{S}^n$ given by its intersection with $\mathbb{S}^n$. Therefore, $N: M_0 \times I \to \mathbb{S}^n$ is a local diffeomorphism (by the assumption that the shape operator of f has rank n everywhere) with the property that the image by N of any leaf of E_0 is a small hypersphere of $\mathbb{S}^n$. Thus, E_0 is a spherical distribution with respect to the metric induced by N. Moreover, by condition (ii), the images by N of the integral curves of E_1 are orthogonal trajectories of the foliation of $\mathbb{S}^n$ given by the images of the leaves of E_0. In other words, the net $\mathcal{E}$ is also an orthogonal net with respect to the metric induced by N. It follows from Corollary 5 that N is a Ribaucour partial tube over a unit-speed curve $\beta: I \to \mathbb{S}^n$.

Now, the Gauss parametrization allows to recover f in terms of N and the support function γ by means of (25). Since the second fundamental form of f is adapted to $\mathcal{E}$, then Hess γ must also be adapted to $\mathcal{E}$ by (27). Thus, γ must be given by (24), as shown in the proof of the direct statement. □

For $n = 2$, Theorem 4 reads as follows.

Corollary 7 *Let $N: J \times I \to \mathbb{S}^2$ be a Ribaucour partial tube over a unit-speed curve $\beta: I \to \mathbb{S}^2$ and let $ds^2 = v_1^2 du_1^2 + v_2^2 du_2^2$ be the metric induced by N. Given $U \in C^\infty(J)$ and $V \in C^\infty(I)$, let $\gamma \in C^\infty(J \times I)$ be given by*

$$\gamma(u_1, u_2) = v_1(u_1, u_2)\left(U(u_1) + \int_0^{u_2} \frac{V(\tau) v_2(u_1, \tau)}{v_1(u_1, \tau)} d\tau\right).$$

Then, the map $f: J \times I \to \mathbb{R}^3$ given by

$$f(u_1, u_2) = \gamma(u_1, u_2) N(u_1, u_2) + \frac{1}{v_1^2} \frac{\partial \gamma}{\partial u_1} \frac{\partial N}{\partial u_1} + \frac{1}{v_2^2} \frac{\partial \gamma}{\partial u_2} \frac{\partial N}{\partial u_2}$$

defines, on the open subset of its regular points, a surface parametrized by lines of curvature whose u_1-lines of curvature are planar.

Conversely, any surface in $\mathbb{R}^3$ free of flat points whose lines of curvature correspondent to one of its principal curvatures are planar can be locally parametrized in this way.

6.2 The General Case

We now address the general problem of describing all hypersurfaces $f\colon M^n = M_0^{n-1} \times I \to \mathbb{R}^{n+1}$ of Enneper type.

Let $N\colon M^n \to \mathbb{S}^n$ be the Gauss map of a hypersurface of Enneper type. By conditions (i) and (ii), the product net $\mathcal{E} = (E_0, E_1)$ of M^n is orthogonal also with respect to the metric $d\sigma^2$ induced by N. It follows from Lemma 2.1 in [16] that there exists a one-parameter family $\{g_s\}$ of metrics on M_0^{n-1} indexed on I and a family $\{g_{p_0}\}$ of metrics on I indexed on M_0, such that $d\sigma^2(p_0, t) = \pi_0^* g_s(p_0) + \pi_1^* g_{p_0}(t)$ for all $(p_0, t) \in M_0^{n-1} \times I = M^n$, where π_0 and π_1 are the projections onto M_0^{n-1} and I, respectively. But since $I \subset \mathbb{R}$ is an interval, there exists $\rho \in C^\infty(M)$ such that

$$d\sigma^2 = g_s + \rho^2 ds^2, \tag{28}$$

where ds^2 is the standard metric on I. In particular, $\rho^{-1}\partial_s$ spans E_1 and has unit length with respect to $d\sigma^2$.

Let us assume that, for each $s_0 \in I$, the image by f of the leaf $s = s_0$ of E_0 is contained in a hypersphere $\mathbb{S}^n(\gamma(s_0), R(s_0))$ of $\mathbb{R}^{n+1}$ with center $\gamma(s_0) \in \mathbb{R}^{n+1}$ and radius $R(s_0)$. The position vector $\zeta(x, s) = f(x, s) - \gamma(s)$ of $\mathbb{S}^n(\gamma(s), R(s))$ at $f(x, s)$ with respect to $\gamma(s)$ can be written as $\zeta = R\cos\theta N + R\sin\theta f_*\delta$, where δ is a unit vector field (with respect to the metric induced by f) spanning E_1 and $\theta \in C^\infty(M)$ is the angle between ζ and N, which depends only on s by Lemma 4. After changing δ by $-\delta$, if necessary, we can assume that $f_*\delta = \rho^{-1}N_*\partial_s$. Thus, we can write

$$f = \gamma + \alpha N + \beta\rho^{-1}N_*\partial_s, \tag{29}$$

where $\alpha = \alpha(s) = R\cos\theta$ and $\beta = \beta(s) = R\sin\theta$. Now, we impose N to be normal to f. Since α and β depend only on s, the condition $0 = \langle f_*T, N\rangle$ is identically satisfied. On the other hand, $0 = \langle f_*\partial_s, N\rangle$ if and only if

$$\langle \gamma', N\rangle + \alpha' = \beta\rho, \tag{30}$$

where the prime stands for derivative with respect to s. We have thus proved the converse statement of the following result.

Theorem 5 *Let $N\colon M^n \to \mathbb{S}^n$ be a local diffeomorphism of a product manifold $M^n = M_0^{n-1} \times I$ such that the product net $\mathcal{E} = (E_0, E_1)$ of M^n is orthogonal with respect to the metric $d\sigma^2$ induced by N (equivalently, N is a local diffeomorphism whose induced metric $d\sigma^2$ is given as in* (28) *for some $\rho \in C^\infty(M)$ and for some smooth family of metrics g_s on M_0^{n-1} indexed on I). If there exist a smooth curve $\gamma\colon I \to \mathbb{R}^{n+1}$ and $\alpha, \beta \in C^\infty(I)$ such that* (30) *holds, then the map $f\colon M^n \to$*

$\mathbb{R}^{n+1}$ given by (29) *parametrizes a hypersurface of Enneper type with respect to $\mathcal{E}$ having N as a Gauss map.*

Conversely, if $f: M^n = M_0^{n-1} \times I \to \mathbb{R}^{n+1}$ is a hypersurface of Enneper type with respect to the product net $\mathcal{E} = (E_0, E_1)$ of M^n having $N: M^n \to \mathbb{S}^n$ as a Gauss map, then there exist a smooth curve $\gamma: I \to \mathbb{R}^{n+1}$ and $\alpha, \beta \in C^\infty(I)$ satisfying (30) *such that f is given by* (29).

Proof If $f: M^n \to \mathbb{R}^{n+1}$ is given by (29) in terms of N and (γ, α, β), then (30) is precisely the condition for N to be a Gauss map of f. Moreover, if that condition is satisfied, then

$$\|f - \gamma\|^2 = \alpha^2 + \beta^2 \text{ and } \left\langle \frac{f-\gamma}{\|f-\gamma\|}, N \right\rangle = \frac{\alpha}{\sqrt{\alpha^2+\beta^2}}. \tag{31}$$

The preceding equations show that the image by f of each leaf of E_0 is contained in a hypersphere of $\mathbb{R}^{n+1}$ that intersects $f(M)$ at a constant angle. By Lemma 4, the distribution E_0 is invariant under the shape operator of f, that is, the second fundamental form of f is adapted to the net $\mathcal{E} = (E_0, E_1)$. Since the net $\mathcal{E}$ is orthogonal with respect to the metric induced by N, this implies that it is also orthogonal with respect to the metric induced by f. Thus, f is a hypersurface of Enneper type with respect to $\mathcal{E}$ having N as a Gauss map. □

Given a hypersurface of Enneper type $f: M^n = M_0^{n-1} \times I \to \mathbb{R}^{n+1}$ with respect to the product net $\mathcal{E} = (E_0, E_1)$ of M^n, we determine next all hypersurfaces of Enneper type with respect to $\mathcal{E}$ sharing the same Gauss map with f.

Proposition 4 *Let $f: M^n = M_0^{n-1} \times I \to \mathbb{R}^{n+1}$ be a hypersurface of Enneper type with respect to the product net $\mathcal{E} = (E_0, E_1)$ of M^n. Assume that its Gauss map $N: M^n \to \mathbb{S}^n$ is a local diffeomorphism whose induced metric $d\sigma^2$ is given by* (28) *and that f is parametrized by* (29) *in terms of N, a smooth regular curve $\gamma: I \to \mathbb{R}^{n+1}$ and $\alpha, \beta \in C^\infty(I)$ satisfying* (30). *Suppose also that there does not exist any leaf $s = s_0$ of E_0 whose image by f is (an open piece of) a round $(n-1)$-dimensional sphere. Then, any other hypersurface $\bar{f}: M^n \to \mathbb{R}^{n+1}$ of Enneper type with respect to $\mathcal{E}$ having N as a Gauss map is parametrized by* (29) *in terms of a smooth curve $\bar{\gamma}: I \to \mathbb{R}^{n+1}$ and $\bar{\alpha}, \bar{\beta} \in C^\infty(I)$ which are related to γ, α, and β by*

$$\bar{\beta} = \lambda\beta, \ \ \bar{\alpha}' = \lambda\alpha' \ \ \textit{and} \ \ \bar{\gamma}' = \lambda\gamma'$$

for some $\lambda \in C^\infty(I)$.

Proof Let $\bar{f}: M^n \to \mathbb{R}^{n+1}$ be another hypersurface of Enneper type with respect to $\mathcal{E}$ sharing the same Gauss map N with f. By Theorem 5, it can be parametrized by (29) in terms of a smooth curve $\bar{\gamma}: I \to \mathbb{R}^{n+1}$ and $\bar{\alpha}, \bar{\beta} \in C^\infty(I)$ satisfying

$$\langle \bar{\gamma}', N \rangle + \bar{\alpha}' = \bar{\beta}\rho. \tag{32}$$

Notice that if β vanishes at some $s_0 \in I$, since γ' is nowhere vanishing by assumption, then (30) implies that the image by N of the leaf $s = s_0$ of E_0 is contained in an affine hyperplane of $\mathbb{R}^{n+1}$ (hence is a $(n-1)$-dimensional round hypersphere of $\mathbb{S}^n$). Hence, also the image by f of the leaf $s = s_0$ of E_0 is contained in an affine hyperplane of $\mathbb{R}^{n+1}$, and therefore, it is an open piece of an $(n-1)$-dimensional round hypersphere given by its intersection with the hypersphere $\mathbb{S}^n(\gamma(s_0), R(s_0))$ containing such image, contradicting our assumption. Thus, β is nowhere vanishing and we can write $\bar{\beta} = \lambda\beta$ for some $\lambda \in C^\infty(I)$. Comparing (32) with (30) yields

$$\langle \bar{\gamma}' - \lambda\gamma', N\rangle + \bar{\alpha}' - \lambda\alpha' = 0. \tag{33}$$

If $\bar{\gamma}' - \lambda\gamma'$ was nonzero for some $s_0 \in I$, then, arguing as before, we would conclude that the image by f of the leaf $s = s_0$ of E_0 would be an open piece of an $(n-1)$-dimensional round hypersphere, a contradiction. Thus, $\bar{\gamma}' - \lambda\gamma'$, and hence also $\bar{\alpha}' - \lambda\alpha'$ by (33), must vanish everywhere. □

We are now able to show how any hypersurface of Enneper type in $\mathbb{R}^{n+1}$ satisfying the assumptions of Proposition 4 can be constructed by means of a hypersurface of Enneper type with extrinsically planar leaves.

Theorem 6 *Let $f: M^n = M_0^{n-1} \times I \to \mathbb{R}^{n+1}$ be a hypersurface given as in Theorem 4, let $\tilde{f} = \mathcal{I} \circ f: M^n = M_0^{n-1} \times I \to \mathbb{R}^{n+1}$ be its composition with an inversion with respect to a hypersphere of unit radius centered at the origin, and let $\tilde{N}: M^n \to \mathbb{S}^n$ be the Gauss map of $\tilde{f}$. Let $\tilde{f}$ be parametrized by (29) in terms of $\tilde{N}$ and a triple $(\tilde{\gamma}, \tilde{\alpha}, \tilde{\beta})$ satisfying (30), where $\tilde{\alpha}, \tilde{\beta} \in C^\infty(I)$ and $\tilde{\gamma}: I \to \mathbb{R}^{n+1}$ is a smooth curve. Define a new triple $(\bar{\gamma}, \bar{\alpha}, \bar{\beta})$ by*

$$\bar{\beta} = \lambda\tilde{\beta}, \;\; \bar{\alpha}' = \lambda\tilde{\alpha}' \;\; \textit{and} \;\; \bar{\gamma}' = \lambda\tilde{\gamma}' \tag{34}$$

for some $\lambda \in C^\infty(I)$. Then, the hypersurface $\bar{f}: M^n = M_0^{n-1} \times I \to \mathbb{R}^{n+1}$ parametrized by (29) in terms of $\tilde{N}$ and $(\bar{\gamma}, \bar{\alpha}, \bar{\beta})$ is of Enneper type.

Conversely, any hypersurface of Enneper type in $\mathbb{R}^{n+1}$ satisfying the assumptions of Proposition 4 can be constructed as above.

Proof Since f satisfies conditions (i), (ii), and (iii'') above, it is clear that $\tilde{f} = \mathcal{I} \circ f$ satisfies conditions (i), (ii), and (iii') and hence is a hypersurface of Enneper type. If the triple $(\tilde{\gamma}, \tilde{\alpha}, \tilde{\beta})$ satisfies (30), then the same holds for the new triple $(\bar{\gamma}, \bar{\alpha}, \bar{\beta})$ defined by (34). Then, the hypersurface $\bar{f}: M^n = M_0^{n-1} \times I \to \mathbb{R}^{n+1}$ parametrized by (29) in terms of $\tilde{N}$ and $(\bar{\gamma}, \bar{\alpha}, \bar{\beta})$ is of Enneper type by Theorem 5.

To prove the converse, let $\bar{f}: M^n = M_0^{n-1} \times I \to \mathbb{R}^{n+1}$ be a hypersurface of Enneper type with respect to the product net $\mathcal{E} = (E_0, E_1)$ of M^n, parametrized by (29) in terms of its Gauss map $\bar{N}: M^n \to \mathbb{S}^n$, a smooth regular curve $\bar{\gamma}: I \to \mathbb{R}^{n+1}$ and $\bar{\alpha}, \bar{\beta} \in C^\infty(I)$ satisfying (30). Given $\lambda \in C^\infty(I)$, let $\tilde{\gamma}: I \to \mathbb{R}^{n+1}$ and

$\tilde{\alpha}, \tilde{\beta} \in C^{\infty}(I)$ be related to $\bar{\gamma}, \bar{\alpha}, \bar{\beta}$ by

$$\lambda\bar{\beta} = \tilde{\beta}, \ \ \lambda\bar{\alpha}' = \tilde{\alpha}', \ \ \lambda\bar{\gamma}' = \tilde{\gamma}'.$$

One can choose λ so that

$$\|\tilde{\gamma}\|^2 = \tilde{\alpha}^2 + \tilde{\beta}^2. \tag{35}$$

In fact, an argument entirely similar to that used by G. Darboux in pages 246–252 of [12], vol. IV, shows that there exist infinitely many solutions to the equation:

$$\left\| \int \lambda\bar{\gamma}' \right\|^2 = \lambda^2\bar{\beta}^2 + \left(\int \lambda\bar{\alpha}' \right)^2.$$

Let $\tilde{f}\colon M^n \to \mathbb{R}^{n+1}$ be the hypersurface given by (29) in terms of $\bar{N}$ and $(\tilde{\gamma}, \tilde{\alpha}, \tilde{\beta})$. By Theorem 5, $\tilde{f}$ also satisfies conditions (i), (ii), and (iii') and has $\bar{N}$ as a Gauss map. Moreover, Eqs. (31) and (35) imply that the hyperspheres that contain the images by $\tilde{f}$ of the leaves of E_0 all pass through the origin. Therefore, the composition $f = \mathcal{I} \circ \tilde{f}$ of $\tilde{f}$ with an inversion with respect to a hypersphere of unit radius centered at the origin satisfies conditions (i), (ii), and (iii'') and hence is given as in Theorem 4. □

6.3 *On Some Special Hypersurfaces of Enneper Type*

In this subsection, we discuss some special hypersurfaces $f\colon M^n = M_0^{n-1} \times I \to \mathbb{R}^{n+1}$ of Enneper type with respect to the product net $\mathcal{E} = (E_0, E_1)$ of M^n, in particular those with the property that the hyperspheres containing the images by f of the leaves of E_0 are concentric, which are ruled out in the converse statement of Theorem 6.

First, recall that there exists a conformal diffeomorphism $\Phi\colon \mathbb{S}^{n-1}\times\mathbb{R} \to \mathbb{R}^n\setminus\{0\}$ given by $(x, t) \mapsto e^t x$. Similarly, there is also a conformal diffeomorphism between $\mathbb{H}^n \times \mathbb{S}^1$ and $\mathbb{R}^{n+1} \setminus \mathbb{R}^{n-1}$ given as follows. Let $e_0, e_1, \ldots, e_{n-1}, e_n$ be a pseudo-orthonormal basis of the Lorentzian space $\mathbb{R}_1^{n+1}$ satisfying $\langle e_0, e_0\rangle = 0 = \langle e_n, e_n\rangle$, $\langle e_0, e_n\rangle = -1/2$ and $\langle e_i, e_j\rangle = \delta_{ij}$ for $1 \le i \le n-1$ and $0 \le j \le n$. Then, the map $\bar{\Phi}\colon \mathbb{H}^n \times \mathbb{S}^1 \subset \mathbb{R}_1^{n+1} \times \mathbb{R}^2 \to \mathbb{R}^{n+1} \setminus \mathbb{R}^{n-1}$ given by

$$\bar{\Phi}(x_0e_0 + \ldots + x_ne_n, (y_1, y_2)) = \frac{1}{x_0}(x_1, \ldots, x_{n-1}, y_1, y_2) \tag{36}$$

is a conformal diffeomorphism. Composing $\bar{\Phi}$ with the isometric covering map

$$\pi : \mathbb{H}^n \times \mathbb{R} \to \mathbb{H}^n \times \mathbb{S}^1 : (x, t) \mapsto (x, (\cos t, \sin t))$$

gives rise to a conformal covering map $\Phi\colon \mathbb{H}^n \times \mathbb{R} \to \mathbb{R}^{n+1} \setminus \mathbb{R}^{n-1}$ given by

$$\Phi(x_0 e_0 + \ldots + x_n e_n, t) = \frac{1}{x_0}(x_1, \ldots, x_{n-1}, \cos t, \sin t). \tag{37}$$

Now, let $g\colon M^{n-1} \to \mathbb{Q}^n_\epsilon$ be a hypersurface, where $\mathbb{Q}^n_\epsilon$ stands for $\mathbb{S}^n$ if $\epsilon = 1$, $\mathbb{R}^n$ if $\epsilon = 0$, and $\mathbb{H}^n$ if $\epsilon = -1$, and let $g_s\colon M^{n-1} \to \mathbb{Q}^n_\epsilon \subset \mathbb{R}^{n+|\epsilon|}_\mu$, where $\mu = 0$ if $\epsilon = 0$ or $\epsilon = 1$, and $\mu = 1$ if $\epsilon = -1$, be the family of its parallel hypersurfaces, that is,

$$g_s(x) = C_\epsilon(s) g(x) + S_\epsilon(s) N(x),$$

where N is a unit normal vector field to g

$$C_\epsilon(s) = \begin{cases} \cos s, & \text{if } \epsilon = 1 \\ 1, & \text{if } \epsilon = 0 \\ \cosh s, & \text{if } \epsilon = -1 \end{cases} \quad \text{and} \quad S_\epsilon(s) = \begin{cases} \sin s, & \text{if } \epsilon = 1 \\ s, & \text{if } \epsilon = 0 \\ \sinh s, & \text{if } \epsilon = -1. \end{cases}$$

Define

$$F\colon M^n = M^{n-1} \times I \to \mathbb{Q}^n_\epsilon \times \mathbb{R} \subset \mathbb{R}^{n+1+|\epsilon|}_\mu$$

by

$$F(x, s) = g_s(x) + a(s)\frac{\partial}{\partial t} \tag{38}$$

for some smooth function $a\colon I \to \mathbb{R}$ with positive derivative on an open interval $I \subset \mathbb{R}$. Here, we regard F as taking values in the underlying flat space.

In the next statement, we denote by Φ either the conformal diffeomorphism $\Phi\colon \mathbb{Q}^n_\epsilon \times \mathbb{R} \to \mathbb{R}^{n+1} \setminus \{0\}$ if $\epsilon = 1$, the conformal covering map $\Phi\colon \mathbb{Q}^n_\epsilon \times \mathbb{R} \to \mathbb{R}^{n+1} \setminus \mathbb{R}^{n-1}$ if $\epsilon = -1$, or the standard isometry $\Phi\colon \mathbb{Q}^n_\epsilon \times \mathbb{R} \to \mathbb{R}^{n+1}$ if $\epsilon = 0$.

Theorem 7 *Let $f\colon M^n = M^{n-1} \times I \to \mathbb{R}^{n+1}$ be a hypersurface of Enneper type with respect to the product net $\mathcal{E} = (E_0, E_1)$ of M^n. Assume that the images by f of the leaves of E_0 are contained in either:*

(a) Concentric hyperspheres
(b) Parallel affine hyperplanes
(c) Affine hyperplanes intersecting along an affine $(n-1)$-dimensional subspace

Then, assuming M^n simply connected in case (c), one of the following possibilities holds:

(i) f is the cone over a hypersurface $g\colon M^{n-1} \to \mathbb{S}^n$ in case (a), the cylinder over a hypersurface $g\colon M^{n-1} \to \mathbb{R}^n$ in case (b) or a rotation hypersurface

over a hypersurface $g\colon M^{n-1} \to \mathbb{R}^n_+$, the latter regarded as the hypersurface model of $\mathbb{H}^n$, in case (c).

(ii) There exists $F\colon M^n \to \mathbb{Q}^n_\epsilon \times \mathbb{R}$ given by (38) in terms of a hypersurface $g\colon M^{n-1} \to \mathbb{Q}^n_\epsilon$, with $\epsilon = 1$ in case (a), $\epsilon = 0$ in case (b) and $\epsilon = -1$ in case (c), such that $f = \Phi \circ F$.

Proof Let $\Psi\colon \mathbb{R}^{n+1} \setminus \mathbb{R}^{k-1} \to \mathbb{H}^k \times \mathbb{S}^{n-k+1} \subset \mathbb{R}^{k+1}_1 \times \mathbb{R}^{n-k+2} = \mathbb{R}^{n+3}_1$ be the map given by

$$\Psi(y_1, \ldots, y_{n+1})$$
$$= \frac{1}{\sqrt{y_k^2 + \cdots + y_{n+1}^2}} \left(e_0 + \sum_{i=1}^{k-1} y_i e_i + \left(\sum_{i=1}^{n+1} y_i^2 \right) e_k, (y_k, \ldots, y_{n+1}) \right),$$

where $\mathbb{R}^{k-1} = \{(y_1, \ldots, y_{n+1}) \in \mathbb{R}^{n+1} : y_k = \cdots = y_{n+1} = 0\}$ and $e_0, \ldots, , e_k$ is a pseudo-orthonormal basis of $\mathbb{R}^{k+1}_1$ with $\langle e_0, e_0 \rangle = 0 = \langle e_k, e_k \rangle$, $\langle e_0, e_k \rangle = -1/2$ and $\langle e_i, e_j \rangle = \delta_{ij}$ for $1 \le i \le k-1$ and $0 \le j \le k$. It is a conformal diffeomorphism whose conformal factor $\varphi \in C^\infty(\mathbb{R}^{n+1} \setminus \mathbb{R}^{k-1})$ is $\varphi(y_1, \ldots, y_{n+1}) = (\sum_{i=k}^{n+1} y_i^2)^{1/2}$. For $k = n$, Ψ is the inverse of the conformal diffeomorphism $\bar{\Phi}\colon \mathbb{H}^n \times \mathbb{S}^1 \to \mathbb{R}^{n+1} \setminus \mathbb{R}^{n-1}$ given by (36). Notice that Ψ takes each half-space of a k-dimensional subspace $\mathbb{R}^k$ containing $\mathbb{R}^{k-1}$ onto a slice $\mathbb{H}^k \times \{x\}$ of $\mathbb{H}^k \times \mathbb{S}^{n-k+1}$, while $(n-k+1)$-dimensional spheres centered at $\mathbb{R}^{k-1}$ lying in subspaces $\mathbb{R}^{n-k+2}$ orthogonal to $\mathbb{R}^{k-1}$ are mapped onto slices $\{x\} \times \mathbb{S}^{n-k+1}$ of $\mathbb{H}^k \times \mathbb{S}^{n-k+1}$.

Assume first that condition (*a*) is satisfied. Let Ψ be the diffeomorphism defined above for $k = 1$, and let $\hat{\Psi}$ be its composition with the isometry:

$$(y^{-1} e_0 + y e_1, x) \in \mathbb{H}^1 \times \mathbb{S}^n \ (y > 0) \mapsto (x, \log y) \in \mathbb{S}^n \times \mathbb{R}.$$

We can assume that the hyperspheres containing the images by f of the leaves of E_0 are centered at the origin. Then, the images by $F = \hat{\Psi} \circ f\colon M^n \to \mathbb{S}^n \times \mathbb{R}$ of the leaves of E_0 are contained in the slices $\mathbb{S}^n \times \{t\}$, $t \in \mathbb{R}$. This means that the height function $(x, s) \mapsto \langle F(x, s), \partial_t \rangle$ depends only on s, where ∂_t is a unit vector field tangent to the factor $\mathbb{R}$. Differentiating with respect to $X \in E_0$ gives $0 = \langle F_* X, \partial_t \rangle = \langle F_* X, F_* T \rangle$, where $\partial_t = F_* T + \langle \partial_t, N \rangle N$. Since $\hat{\Psi}$ is a conformal diffeomorphism, the metrics induced by f and F are conformal; hence, T spans E_1. Moreover, since conformal diffeomorphisms preserve principal directions and the integral curves of E_1 are lines of curvature of f, it follows that T is a principal direction of F.

Now, assume that condition (*b*) holds. Denoting by $\Psi\colon \mathbb{R}^{n+1} \to \mathbb{R}^n \times \mathbb{R} \to \mathbb{R}^{n+1}$ the standard isometry, the images under $F = \Psi \circ f$ of the leaves of E_0 are contained in slices $\mathbb{R}^n \times \{t\}$, $t \in \mathbb{R}$. Arguing as in the preceding paragraph, this means that the tangent component of the unit vector field tangent to the factor $\mathbb{R}$ is a principal direction of F.

Finally, suppose that condition (c) holds and let Ψ be the diffeomorphism defined in the preceding paragraph for $k = n$. Then, the images by $\hat{F} = \Psi \circ f : M^n \to \mathbb{H}^n \times \mathbb{S}^1$ of the leaves of E_0 are contained in slices $\mathbb{H}^n \times \{x\}$, $x \in \mathbb{S}^1$. Let $F : M^n \to \mathbb{H}^n \times \mathbb{R}$ be such that $\hat{F} = \pi \circ F$, where $\pi : \mathbb{H}^n \times \mathbb{R} \to \mathbb{H}^n \times \mathbb{S}^1$ is the covering map $(x, t) \mapsto (x, (\cos t, \sin t))$. Then, the images by F of the leaves of E_0 are contained in slices $\mathbb{H}^n \times \{t\}$, $t \in \mathbb{R}$. Arguing as in case (a), we conclude that T is a principal direction of F, where $\partial_t = F_* T + \langle \partial_t, N \rangle N$, with ∂_t a unit vector field tangent to $\mathbb{R}$.

In either of the preceding cases, it follows from Theorem 1 in [17] that the map F is either a vertical cylinder given by $F(x, s) = g(x) + s\frac{\partial}{\partial t}$, over a hypersurface $g : M^{n-1} \to \mathbb{Q}^n_\epsilon$, or it is given by (38) in terms of such a hypersurface, with $\epsilon = 1$ in case (a), $\epsilon = 0$ in case (b), and $\epsilon = -1$ in case (c). In any case, we have $f = \Phi \circ F$, and the statement follows. □

Surfaces of Enneper type in $\mathbb{R}^3$ satisfying condition (c) in the preceding theorem are known in the literature as *Joachimsthal surfaces*. These are the surfaces whose lines of curvature corresponding to one of the principal curvatures are contained in planes that intersect along a common line, whereas the lines of curvature corresponding to the other principal curvature lie on spheres centered on that line. The following consequence of Theorem 7 shows how any such surface arises.

Corollary 8 *Let $\gamma : J \subset \mathbb{R} \to \mathbb{H}^2$ be a unit-speed curve and let $\gamma_s : J \subset \mathbb{R} \to \mathbb{H}^2 \subset \mathbb{R}^3_1$ be the family of its parallel curves, that is,*

$$\gamma_s(t) = \cosh(s)\gamma(t) + \sinh(s)\gamma(t) \wedge \gamma'(t).$$

Define $F : J \times I \to \mathbb{H}^2 \times \mathbb{R} \subset \mathbb{R}^4_1$ by

$$F(t, s) = \gamma_s(t) + a(s)\frac{\partial}{\partial t},$$

where $I \subset \mathbb{R}$ is an open interval and $a \in C^\infty(I)$ has positive derivative. Then, on the subset $M^2 \subset J \times I$ of its regular points, the map $f = \Phi \circ F : M^2 \to \mathbb{R}^3 \setminus \mathbb{R}$, where $\Phi : \mathbb{H}^2 \times \mathbb{R} \to \mathbb{R}^3 \setminus \mathbb{R}$ is the conformal covering map given by (37)*, defines a Joachimsthal surface.*

Conversely, any Joachimsthal surface in $\mathbb{R}^3$ can either be parametrized in this way or is a rotation surface, in which case it is a composition $f = \Phi \circ F$, where $F : J \times I \to \mathbb{H}^2 \times \mathbb{R} \subset \mathbb{R}^4_1$ is a vertical cylinder, given by

$$F(t, s) = \gamma(t) + s\frac{\partial}{\partial t},$$

over a unit-speed curve $\gamma : J \subset \mathbb{R} \to \mathbb{H}^2$.

Acknowledgments Ruy Tojeiro was partially supported by Fapesp grant 2022/16097-2 and CNPq grant 303002/2017-4.

References

1. L. Bianchi, "Lezioni di Geometria Différenziale", Bologna, 1902.
2. A. I. Bobenko, T. Hoffmann and A. O. Sageman-Furnas, *Compact Bonnet pairs: isometric tori with the same curvatures.* (available at arXiv [math.DG] 2110.06335v1.
3. A. I. Bobenko, T. Hoffmann and A. O. Sageman-Furnas, *Isothermic tori with one family of planar curvature lines and area constrained hyperbolic elastica.* (available at arXiv [math.DG] arXiv:2312.14956.
4. T. Cecil, G. Jensen, *Dupin hypersurfaces with three principal curvatures.* Invent. Math. 132 (1998), 121–178. https://doi.org/10.1007/s002220050220.
5. T. Cecil, *Classifications of Dupin Hypersurfaces in Lie Sphere Geometry*, Acta Math. Sci. 44 (2024), 1–36, https://doi.org/10.1007/s10473-024-0101-7.
6. M. Dajczer and R. Tojeiro, "Submanifold Theory beyond an introduction", Springer, New York, 2019, Universitext Series.
7. M. Dajczer and R. Tojeiro, *An extension of the classical Ribaucour transformation.* Proc. London Math. Soc. 85 (2002), 211–232.
8. M. Dajczer and R. Tojeiro, *Commuting Codazzi tensors and the Ribaucour transformation for submanifolds.* Results Math. 44 (2003), 258–278.
9. M. Dajczer and R. Tojeiro, *Hypersurfaces of space forms carrying a totally geodesic foliation.* Geom. Dedicata 205 (2020), 129–146.
10. M. Dajczer, V. Rovenski and R. Tojeiro, *Euclidean hypersurfaces with a totally geodesic foliation of codimension one*. Geom. Dedicata 176 (2015), 215–224.
11. M. Dajczer, L. Florit and R. Tojeiro, *Reducibility of Dupin submanifolds.* Illinois J. Math. 49 (2005), 759–791.
12. G. Darboux, "Théorie genérale des surfaces", Chelsea Publishing Company, Bronx, New York, 1972, third edition.
13. U. Pinkall and I. Sterling, *On the classification of constant mean curvature tori.* Ann. of Math. 130 (1989), 407–451.
14. J. P. dos Santos and R. Tojeiro, *Cyclic conformally flat hypersurfaces revisited.* Matemática Contemporânea 49 (2022), 188–211 (special volume in honor of Renato Tribuzy).
15. R. Tojeiro, *Conformal immersions of warped products.* Geom. Dedicata 128 (2007), 17–31.
16. R. Tojeiro, *A decomposition theorem for immersions of product manifolds.* Proc. Edinb. Math. Soc. 59 (2016), 247–269.
17. R. Tojeiro, *On a class of hypersurfaces in* $\mathbb{S}^n \times \mathbb{R}$ *and* $\mathbb{H}^n \times \mathbb{R}$, Bulletin Braz. Math. Soc. 41 (2010), 199–209.
18. H. C. Wente, *Counterexample to a conjecture of H. Hopf.* Pacific J. Math. (1986), 193–243.
19. H. C. Wente, *Constant Mean Curvature Immersions of Enneper Type*, Memoirs of the A.M.S. v. 100 (1992), number 478.
20. H.C. Wente, *Twisted Tori of Constant Mean Curvature in* $\mathbb{R}^3$, Seminar on New Results in Nonlinear Partial Differential Equations, Aspects of Mathematics, Vieweg+Teubner Verlag, 1987.

Conformal Immersions of Product Manifolds Endowed with Polar Metrics

Sergio Chion and Ruy Tojeiro

Dedicated to Professor Marcos Dajczer on his 75th birthday.

Abstract In our previous article (Chion and Tojeiro, Ribaucour Partial Tubes and Hypersurfaces of Enneper Type. Advances in Submanifold theory and related topics, Latin American Mathematical Series, UFSCar Subseries, Springer, 2025), we introduced the concept of a Ribaucour partial tube in a space form and proved that it can be characterized as an immersion of a product of two manifolds such that the distributions given by the tangent spaces of the factors are orthogonal to each other with respect to the induced metric and are invariant under all shape operators, and one of them is spherical. In this article, we prove a de Rham-type theorem that provides conditions under which a certain Riemannian manifold is locally isometric to a product manifold whose metric is conformal to a polar metric and then use the characterization above of Ribaucour partial tubes to derive a general decomposition theorem for immersions of product manifolds which extends various related results.

1 Introduction

This article deals with the problem of investigating the geometry of an isometric immersion with high codimension by trying to "decompose it into simpler components." This problem finds parallels in various branches of mathematics, a basic

S. Chion
CENTRUM Católica Graduate Business School, Pontificia Universidad Católica del Perú, Lima, Peru
e-mail: sjchiona@pucp.edu.pe

R. Tojeiro (✉)
Instituto de Ciências Matemáticas e de Computação, Universidade de São Paulo, São Carlos, Brazil
e-mail: tojeiro@icmc.usp.br

© The Author(s), under exclusive license to Springer Nature Switzerland AG 2025

R. Tojeiro (ed.), *Advances in Submanifold Theory and Related Topics*,
Latin American Mathematics Series – UFSCar subseries,
https://doi.org/10.1007/978-3-032-00918-0_7

example being the decomposition of an integer into prime factors. In the realm of differential geometry, from an intrinsic point of view, it has led to the development of several de Rham-type theorems, which provide conditions under which a certain Riemannian manifold is (locally or globally) isometric to a product manifold whose metric is of a specific type. See, for instance, [2, 4–6], and [8] for results of this type.

We first derive a de Rham-type theorem that establishes conditions for a Riemannian manifold to be locally isometric to a product manifold whose metric is conformal to a polar metric (Theorem 2). Recall that a metric g on a product manifold $M = \prod_{i=0}^{r} M_i$ is said to be *polar* if there exists a metric g_0 on M_0 and a family of metrics on M_a, $1 \leq a \leq r$, smoothly parametrized by M_0, such that

$$g = \pi_0^* g_0 + \sum_{a=1}^{r} \pi_a^*(g_a \circ \pi_0),$$

where $\pi_i : M \to M_i$ is the projection for $0 \leq i \leq r$. Polar metrics include as special cases the warped product of metrics $g_0, \ldots, g_r$ on $M_0, \ldots, M_r$, respectively, with smooth warping functions $\rho_a : M_0 \to \mathbb{R}_+$, $1 \leq a \leq r$, that is, metrics given by

$$g = \pi_0^* g_0 + \sum_{a=1}^{r} (\rho_a \circ \pi_0)^2 \pi_a^* g_a.$$

The Riemannian product of $g_0, \ldots, g_r$ is the particular case in which the warping functions ρ_a, $1 \leq a \leq r$, are identically one. Warped (respectively, Riemannian) product metrics correspond to polar metrics for which all metrics $g_a(x_0)$ on M_a, $1 \leq a \leq r$, $x_0 \in M_0$, are homothetical (respectively, isometric) to a fixed Riemannian metric. Our result extends previous results in [2, 6, 8], and [9] on warped product metrics, metrics conformal to Riemannian and warped product metrics, and polar metrics, respectively.

From an extrinsic perspective, various decomposition theorems for immersions of product manifolds have been obtained under the assumption that the tangent spaces to the factors are invariant by all shape operators. The first theorem along this line was Moore's basic result characterizing extrinsic products of immersions as isometric immersions of Riemannian product manifolds that satisfy that condition. Moore's theorem has been subsequently generalized in [3, 7, 9], and [6] to cover isometric immersions of product manifolds with more general types of metrics, namely, warped product metrics, metrics conformal to Riemannian product and warped product metrics, and polar metrics, respectively.

In this article, we use the notion of a Ribaucour partial tube to extend those results further to the broader class of metrics conformal to polar metrics. Specifically, we provide a comprehensive description of all conformal immersions of a product manifold equipped with a polar metric under the assumption that the tangent spaces of the factors are invariant by all shape operators (Theorem 3).

2 A De Rham-Type Theorem

This section aims to prove an intrinsic de Rham-type theorem that characterizes Riemannian manifolds that are locally conformal to product manifolds endowed with polar metrics.

We use the following terminology introduced in [5].

A *net* $\mathcal{E} = (E_i)_{i=0,\dots,r}$ on a differentiable manifold M is a decomposition of its tangent bundle $TM = \oplus_{i=0}^r E_i$ as a Whitney sum of integrable distributions. If M is a Riemannian manifold, the net $\mathcal{E} = (E_i)_{i=0,\dots,r}$ is said to be an *orthogonal net* if the distributions E_i are mutually orthogonal. For an isometric immersion $f: M \to \mathbb{R}^m$ of a Riemannian manifold M equipped with a net $\mathcal{E} = (E_i)_{i=0,\dots,r}$, the second fundamental form α_f of f is said to be *adapted to the net* $\mathcal{E}$ if $\alpha_f(X_i, X_j) = 0$ whenever $X_i \in \Gamma(E_i)$ and $X_j \in \Gamma(E_j)$, with $0 \le i \ne j \le r$.

In a product manifold $M = \prod_{i=0}^r M_i$, there exists a natural net $\mathcal{E} = (E_i)_{i=0,\dots,r}$, known as its *product net*, given by the tangent bundles of its factors, that is, $E_i(x) = \tau_i^x{}_* T_{x_i} M_i$ for all $x = (x_0, \dots, x_r) \in M$ and $0 \le i \le r$, where $\tau_i^x : M_i \to M$ stands for the inclusion of M_i into M given by $\tau_i^x(y_i) = (x_0, \dots, x_{i-1}, y_i, x_{i+1}, \dots, x_r)$. For each $0 \le i \le r$, we denote

$$M_{\perp_i} = \prod_{\substack{j=0 \\ j \ne i}}^r M_j$$

and define the projection $\pi_{\perp_i} : M \to M_{\perp_i}$ by

$$\pi_{\perp_i}(x_0, \dots, x_r) = (x_0, \dots, x_{i-1}, x_{i+1}, \dots, x_r).$$

Given $x_i \in M_i$, the map $\mu_{x_i} : M_{\perp_i} \to M$ stands for the inclusion:

$$\mu_{x_i}(y_0, \dots, y_{i-1}, y_{i+1}, \dots, y_r) = (y_0, \dots, y_{i-1}, x_i, y_{i+1}, \dots, y_r).$$

A distribution E on a Riemannian manifold is *umbilical* if there exists $Z \in \Gamma(E^\perp)$ such that

$$(\nabla_T S)_{E^\perp} = \langle T, S \rangle Z$$

for all $T, S \in \Gamma(E)$, where the subscript $E^\perp$ denotes the orthogonal projection onto $E^\perp$. Umbilical distributions are integrable and the leaves are totally umbilical submanifolds with mean curvature vector field Z. If, in addition, Z satisfies

$$(\nabla_T Z)_{E^\perp} = 0$$

for any $T \in \Gamma(E)$, then E is said to be *spherical*, in which case its leaves are extrinsic spheres.

Proposition 4 in [6] shows that a Riemannian metric on a product manifold $M = \prod_{i=0}^r M_i$ is polar if and only if the product net $\mathcal{E} = (E_i)_{i=0,\ldots,r}$ of M is an orthogonal net such that $E_a^\perp$ is totally geodesic for all $1 \le a \le r$. Our first goal is to obtain a similar characterization for metrics on a product manifold that are *conformal* to a polar metric. Recall that two Riemannian metrics g_1 and g_2 on a manifold M are conformal if there exists a positive function $\lambda \in C^\infty(M)$ such that $g_2 = \lambda^2 g_1$. The function λ is called the *conformal factor* of g_2 with respect to g_1.

Theorem 1 *A Riemannian metric on a product $M = \prod_{i=0}^r M_i$ of connected manifolds is conformal to a polar metric if and only if the product net $\mathcal{E} = (E_i)_{i=0,\ldots,r}$ of M is an orthogonal net such that $E_a^\perp$ is totally umbilical for all $1 \le a \le r$.*

Proof Consider a polar metric g_1 on $M = \prod_{i=0}^r M_i$ and let g_2 be a metric conformal to g_1 with conformal factor $\lambda \in C^\infty(M)$. It is well known that the Levi-Civita conections ∇^1 and ∇^2 associated with g_1 and g_2, respectively, are related by

$$\nabla^2_X Y = \nabla^1_X Y + \frac{1}{\lambda}\left(Y(\lambda)X + X(\lambda)Y - g_2(X,Y)\mathrm{grad}_2\lambda\right), \tag{1}$$

where grad_2 denotes the gradient with respect to g_2. Since $E_a^\perp$ is a totally geodesic distribution with respect to g_1 for $1 \le a \le r$, it follows from (1) that $E_a^\perp$ is totally umbilical with respect to g_2 with mean curvature vector field $-(\mathrm{grad}_2 \log \lambda)_{E_a}$.

To prove the converse statement, for each $1 \le a \le r$, decompose $M = M_a \times M_{\perp_a}$. Fix $p = (p_0, \ldots, p_r) \in M$ and endow $M_{\perp_a}$ with the metric $g_{\perp_a} = \mu_{p_a}^* g$. Given any $x = (x_0, \ldots, x_r)$, denote

$$p^a = (x_0, \ldots, x_{a-1}, p_a, x_{r+1}, \ldots, x_r), \quad x^a = (p_0, \ldots, p_{a-1}, x_a, p_{a+1}, \ldots, p_r)$$

and

$$x^{0,a} = (x_0, p_1, \ldots, p_{a-1}, x_a, p_{a+1}, \ldots, p_r).$$

Since $E_a^\perp$ is totally umbilical, Proposition 1 in [4] guarantees the existence of a positive $\lambda_a \in C^\infty(M)$ such that

$$\mu_{x_a}^* g = \lambda_{x_a}^2 g_{\perp_a} = \lambda_{x_a}^2 \mu_{p_a}^* g, \quad \text{for } 1 \le a \le r,$$

where $\lambda_{x_a} = \lambda_a \circ \mu_{x_a}$. Thus, for all $0 \le b \ne a \le r$ and $X_b, Y_b \in T_{x_b}M_b$, we have

$$g(x_0, \ldots, x_r)(\tau_{b*}^x X_b, \tau_{b*}^x Y_b) = \lambda_{x_a}^2(\pi_{\perp_a}(x)) g(p^a)(\tau_{b*}^{p^a} X_b, \tau_{b*}^{p^a} Y_b). \tag{2}$$

For a fixed $1 \leq a \leq r$, applying (2) for each $1 \leq b \neq a \leq r$, we obtain

$$\begin{aligned} & g(x_0, \ldots, x_r)(\tau_{a*}^x X_a, \tau_{a*}^x Y_a) \\ & \quad = \lambda_{x_r}^2(x_0, \ldots, x_{r-1}) \cdots \lambda_{x_{a+1}}^2(x_0, \ldots, x_a, p_{a+2}, \ldots, p_r) \cdots \\ & \quad \cdots \lambda_{x_{a-1}}^2(x_0, \ldots, x_{a-2}, x_a, p_{a+1}, \ldots, p_r) \cdots \\ & \quad \cdots \lambda_{x_1}^2(x_0, p_2, \ldots, p_{a-1}, x_a, p_{a+1}, \ldots, p_r) \cdots \\ & \quad \cdots g(x^{0,a})(\tau_a^{x^{0,a}}{}_* X_a, \tau_a^{x^{0,a}}{}_* Y_a). \end{aligned} \tag{3}$$

On the other hand, for $b = 0$, we can use (2) recursively, following the order given by an arbitrary permutation $s : \{1, \ldots, r\} \to \{1, \ldots, r\}$. This yields

$$\begin{aligned} & g(x_0, \ldots, x_r)(\tau_{0*}^x X_0, \tau_{0*}^x Y_0) \\ & \quad = \lambda_{x_{s(1)}}^2(x_0, \ldots, x_{s(1)-1}, x_{s(1)+1}, \ldots, x_r) \cdots \\ & \quad \cdots \lambda_{x_{s(r)}}^2(x_0, p_1, \ldots, p_{s(r)-1}, p_{s(r)+1}, \ldots, p_r) g(x^0)(\tau_0^{x^0}{}_* X_0, \tau_0^{x^0}{}_* Y_0). \end{aligned} \tag{4}$$

Since the permutation s in (4) is arbitrary, for any two such permutations s, t, we have

$$\begin{aligned} & \lambda_{x_{s(1)}}^2(x_0, \ldots, x_{s(1)-1}, x_{s(1)+1}, \ldots, x_r) \cdots \\ & \lambda_{x_{s(r)}}^2(x_0, p_1, \ldots, p_{s(r)-1}, p_{s(r)+1}, \ldots, p_r) \\ & = \lambda_{x_{t(1)}}^2(x_0, \ldots, x_{t(1)-1}, x_{t(1)+1}, \ldots, x_r) \cdots \\ & \lambda_{x_{t(r)}}^2(x_0, p_1, \ldots, p_{t(r)-1}, p_{t(r)+1}, \ldots, p_r) \end{aligned} \tag{5}$$

Consider the metric $g_0 = \tau_0^{p*} g$ on M_0. For each $1 \leq a \leq r$ and $x_0 \in M_0$, let $g_a(x_0)$ be the metric on M_a given by

$$g_a(x_0)(x_a) = \lambda_{x_a}^{-2}(x_0, p_1, \ldots, \hat{p}_a, \ldots, p_r)(\tau_a^{x^0 *} g)(x_a).$$

The proof will be completed as soon as we show that

$$g = \varphi^2(\pi_0^* g_0 + \sum_{b=1}^{r} \pi_b^*(g_b \circ \pi_0)), \tag{6}$$

where

$$\varphi^2(x_0, \ldots, x_r) = \lambda_{x_r}^2(x_0, \ldots, x_{r-1}) \cdots \lambda_{x_1}^2(x_0, p_2, \ldots, p_r).$$

Using (4) for the permutation $s(\kappa) = r + 1 - \kappa$, we obtain

$$\begin{aligned} & g(x)(\tau_{0*}^x X_0, \tau_{0*}^x Y_0) \\ & \quad = \lambda_{x_r}^2(x_0, \ldots, x_{r-1}) \cdots \lambda_{x_1}^2(x_0, p_2, \ldots, p_r) g(x^0)(\tau_0^{x^0}{}_* X_0, \tau_0^{x^0}{}_* Y_0), \end{aligned}$$

while from the definition of g_0, we have

$$\begin{aligned}\varphi^2(\pi_0^* g_0)(x)(\tau_{0*}^x X_0, \tau_{0*}^x Y_0) &= \varphi^2(x) g_0(x_0)(X_0, Y_0)\\ &= \varphi^2(x) g(x^0)(\tau_{0*}^p X_0, \tau_{0*}^p Y_0)\\ &= \varphi^2(x) g(x^0)(\tau_{0*}^{x^0} X_0, \tau_{0*}^{x^0} Y_0).\end{aligned}$$

Thus, (6) holds for pairs of vectors that belong to E_0. Since the product net is orthogonal, we only have to show that (6) holds for pairs of vectors in E_a for any $1 \le a \le r$. From the definition of the metric $g_a(x_0)$ for $x_0 \in M_0$, we have

$$\begin{aligned}\pi_a^*(g_a \circ \pi_0)(x)(\tau_{a*}^x X_a, \tau_{a*}^x Y_a) &= g_a(x_0)(x_a)(X_a, Y_a)\\ &= \lambda_{x_a}^{-2}(x_0, p_1, \dots, \hat{p}_a, \dots, p_r)(\tau_a^{x^0 *} g)(x_a)(X_a, Y_a)\\ &= \lambda_{x_a}^{-2}(x_0, p_1, \dots, \hat{p}_a, \dots, p_r) g(x^{0,a})(\tau_{a*}^{x^0} X_a, \tau_{a*}^{x^0} Y_a)\\ &= \lambda_{x_a}^{-2}(x_0, p_1, \dots, \hat{p}_a, \dots, p_r) g(x^{0,a})(\tau_{a*}^{x^{0,a}} X_a, \tau_{a*}^{x^{0,a}} Y_a),\end{aligned}$$

whereas from the definition of φ and (5), we obtain

$$\begin{aligned}\varphi^2(x) = {}& \lambda_{x_r}^2(x_0, \dots, x_{r-1}) \cdots \lambda_{x_{a+1}}^2(x_0, \dots, x_a, p_{a+2}, \dots, p_r) \cdots\\ & \cdots \lambda_{x_{a-1}}^2(x_0, \dots, x_{a-2}, x_a, p_{a+1}, \dots, p_r) \cdots\\ & \cdots \lambda_{x_1}^2(x_0, p_2, \dots, p_{a-1}, x_a, p_{a+1}, \dots, p_r) \lambda_{x_a}^2(x_0, p_1, \dots, \hat{p}_a, \dots, p_r).\end{aligned}$$

The last two equalities give

$$\begin{aligned}&\varphi^2(x)(\pi_a^*(g_a \circ \pi_0)(x)(\tau_{a*}^x X_a, \tau_{a*}^x Y_a))\\ &= \lambda_{x_r}^2(x_0, \dots, x_{r-1}) \cdots \lambda_{x_{a+1}}^2(x_0, \dots, x_a, p_{a+2}, \dots, p_r) \cdots\\ &\cdots \lambda_{x_{a-1}}^2(x_0, \dots, x_{a-2}, x_a, p_{a+1}, \dots, p_r) \cdots\\ &\cdots \lambda_{x_1}^2(x_0, p_2, \dots, p_{a-1}, x_a, p_{a+1}, \dots, p_r) g(x^{0,a})(\tau_{a*}^{x^{0,a}} X_a, \tau_{a*}^{x^{0,a}} Y_a),\end{aligned}$$

thus showing (6) and completing the proof. □

The following additional fact will be needed in the proof of Theorem 3 below.

Proposition 1 *If g_1 is a polar metric on a product manifold $M = \prod_{i=0}^r M_i$ with product net $\mathcal{E} = (E_i)_{i=0,\dots,r}$ and g_2 is conformal to g_1 with conformal factor $\lambda \in C^\infty(M)$, then E_0 is a spherical distribution if and only if Hess λ is adapted to the net $(E_0, E_0^\perp)$.*

Proof Since E_0 is totally geodesic with respect to g_1, by (1) it is umbilical with mean curvature vector field $\eta = -(\operatorname{grad}\log\lambda)_{E_0^\perp}$ with respect to g_2. Now, for $X_0 \in \Gamma(E_0)$ and $Y_j \in \Gamma(E_j) \subset \Gamma(E_0^\perp)$, $j \neq 0$, we have

$$\begin{aligned} -\left\langle \nabla_{X_0}\eta, Y_j \right\rangle &= \left\langle \nabla_{X_0}\operatorname{grad}\log\lambda, Y_j \right\rangle - \left\langle \nabla_{X_0}(\operatorname{grad}\log\lambda)_{E_0}, Y_j \right\rangle \\ &= \left\langle \nabla_{X_0}(\lambda^{-1}\operatorname{grad}\lambda), Y_j \right\rangle - \langle X_0, \operatorname{grad}\log\lambda\rangle \left\langle Y_j, -(\operatorname{grad}\log\lambda)_{E_0^\perp} \right\rangle \\ &= -\lambda^{-2}X_0(\lambda)Y_j(\lambda) + \lambda^{-1}\operatorname{Hess}\lambda(X_0, Y_j) + \lambda^{-2}X_0(\lambda)Y_j(\lambda) \\ &= \lambda^{-1}\operatorname{Hess}\lambda(X_0, Y_j), \end{aligned}$$

and the statement follows. □

Given a net $\mathcal{F} = (F_i)_{i=0,\ldots,r}$ on a manifold M, a diffeomorphism $\Psi\colon \prod_{i=0}^r M_i \to M$ from a product manifold with product net $\mathcal{E} = (E_i)_{i=0,\ldots,r}$ is called a *product representation* of $\mathcal{F}$ if $\Psi_* E_i(p) = F_i(\Psi(p))$ for $0 \leq i \leq r$. Combining Theorem 1 in [5] and Theorem 1 yields the following de Rham-type result.

Theorem 2 *Let M be a Riemannian manifold and let $\mathcal{E} = (E_i)_{i=0,\ldots,r}$ be an orthogonal net such that $E_a^\perp$ is totally umbilical for each $1 \leq a \leq r$. Then, for every point $p \in M$, there exists a local product representation $\Psi\colon \prod_{i=0}^r M_i \to U$ of $\mathcal{E}$, with $p \in U \subset M$, which is conformal with respect to a polar metric on $\prod_{i=0}^r M_i$.*

The decomposition in Theorem 2 is of a local nature. Indeed, an example provided before Theorem 1 in [4] shows that a global representation cannot always be achieved.

Remark 1 Observe that any orthogonal net $\mathcal{E} = (E_0, E_1)$ with only two factors such that E_0 has rank one satisfies the conditions in Theorem 2, since any one-dimensional distribution is umbilical. Moreover, if Z is a vector field spanning E_0 on a simply connected open subset $U \subset M^n$, then $E_1 = E_0^\perp$ is integrable if and only if Z is the gradient of a smooth function $\varphi \in C^\infty(U)$, in which case the leaves of E_1 are the level sets of φ. Therefore, on any Riemannian manifold M, one can find as many such orthogonal nets $\mathcal{E}$ and hence as many local product representations of them that are conformal with respect to a polar metric, as there are smooth functions defined on open subsets of M.

3 An Extrinsic Decomposition Theorem

In this section, we prove a decomposition theorem for immersions of product manifolds endowed with metrics conformal to polar metrics. Besides some of the results of the previous section, the main tool is the notion of Ribaucour partial tubes, introduced in the article [1] in this volume. More precisely, the proof of Theorem 3 below relies on the characterization of Ribaucour partial tubes as the immersions of product manifolds with two factors such that the distributions given by the tangent spaces of the factors are orthogonal to each other with respect to the induced metric and are invariant under all shape operators, and one of them is spherical. We refer the reader to Sections 3 and 4 of [1], where Ribaucour partial tubes are defined and that characterization is derived (see Corollary 1 and Theorem 3 therein).

Theorem 3 *Let $f: M = \prod_{i=0}^r M_i \to \mathbb{R}^m$ be a conformal immersion with conformal factor $\lambda \in C^\infty(M)$ of a product manifold M endowed with a polar metric. Assume that the second fundamental form of f is adapted to the product net $\mathcal{E} = (E_i)_{i=0,\cdots,r}$ of M. If $\dim M_0 = 1$, suppose further that $\operatorname{Hess}\lambda$ is adapted to the net $(E_0, E_0^\perp)$. Then, f is a Ribaucour partial tube over an immersion $\tilde{f}: \tilde{M} = \prod_{a=1}^r M_a \to \mathbb{R}^m$ given in one of the following ways:*

1. *$\tilde{f} = \tau \circ \bar{f}$, where $\bar{f}$ is an extrinsic product of substantial immersions into either $\mathbb{R}^m$ or $\mathbb{S}_c^m$ and τ is, respectively, a conformal transformation of $\mathbb{R}^m$ or a conformal diffeomorphism of $\mathbb{S}_c^m$ (with one point removed) onto $\mathbb{R}^m$.*
2. *$\tilde{f} = \Theta \circ (f_1 \times \bar{f})$, where $\Theta: \mathbb{H}_{-c}^k \times \mathbb{S}_c^{m-k} \to \mathbb{R}^m$ is a conformal diffeomorphism (onto the complement of a $(k-1)$-dimensional subspace), and, after possibly relabeling factors, $f_1: M_1 \to \mathbb{H}_{-c}^k$ is an isometric immersion and $\bar{f}: \prod_{a=2}^r M_i \to \mathbb{S}_c^{m-k}$ is an extrinsic product of substantial isometric immersions.*

Proof First, let us consider the case when $r = 1$. In this case, by Theorem 1, the assumption that the metric induced by f is conformal to a polar metric implies that E_0 is an umbilical distribution. It follows from Lemma 12 of [7] if $\dim M_0 \geq 2$, or from the assumption that $\operatorname{Hess}\lambda$ is adapted to the net $(E_0, E_0^\perp)$ combined with Proposition 1 if $\dim M_0 = 1$, that E_0 is indeed spherical. Since the second fundamental form of f is adapted to the product net of M, the statement in this case is a consequence of Theorem 3 in [1].

If $r \geq 2$ is arbitrary, we can apply the case $r = 1$ just proved to the immersion f considered as a map from $M_0 \times \tilde{M}$ into $\mathbb{R}^m$, where $\tilde{M} = \prod_{a=1}^r M_i$. It follows that f is a Ribaucour partial tube over $\tilde{f}: \tilde{M} \to \mathbb{R}^m$ given by $\tilde{f} = f \circ \mu_{\bar{x}_0}$ for some $\bar{x}_0 \in M_0$, where $\mu_{\bar{x}_0}: \tilde{M} \to M$ is the inclusion given by $\mu_{\bar{x}_0}(\tilde{x}) = (\bar{x}_0, \tilde{x})$. The metric $\tilde{g}$ on $\tilde{M}$ induced by $\mu_{\bar{x}_0}$ from the metric g of M is

$$\tilde{g} = \mu_{\bar{x}_0}^* g = (\lambda \circ \mu_{\bar{x}_0})^2 \mu_{\bar{x}_0}^*\big(\pi_0^* g_0 + \sum_{a=1}^r \pi_a^*(g_a \circ \pi_0)\big) = (\lambda \circ \mu_{\bar{x}_0})^2 \sum_{a=1}^r \tilde{\pi}_a^* g_a(\bar{x}_0),$$

where $\tilde{\pi}_a: \tilde{M} \to M_a$ is the projection. Hence, $\tilde{g}$ is conformal to a Riemannian product metric.

We now claim that the second fundamental form of $\tilde{f} = f \circ \mu_{\bar{x}_0}$ is adapted to the product net on $\tilde{M}$. We have

$$\begin{aligned}\alpha^{\tilde{f}}(\tau_a^{\tilde{x}}{}_*X_a, \tau_b^{\tilde{x}}{}_*X_b) &= \alpha^f(\tau_a^{(\bar{x}_0,\tilde{x})}{}_*X_a, \tau_b^{(\bar{x}_0,\tilde{x})}{}_*X_b) + f_*\alpha^{\mu}_{\bar{x}_0}(\tau_a^{\tilde{x}}{}_*X_a, \tau_b^{\tilde{x}}{}_*X_b)\\ &= f_*\alpha^{\mu}_{\bar{x}_0}(\tau_a^{\tilde{x}}{}_*X_a, \tau_b^{\tilde{x}}{}_*X_b),\end{aligned}$$

since the second fundamental form of f is adapted to the product net of M. Because $E_b^{\perp}$ is umbilic, as follows from Theorem 1 and the assumption that the metric induced by f is conformal to a polar metric, we have

$$\left\langle \nabla_{\tau_a^{\tilde{x}}{}_*X_a} \tau_b^{\tilde{x}}{}_*X_b, \tau_0^{\tilde{x}}{}_*X_0 \right\rangle = -\left\langle \tau_b^{\tilde{x}}{}_*X_b, \nabla_{\tau_a^{\tilde{x}}{}_*X_a} \tau_0^{\tilde{x}}{}_*X_0 \right\rangle = 0.$$

Thus, $\alpha^{\mu}_{\bar{x}_0}(\tau_a^{\tilde{x}}{}_*X_a, \tau_b^{\tilde{x}}{}_*X_b) = 0$, and our claim follows.

To complete the proof, it remains to show that $\tilde{f}\colon \prod_{a=1}^r M_i \to \mathbb{R}^m$ is given as in the statement. But this follows from Theorem 5 in [7], where conformal immersions of a Riemannian product whose second fundamental forms are adapted to the product net of the manifold have been classified. □

Corollary 1 *Let $f\colon M^m = \prod_{i=0}^r M_i \to \mathbb{R}^m$ be a conformal local diffeomorphism with conformal factor $\lambda \in C^\infty(M)$ of a product manifold M endowed with a polar metric. If $n_0 = \dim M_0 = 1$, suppose further that Hess λ is adapted to the net $(E_0, E_0^{\perp})$. Then, f is a Ribaucour partial tube over an immersion $\tilde{f}\colon \tilde{M} = \prod_{a=1}^r M_a \to \mathbb{R}^m$ given as in Theorem 3, with each of the immersions f_a, $1 \le a \le r$, having flat normal bundle and $f_0\colon M_0^{m_0} \to \mathbb{R}^{m_0}$ being a local isometry.*

Acknowledgments Ruy Tojeiro was partially supported by Fapesp grant 2022/16097-2 and CNPq grant 303002/2017-4.

References

1. S. Chion and R. Tojeiro, *Ribaucour partial tubes and hypersurfaces of Enneper type.* Advances in Submanifold theory and related topics, Latin American Mathematical Series, UFSCar Subseries, Springer, 2025.
2. S. Hiepko, *Eine innere Kennzeichnung der verzerrten Produkte*, Math. Ann. **241** (1979), 209–215.
3. Nölker, S., *Isometric immersions of warped products*, Differential Geom. Appl. 6 (1996), 1–30.
4. R. Ponge and H. Reckziegel, *Twisted products in pseudo-riemannian geometry.* Geom. Dedicata. **48**, (1993), 15–25.
5. H. Reckziegel, and M. Schaaf, *De Rham decompositions of netted manifolds.* Result. Math. **35**, (19999), 175–191.
6. R. Tojeiro, *A decomposition theorem for immersions of product manifolds.* Edinb. Math. Soc. **59**, (2016), 247–269.

7. R. Tojeiro, *Isothermic submanifolds of Euclidean space.* J. Reine Angew. Math. **598**, (2006), 1–24.
8. R. Tojeiro, *Conformal de Rham decomposition of Riemannian manifolds*, Houston J. Math. **32** (2006), 725-743.
9. R. Tojeiro, *Conformal immersions of warped products*, Geom. Dedicata 128 (2007), 17–31.

Compact Flat Manifolds and Reducibility

Andrzej Derdzinski and Paolo Piccione

Dedicated to Professor Marcos Dajczer on his 75th birthday.

Abstract Hiss and Szczepański proved in 1991 that the holonomy group of any compact flat Riemannian manifold, of dimension at least two, acts reducibly on the rational span of the Euclidean lattice associated with the manifold via the first Bieberbach theorem. Geometrically, their result states that such a manifold must admit a nonzero proper parallel distribution with compact leaves. We study algebraic and geometric properties of the sublattice-spanned holonomy-invariant rational vector subspaces that exist due to the above theorem, and of the resulting compact-leaf foliations of compact flat manifolds. The class consisting of the former subspaces, in addition to being closed under spans and intersections, also turns out to admit (usually nonorthogonal) complements. As for the latter foliations, we provide descriptions, first, and foremost, of the intrinsic geometry of their generic leaves in terms of that of the original flat manifold and, secondly, as an essentially obvious afterthought, of the leaf-space orbifold. The general conclusions are then illustrated by examples in the form of generalized Klein bottles.

A. Derdzinski (✉)
Department of Mathematics, The Ohio State University, Columbus, OH, USA
e-mail: andrzej@math.ohio-state.edu

P. Piccione
Departamento de Matemática, Instituto de Matemática e Estatística, Universidade de São Paulo, São Paulo, SP, Brazil

Department of Mathematics, School of Sciences, Great Bay University, Dongguan, Guangdong, China
e-mail: paolo.piccione@usp.br; paolo.piccione@gbu.edu.cn

© The Author(s), under exclusive license to Springer Nature Switzerland AG 2025
R. Tojeiro (ed.), *Advances in Submanifold Theory and Related Topics*,
Latin American Mathematics Series – UFSCar subseries,
https://doi.org/10.1007/978-3-032-00918-0_8

1 Introduction

As shown by Hiss and Szczepański [10, the corollary in Sect. 1], on any compact flat Riemannian manifold $\mathcal{M}$ with $\dim \mathcal{M} = n \geq 2$, there exists a parallel distribution D of dimension k, where $0 < k < n$, such that the leaves of D are all compact. In the appendix, we reproduce the original algebraic phrasing of their result and mention a stronger version of it, established more recently by Lutowski [12], which implies that—unless $\mathcal{M}$ is a flat torus—there exist at least two distributions D with the above properties, having nonequivalent irreducible holonomy representations.

The present paper deals with geometric consequences of Hiss and Szczepański's theorem. We do not pursue analogous ramifications of Lutowski's generalization.

Our main results are Theorems 1, 2, 3, 4, and 5 and Corollary 2.

Theorem 1 describes geometries of the individual leaves $\mathcal{M}'$ of a distribution D on $\mathcal{M}$ having the properties mentioned above, in terms of the short exact sequence $L \to \Pi \to H$ formed by the lattice L, holonomy group H and Bieberbach group Π associated with $\mathcal{M}$, and its analog $L' \to \Pi' \to H'$ for $\mathcal{M}'$. Specifically, according to our Theorem 1(ii), Π' (or, L') may be treated as a subgroup of Π (or, of a certain Euclidean vector space $\mathcal{V}$) and H' as a homomorphic image of a subgroup of H.

In Theorem 2, we establish the existence, on every compact flat Riemannian manifold $\mathcal{M}$ of dimension $n \geq 2$, of *two* proper parallel distributions D and $\hat{D}$ with compact leaves, complementary to each other in the sense that $T\mathcal{M} = D \oplus \hat{D}$.

Corollary 2 states that, in any manifold $\mathcal{M}$ as above, the class of parallel distributions with compact leaves is closed under spans and intersections.

Theorem 3 addresses the particularly simple form of the sequence $L' \to \Pi' \to H'$ described by Theorem 1(ii), arising in the case of leaves $\mathcal{M}'$ which we call *generic*. The union of all generic leaves is an open dense subset of $\mathcal{M}$, they all have the same triple L', Π', H', and they are mutually isometric. When all leaves of D happen to be generic, they form a locally trivial bundle with compact flat manifolds serving both as the base and the fiber (the *fibration case*).

Theorems 4 and 5 describe the intersection numbers of the leaves of the two mutually complementary foliations resulting from Theorem 2.

Aside from the holonomy group H' of each individual leaf $\mathcal{M}'$ of D, forming a part of its intrinsic (submanifold) geometry, $\mathcal{M}'$ also gives rise to two "extrinsic" holonomy groups, one arising since $\mathcal{M}'$ is a leaf of the foliation $F_{\mathcal{M}}$ of $\mathcal{M}$ tangent to D and the other coming from the normal connection of $\mathcal{M}'$. Due to the flatness of the normal connection, the two extrinsic holonomy groups coincide and are trivial for all generic leaves. In Sect. 12, we briefly discuss the leaf space $\mathcal{M}/F_{\mathcal{M}}$, pointing out that (not surprisingly!) $\mathcal{M}/F_{\mathcal{M}}$ is a flat compact orbifold which, in the fibration case mentioned above, constitutes the base manifold of the bundle.

Our results are illustrated by examples (generalized Klein bottles, Sect. 15), where both the fibration and non-fibration cases occur, depending on the choice of D.

In Sect. 16, we discuss certain analogs of the correspondence between Bieberbach groups and compact flat manifolds: one provided by almost-Bieberbach groups and infra-nilmanifolds and the other by the group $\mathrm{Spin}(m, 1)$ acting on the orthonormal-frame bundle of the hyperbolic m-space, leading to quotient manifolds that include some compact locally symmetric pseudo-Riemannian Einstein manifolds.

An earlier version [7] of this paper is cited by [2] and therefore still available on the arXiv. The presentation in [7] was—as we eventually realized—rather far from reader-friendly, which prompted us to thoroughly rewrite the whole text.

2 Preliminaries

Manifolds, mappings (except in Lemma 2), and tensor fields, such as bundle and covering projections, submanifold inclusions, and Riemannian metrics, are by definition of class C^∞. Submanifolds need not carry the subset topology, and a manifold may be disconnected (although, being required to satisfy the second countability axiom, it must have at most countably many connected components). Connectedness/compactness of a submanifold always refers to its own topology and implies the same for its underlying set within the ambient manifold. Thus, a compact submanifold is always endowed with the subset topology.

By a *distribution* on a manifold $\mathcal{N}$, we mean, as usual, a (smooth) vector subbundle D of the tangent bundle $T\mathcal{N}$. An *integral manifold* of D is any submanifold $\mathcal{L}$ of $\mathcal{N}$ with $T_x\mathcal{L} = D_x$ whenever $x \in \mathcal{L}$. The maximal connected integral manifolds of D will also be referred to as the *leaves* of D. If D is integrable, its leaves form the foliation associated with D. We call D *projectable* under a mapping $\psi : \mathcal{N} \to \hat{\mathcal{N}}$ onto a distribution $\hat{D}$ on the target manifold $\hat{\mathcal{N}}$ if $d\psi_x(D_x) = \hat{D}_{\psi(x)}$ for all $x \in \mathcal{N}$.

Remark 1 The following facts are well known: (c) is the compact case of Ehresmann's fibration theorem [8, Corollary 8.5.13]; (b) follows from (c). For (a), see [11, pp. 43–44 and 61–62]—note that finiteness trivially implies proper discontinuity:

a. Free diffeomorphic actions of finite groups on manifolds are properly discontinuous, leading to covering projections onto the resulting quotient manifolds.
b. Any locally diffeomorphic mapping from a compact manifold into a connected manifold is a (surjective) finite covering projection.
c. More generally, the phrases "locally diffeomorphic mapping" and "finite covering projection" in (b) may be replaced with *submersion* and *fibration*.

Lemma 1 *Given manifolds $\hat{\mathcal{M}}$ and $\mathcal{M}$ with distributions $\hat{D}$ and D, let $\hat{D}$ be projectable onto D under a locally diffeomorphic surjective mapping $\psi : \hat{\mathcal{M}} \to \mathcal{M}$:*

i. *The ψ-image of any leaf of $\hat{D}$ is a connected integral manifold of D.*
ii. *Integrability of $\hat{D}$ implies that of D.*
iii. *For any compact leaf $\mathcal{L}$ of $\hat{D}$, the image $\mathcal{L}' = \psi(\mathcal{L})$ is a compact leaf of D, and the restriction $\psi : \mathcal{L} \to \mathcal{L}'$ constitutes a covering projection.*

iv. *If the leaves of* $\hat{D}$ *are all compact, so are those of* D.

Proof Assertion (i) is immediate from the definitions of a leaf and projectability, while (i) yields (ii) as integrability amounts to the existence of an integral manifold through every point. Remark 1(b) and (i) give (iii). Now, (iv) follows. □

Lemma 2 *Suppose that* F *is a mapping from a manifold* $\mathcal{W}$ *into any set. If for every* $x \in \mathcal{W}$ *there exists a diffeomorphic identification of a neighborhood* $\mathcal{B}_x$ *of* x *in* $\mathcal{W}$ *with a unit open Euclidean ball centered at* 0 *under which* x *corresponds to* 0 *and* F *becomes constant on each open straight-line interval of length* 1 *in the open Euclidean ball having* 0 *as an endpoint, then* F *is locally constant on some open dense subset of* $\mathcal{W}$.

Proof We use induction on $n = \dim \mathcal{W}$. The case $n = 1$ being trivial, let us suppose that the assertion holds in dimension $n-1$ and consider a mapping F from an n-dimensional manifold $\mathcal{W}$, satisfying our hypothesis, along with an embedded open Euclidean ball $\mathcal{B}_x \subseteq \mathcal{M}$ "centered" at a fixed point x, as in the statement of the lemma. The constancy of F along the fibers of the normalization projection $\mu : \mathcal{B}_x \smallsetminus \{x\} \to \mathcal{S}$ onto the unit $(n-1)$-sphere $\mathcal{S}$ gives rise to a mapping G with the domain $\mathcal{S}$ and $F = G \circ \mu$. Let us now fix $s \in \mathcal{S}$, any $y \in \mathcal{B}_x \smallsetminus \{x\}$ with $\mu(y) = s$, and an embedded open Euclidean ball $\mathcal{B}_y$ "centered" at y, such that F is constant on each radial open interval in $\mathcal{B}_y$. The obvious submersion property of μ allows us to pass from $\mathcal{B}_y$ to a smaller concentric ball and then choose a codimension-one open Euclidean ball $\mathcal{B}'_y$ arising as a union of radial intervals within this smaller version of $\mathcal{B}_y$, for which $\mu : \mathcal{B}'_y \to \mathcal{S}$ is an embedding. The assumption of the lemma thus holds when $\mathcal{W}$ and F are replaced by $\mathcal{S}$ and G, leading to the local constancy of G (and F) on a dense open set in $\mathcal{S}$ (and, respectively, in $\mathcal{B}_x \smallsetminus \{x\}$). Since the union of the latter sets over all x is obviously dense in $\mathcal{W}$, our claim follows. □

We have the following well-known consequence of the inverse mapping theorem combined with the Gauss lemma for submanifolds.

Lemma 3 *Given a compact submanifold* $\mathcal{M}'$ *of a Riemannian manifold* $\mathcal{M}$, *every sufficiently small* $\delta \in (0, \infty)$ *has the following properties:*

a. *The normal exponential mapping restricted to the radius* δ *open-disk subbundle* $\mathcal{N}_\delta$ *of the normal bundle of* $\mathcal{M}'$ *constitutes a diffeomorphism* $\mathrm{Exp}^\perp : \mathcal{N}_\delta \to \mathcal{M}'_\delta$ *onto the open submanifold* $\mathcal{M}'_\delta$ *of* $\mathcal{M}$ *equal to the preimage of* $[0, \delta)$ *under the function* $\mathrm{dist}(\mathcal{M}', \cdot)$ *of metric distance from* $\mathcal{M}'$.
b. *Each* $x \in \mathcal{M}'_\delta$ *has a unique point* $y \in \mathcal{M}'$ *nearest to* x, *which is simultaneously the unique point* y *of* $\mathcal{M}'$ *joined to* x *by a geodesic in* $\mathcal{M}'_\delta$ *normal to* $\mathcal{M}'$ *at* y, *and the resulting assignment* $\mathcal{M}'_\delta \ni x \mapsto y \in \mathcal{M}'$ *coincides with the composition of the inverse diffeomorphism of* $\mathrm{Exp}^\perp : \mathcal{N}_\delta \to \mathcal{M}'_\delta$ *followed by the normal bundle projection* $\mathcal{N}_\delta \to \mathcal{M}'$.
c. *The* $\mathrm{Exp}^\perp$ *images of length* δ *radial line segments emanating from the zero vectors in the fibers of* $\mathcal{N}_\delta$ *coincide with the length* δ *minimizing geodesic segments in* $\mathcal{M}'_\delta$ *emanating from* $\mathcal{M}'$ *and normal to* $\mathcal{M}'$. *They are also normal to*

all the levels of $\mathrm{dist}(\mathcal{M}', \cdot)$ *in* $\mathcal{M}'_\delta$ *and realize the minimum distance between any two such levels within* $\mathcal{M}'_\delta$.

Lemma 4 *In a complete metric space, any countable union of closed sets with empty interiors has an empty interior.*

Proof This is Baire's theorem [9, p. 187] stating, equivalently, that the intersection of countably many dense open subsets is dense. □

3 Free Abelian Groups

The following well-known facts, cf. [1, p. 2], are gathered here for easy reference.

For a finitely generated Abelian group G, being torsion-free amounts to being free, in the sense of having a $\mathbb{Z}$-basis, by which one means an ordered n-tuple $e_1, \ldots, e_n$ of elements of G such that every $x \in G$ can be uniquely expressed as an integer combination of $e_1, \ldots, e_n$. The integer $n \geq 0$, also denoted by $\dim_{\mathbb{Z}} G$, is an algebraic invariant of G, called its *rank*, or *Betti number*, or $\mathbb{Z}$*-dimension*.

Any finitely generated Abelian group G is isomorphic to the direct sum of its (necessarily finite) torsion subgroup S and the free group G/S. We then set $\dim_{\mathbb{Z}} G = \dim_{\mathbb{Z}}[G/S]$. A subgroup G' (or, a homomorphic image G') of such G, in addition to being again finitely generated and Abelian, also satisfies the inequality $\dim_{\mathbb{Z}} G' \leq \dim_{\mathbb{Z}} G$, strict unless G/G' is finite (or, respectively, the homomorphism in question has a finite kernel).

Lemma 5 *A subgroup* G' *of a finitely generated free Abelian group* G *constitutes a direct summand of* G *if and only if the quotient group* G/G' *is torsion-free.*

In fact, more generally, given a surjective homomorphism $\chi : P \to P'$ between Abelian groups P, P' and elements $x_j, y_a \in P$ (with j, a ranging over finite sets), such that x_j and $\chi(y_a)$ happen to form $\mathbb{Z}$-bases of $\mathrm{Ker}\,\chi$ and, respectively, of P', the system consisting of all x_j and y_a is a $\mathbb{Z}$-basis of P (and so P must be free). This is clear as every element of P' (or, of P) then can be uniquely expressed as an integer combination of $\chi(y_a)$ (or, consequently, of x_j and y_a).

Lemma 6 *For each finitely generated subgroup* G *of the additive group of a finite-dimensional real vector space* $\mathcal{V}$*, the intersection* $G \cap \mathcal{V}'$ *with any vector subspace* $\mathcal{V}' \subseteq \mathcal{V}$ *forms a direct-summand subgroup of* G*. Furthermore, the class of direct-summand subgroups of* G *is closed under intersections, finite or not.*

Both claims are obvious from Lemma 5. Next, we have a straightforward exercise:

Lemma 7 *If normal subgroups* G', G'' *of a group* G *intersect trivially and every* $\gamma' \in G'$ *commutes with every* $\gamma'' \in G''$, *then* $G'G''=\{\gamma'\gamma'' : (\gamma', \gamma'')\in G' \times G''\}$ *is a normal subgroup of* G, *and the assignment* $(\gamma', \gamma'') \mapsto \gamma'\gamma''$ *defines an isomorphism* $G' \times G'' \to G'G''$.

4 Lattices and Vector Subspaces

Throughout this section, $\mathcal{V}$ denotes a fixed finite-dimensional real vector space, and

$$\text{subspaces } \mathcal{V}', \mathcal{V}'' \subseteq \mathcal{V} \text{ with } \mathcal{V} = \mathcal{V}' \oplus \mathcal{V}'' \text{ are called } \textit{complementary} \tag{1}$$

to each other. As usual, we define a (full) *lattice* in $\mathcal{V}$ to be any subgroup L of the additive group of $\mathcal{V}$ generated by a basis of $\mathcal{V}$ (which must consequently also be a $\mathbb{Z}$-basis of L). The quotient Lie group $\mathcal{V}/L$ then is a torus, and we use the term *subtori* when referring to its compact connected Lie subgroups.

Definition 1 Given a lattice L in $\mathcal{V}$, by an L-*subspace* of $\mathcal{V}$, we will mean any vector subspace $\mathcal{V}'$ of $\mathcal{V}$ spanned by $L \cap \mathcal{V}'$. One may equivalently require $\mathcal{V}'$ to be the span of just a subset of L, rather than specifically of $L \cap \mathcal{V}'$.

Lemma 8 *For a lattice* L *in* $\mathcal{V}$, *the parallel distribution on* $\mathcal{V}$ *tangent to any prescribed vector subspace* $\mathcal{V}'$ *projects onto a parallel distribution* D *on the torus group* $\mathcal{V}/L$. *The leaves of* D *must be either all compact or all noncompact, and they are compact if and only if* $\mathcal{V}'$ *is an* L-*subspace, in which case the leaf of* D *through zero is a subtorus of* $\mathcal{V}/L$.

Proof The projectability assertion is obvious from the general fact, here applied to the projection $\mathcal{V} \to \mathcal{V}/L$, that projectability of distributions under covering projections amounts to their deck transformation invariance. The first claim about the leaves of D follows as the leaves are one another's translation images. For the second, let $\mathcal{N}$ be the leaf of D through zero. Requiring $\mathcal{V}'$ to be (or, not to be) an L-subspace makes $L \cap \mathcal{V}'$, by Lemma 6, a direct-summand subgroup of L spanning $\mathcal{V}'$ or, respectively, yields the existence of a nonzero linear functional f on $\mathcal{V}'$, the kernel of which contains $L \cap \mathcal{V}'$. In the former case, $\mathcal{N}$ is a factor of a product-of-tori decomposition of $\mathcal{V}/L$; in the latter, f descends to an unbounded function on $\mathcal{N}$. □

Lemma 9 *Given a lattice* L *in* $\mathcal{V}$, *the span and intersection of any family of* L-*subspaces are* L-*subspaces. The same is true if one replaces the phrase "*L-*subspaces" with "*H-*invariant* L-*subspaces" for any fixed group* H *of linear automorphisms of* $\mathcal{V}$ *sending* L *into itself.*

Proof The assertion about spans follows from the case of two L-subspaces, obvious in turn due to the second sentence of Definition 1. Next, the intersection of the family of subtori in $\mathcal{V}/L$, arising via Lemma 8 from the given family of L-subspaces, constitutes a compact Lie subgroup of $\mathcal{V}/L$, so that it is the union of finitely many cosets of a subtorus $\mathcal{N}$. Since subtori are totally geodesic relative to the translation-invariant flat affine connection on $\mathcal{V}/L$, while the projection $\mathcal{V} \to \mathcal{V}/L$ is locally diffeomorphic, the tangent space of $\mathcal{N}$ at zero equals the intersection of the tangent spaces of the subtori forming the family, and each tangent space corresponds to an L-subspace from our family. The conclusion is now immediate from Lemma 8. □

Remark 2 For a lattice L in $\mathcal{V}$ generated by a basis $e_1, \ldots, e_n$ of $\mathcal{V}$, the translational action of L on $\mathcal{V}$ has an obvious compact *fundamental domain* (a compact subset of $\mathcal{V}$ intersecting all orbits of L): the parallelepiped formed by all the combinations $t_1e_1 + \ldots + t_ne_n$ with $t_1, \ldots, t_n$ ranging over $[0, 1]$.

The next lemma is immediate from the first part of Lemma 6 and the well-known fact [3, Chap. VII, Théorème 2] that lattices in $\mathcal{V}$ are precisely the same as discrete subgroups of $\mathcal{V}$, spanning $\mathcal{V}$.

Lemma 10 *For a lattice L in $\mathcal{V}$, a vector subspace $\mathcal{V}' \subseteq \mathcal{V}$, and $L' = L \cap \mathcal{V}'$:*

a. *L' is a lattice in the vector subspace spanned by it.*
b. *L' constitutes a direct-summand subgroup of L.*

Lemma 11 *Let $\mathcal{W}$ be the rational vector subspace of a finite-dimensional real vector space $\mathcal{V}$, spanned by a fixed lattice L in $\mathcal{V}$. Then, the four sets consisting, respectively, of all:*

i. *L-subspaces $\mathcal{V}'$ of $\mathcal{V}$*
ii. *Direct-summand subgroups L' of L*
iii. *Rational vector subspaces $\mathcal{W}'$ of $\mathcal{W}$*
iv. *Subtori $\mathcal{N}'$ of the torus group $\mathcal{V}/L$, that is, its compact connected Lie subgroups*

stand in mutually consistent, natural bijective correspondences with one another, obtained by declaring $\mathcal{V}'$ to be the real span of both L' and $\mathcal{W}'$ as well as the identity component of the preimage of $\mathcal{N}'$ under the projection homomorphism $\mathcal{V} \to \mathcal{V}/L$. Furthermore, $\mathcal{W}'$ equals $\mathcal{W} \cap \mathcal{V}'$ and, simultaneously, is the rational span of L', while $\mathcal{N}' = \mathcal{V}'/L'$ and $L' = L \cap \mathcal{V}' = L \cap \mathcal{W}'$. Finally, $\dim_{\mathbb{R}} \mathcal{V}' = \dim_{\mathbb{Z}} L' = \dim_{\mathbb{Q}} \mathcal{W}' = \dim \mathcal{N}'$.

"Mutual consistency" means here that the above finite set of bijections is closed under the operations of composition and inverse.

Proof The mappings (ii) →(i) and (iii) →(i), as well as (iv) →(i), defined in the three lines following (iv), are all bijections, with the inverses given by $(L', \mathcal{W}', \mathcal{N}') = (L \cap \mathcal{V}', \mathcal{W} \cap \mathcal{V}', \mathcal{V}'/L')$. Namely, each of the three mappings and their purported inverses takes values in the correct set, and each of the six mapping-inverse compositions is the respective identity. To be specific, the claim about the values follows from Lemma 8 for (iv) →(i) and (i) →(iv) and from Definition 1 and Lemma 6 for (ii) →(i) and (i) →(ii), while it is obvious for (i) →(iii) and, for

(iii) $\to$(i), immediate from Definition 1, since we are free to assume that

$$(L,\, \mathcal{W},\, \mathcal{V}) \;=\; (\mathbb{Z}^n,\, \mathbb{Q}^n,\, \mathbb{R}^n), \qquad \text{where } n = \dim \mathcal{V}, \tag{2}$$

and every rational vector subspace of $\mathbb{Q}^n$ has a basis contained in $\mathbb{Z}^n$. Next, the compositions (ii) $\to$ (i) $\to$ (ii) and (i) $\to$ (ii) $\to$ (i) are the identity mappings—the former due to the fact that $L \cap \operatorname{span}_{\mathbb{R}} L' \subseteq L'$ (which one sees extending a $\mathbb{Z}$-basis of L' to a $\mathbb{Z}$-basis of L)—the opposite inclusion being obvious; the latter, as Definition 1 gives $\mathcal{V}' = \operatorname{span}_{\mathbb{R}}(L \cap \mathcal{V}')$. Similarly for (iii) $\to$ (i) $\to$ (iii) and (i) $\to$ (iii) $\to$ (i), as long as one replaces the letters L and $\mathbb{Z}$ with $\mathcal{W}$ and $\mathbb{Q}$, using (2) and the line following it. Finally, (iv) $\to$ (i) $\to$ (iv) and (i) $\to$ (iv) $\to$ (i) are the identity mappings as a consequence of Lemma 8, and the dimension equalities become obvious if one, again, chooses a $\mathbb{Z}$-basis of L containing a $\mathbb{Z}$-basis of L'. □

In the next theorem, as H is finite, all $A \in H$ must have $\det A = \pm 1$, and so the L-preserving property of H means that $AL = L$ (rather than just $AL \subseteq L$).

Proposition 1 *For a lattice L in a finite-dimensional real vector space $\mathcal{V}$, a finite group H of L-preserving linear automorphisms of $\mathcal{V}$, and an H-invariant L-subspace $\mathcal{V}'$ of $\mathcal{V}$, there exists an H-invariant L-subspace $\mathcal{V}''$ of $\mathcal{V}$, complementary to $\mathcal{V}'$ in the sense of (1).*

Proof Let $\mathcal{W}' = \mathcal{W} \cap \mathcal{V}'$, where $\mathcal{W}$ is the rational span of L (see Lemma 11). Restricted to $\mathcal{W}$, the elements of H act by conjugation on the rational affine space $\mathcal{P}$ of all $\mathbb{Q}$-linear projections $\mathcal{W} \to \mathcal{W}'$ (by which we mean linear operators $\mathcal{W} \to \mathcal{W}'$ equal to the identity on $\mathcal{W}'$). The average of any orbit of the action of H on $\mathcal{P}$ is an H-invariant projection $\mathcal{W} \to \mathcal{W}'$ with a kernel $\mathcal{W}''$ corresponding via Lemma 11 to our required $\mathcal{V}''$. □

Corollary 1 *If $L, \mathcal{V}, H$ satisfy the hypotheses of Proposition 1, then every nonzero H-invariant L-subspace $\mathcal{V}_0'$ of $\mathcal{V}$ can be decomposed into a direct sum of one or more nonzero H-invariant L-subspaces, each of which is minimal in the sense of not containing any further nonzero proper H-invariant L-subspace.*

Proof Induction on the possible values of $\dim \mathcal{V}_0'$. The case $\dim \mathcal{V}_0' = 1$ is trivial. Assuming the claim true for subspaces of dimensions less than $\dim \mathcal{V}_0'$, along with non-minimality of $\mathcal{V}_0'$, we fix a nonzero proper H-invariant L-subspace $\mathcal{V}'$ of $\mathcal{V}$, contained in $\mathcal{V}_0'$, and choose an H-invariant complement $\mathcal{V}''$ of $\mathcal{V}'$, guaranteed to exist by Proposition 1. Since $\mathcal{V}''$ intersects every coset of $\mathcal{V}'$ in $\mathcal{V}$, including cosets within $\mathcal{V}_0'$, the subspace $\mathcal{V}_0' \cap \mathcal{V}''$ is an H-invariant complement of $\mathcal{V}'$ in $\mathcal{V}_0'$, as well as an L-subspace (due to Lemma 9). We may now apply the inductive assumption to both $\mathcal{V}'$ and $\mathcal{V}_0' \cap \mathcal{V}''$. □

For geometric consequences of Corollary 1, see the end of Sect. 9, where we also point out that a decomposition into minimal summands is in general nonunique.

Given a lattice L in a finite-dimensional real vector space $\mathcal{V}$ and an L-subspace $\mathcal{V}'$ of $\mathcal{V}$, the restriction to L of the quotient-space projection $\mathcal{V} \to \mathcal{V}/\mathcal{V}'$ has the kernel $L' = L \cap \mathcal{V}'$, and so it descends to an injective group homomorphism $L/L' \to \mathcal{V}/\mathcal{V}'$, the image of which is a (full) lattice in an $\mathcal{V}/\mathcal{V}'$ (which follows if one uses a $\mathbb{Z}$-basis of L containing a $\mathbb{Z}$-basis of L'). From now on, we will treat L/L' as a subset of $\mathcal{V}/\mathcal{V}'$. The discreteness of the lattice $L/L' \subseteq \mathcal{V}/\mathcal{V}'$ clearly implies the existence of an open subset $\mathcal{U}'$ of $\mathcal{V}$, containing $\mathcal{V}'$ and forming a union of cosets of $\mathcal{V}'$, such that $L \cap \mathcal{U}' = L'$.

5 Affine Spaces

In this section, all the affine and vector spaces are real and finite dimensional, we denote by $\mathrm{End}\mathcal{V}$ the space of linear endomorphisms of a given real vector space $\mathcal{V}$, and scalars stand for the corresponding multiples of identity, so that the identity itself becomes $1 \in \mathrm{End}\,\mathcal{V}$.

For an affine space $\mathcal{E}$ with the translation vector space $\mathcal{V}$, let $\mathrm{Aut}\,\mathcal{E}$ be the group of all affine transformations (automorphisms) of $\mathcal{E}$ and $\mathrm{Aff}\,\mathcal{E}$ the set of all (possibly nonbijective) affine mappings $\mathcal{E} \to \mathcal{E}$. We have the inclusions $\mathrm{Aut}\,\mathcal{E} \subseteq \mathrm{Aff}\,\mathcal{E}$ and $\mathcal{V} \subseteq \mathrm{Aut}\,\mathcal{E}$, the latter expressing the fact that $\mathrm{Aut}\,\mathcal{E}$ contains the normal subgroup consisting of all translations. Any vector subspace $\mathcal{V}'$ of $\mathcal{V}$ gives rise to a foliation of $\mathcal{E}$, with the leaves formed by affine subspaces $\mathcal{E}'$ parallel to $\mathcal{V}'$, meaning

$$\text{the orbits of the translational action of } \mathcal{V}' \text{ on } \mathcal{E}, \tag{3}$$

which will also be referred to as the *cosets* of $\mathcal{V}'$ in $\mathcal{E}$. The resulting leaf (quotient) space $\mathcal{E}/\mathcal{V}'$ constitutes an affine space having the translation vector space $\mathcal{V}/\mathcal{V}'$. Clearly, for cosets $\mathcal{E}'$ and $\mathcal{E}''$ of vector subspaces $\mathcal{V}', \mathcal{V}''$ in a vector space $\mathcal{V}$,

$$\text{the complementarity condition (1) implies that } \mathcal{E}' \cap \mathcal{E}'' \text{ is a one-point set.} \tag{4}$$

A fixed inner product in $\mathcal{V}$ turns $\mathcal{E}$ into a *Euclidean affine space*, with the isometry group $\mathrm{Iso}\,\mathcal{E} \subseteq \mathrm{Aut}\,\mathcal{E}$. If $\delta \in (0, \infty)$, we define the *δ-neighborhood* of an affine subspace $\mathcal{E}'$ of $\mathcal{E}$ to be the set of points in $\mathcal{E}$ lying at distances less that δ from $\mathcal{E}'$. Clearly, the δ-neighborhood of $\mathcal{E}'$ is a union of cosets of a vector subspace $\mathcal{V}'$ of $\mathcal{V}$ (one of them being $\mathcal{E}'$ itself), as well as the preimage, under the projection $\mathcal{E} \to \mathcal{E}/\mathcal{V}'$, of the radius δ open ball centered at the point $\mathcal{E}'$ in the quotient Euclidean affine space $\mathcal{E}/\mathcal{V}'$ (for the obvious inner product on $\mathcal{V}/\mathcal{V}'$).

Given a Euclidean affine space $\mathcal{E}$ with the translation vector space $\mathcal{V}$ and an affine subspace $\mathcal{E}' \subseteq \mathcal{E}$ parallel, as in (3), to a vector subspace $\mathcal{V}' \subseteq \mathcal{V}$, (affine) self-isometries ζ of $\mathcal{E}$ such that $\zeta(x) = x$ for all $x \in \mathcal{E}'$ are in an obvious one-to-one correspondence with linear self-isometries A of the orthogonal complement of $\mathcal{V}'$. In this case, for easy later reference (in the proof of Lemma 14),

$$\text{we will call } \zeta \text{ the affine extension of } A \text{ centered on } \mathcal{E}'. \tag{5}$$

Definition 2 In an affine space $\mathcal{E}$ having the translation vector space $\mathcal{V}$, given an affine mapping $\gamma \in \mathrm{Aff}\,\mathcal{E}$ with the linear part $A \in \mathrm{End}\,\mathcal{V}$, we define the *translational-part coset* of γ to be the subset $b + \hat{\mathcal{V}}$ of $\mathcal{V}$, where $\hat{\mathcal{V}}$ denotes the image of $A - 1$ and $b \in \mathcal{V}$ is the translational part of γ relative to a fixed origin $o \in \mathcal{E}$, in the sense that $\gamma(o + v) = o + Av + b$ for all $v \in \mathcal{V}$. The coset $b + \hat{\mathcal{V}}$ is clearly independent of the choice of an origin o, as a new origin $o + w$ results in the replacement of b with $b + (A - 1)w$.

For an affine transformation $\gamma \in \mathrm{Aut}\,\mathcal{E}$ of an affine space $\mathcal{E}$ with the translation vector space $\mathcal{V}$, and a vector subspace $\mathcal{V}'$ of $\mathcal{V}$, consider this condition:

$$\text{the linear part } A \text{ of } \gamma \text{ leaves } \mathcal{V}' \text{ invariant and descends to the identity transformation of } \mathcal{V}/\mathcal{V}', \text{that is, } (A-1)(\mathcal{V}) \subseteq \mathcal{V}'. \tag{6}$$

Lemma 12 *If, in Lemma 3, $\mathcal{M}'$ is a compact leaf of a parallel distribution D on a complete flat Riemannian manifold $\mathcal{M}$, we get the following additional conclusions:*

a. *Every level of* $\mathrm{dist}(\mathcal{M}', \cdot)$ *in $\mathcal{M}'_\delta$, and $\mathcal{M}'_\delta$ itself, is a union of leaves of D.*
b. *Restrictions of $\mathcal{M}'_\delta \ni x \mapsto y \in \mathcal{M}'$ to leaves of D in $\mathcal{M}'_\delta$ are locally isometric.*
c. *The local inverses of all the above locally isometric restrictions correspond via the diffeomorphism* $\mathrm{Exp}^\perp$ *to all local sections of the normal bundle of $\mathcal{M}'$ obtained by restricting to $\mathcal{M}'$ local parallel vector fields of lengths $r \in [0, \delta)$ that are tangent to $\mathcal{M}$ and normal to $\mathcal{M}'$, with r equal to the value of* $\mathrm{dist}(\mathcal{M}', \cdot)$ *on the leaf in question.*

This trivially follows from the fact the pullback of D to the Euclidean affine space $\mathcal{E}$ constituting the Riemannian universal covering space of $\mathcal{M}$ is a distribution whose integral manifolds are the affine subspaces parallel to $\mathcal{V}'$, in the sense of (3), for some vector subspace $\mathcal{V}'$ of the translation vector space $\mathcal{V}$ of $\mathcal{E}$.

6 Bieberbach Groups and Flat Manifolds

Let $\mathcal{E}$ be a Euclidean affine n-space with the translation vector space $\mathcal{V}$. By a *Bieberbach group* [4, p. 4, Definition 1.7] in $\mathcal{E}$, one means any torsion-free discrete subgroup Π of $\mathrm{Iso}\,\mathcal{E}$ for which there exists a compact fundamental domain (Remark 2). Using the linear-part homomorphism $\alpha : \mathrm{Aut}\,\mathcal{E} \to \mathrm{Aut}\,\mathcal{V} \cong \mathrm{GL}(n, \mathbb{R})$, one defines the *lattice subgroup* L of Π and its *holonomy group* $H \subseteq \mathrm{Iso}\,\mathcal{V} \cong \mathrm{O}(n)$ by

$$L = \Pi \cap \mathcal{V}, \qquad H = \alpha(\Pi). \tag{7}$$

Thus, L is the set of all translations lying in Π (which also makes it the kernel of the restriction $\alpha : \Pi \to H$), and H consists of the linear parts of elements of Π. Note that $L \subseteq \mathcal{V}$ is a (full) lattice in the usual sense [4, p. 17, Theorem 3.1(ii)], as defined in Sect. 4. The relations involving Π, L, and H are conveniently summarized by the short exact sequence

$$L \to \Pi \to H, \quad \text{where the arrows are the inclusion homomorphism and } \alpha. \tag{8}$$

As the normal subgroup L of Π is Abelian, the action of Π on L by conjugation descends to an action on L of the quotient group Π/L, identified via (8) with H.

This last action clearly coincides with the ordinary linear action of H on $\mathcal{V}$, restricted to the lattice $L \subseteq \mathcal{V}$, and so L is H-invariant. (9)

Remark 3 The action of a Bieberbach group Π on the Euclidean affine space $\mathcal{E}$ being always free and properly discontinuous [4, p. 3, Proposition 1.1], the quotient $\mathcal{M} = \mathcal{E}/\Pi$, with the projected metric, forms a compact flat Riemannian manifold, while H must be finite [4, p. 17, Theorem 3.1(i)].

Remark 4 The assignment of $\mathcal{M} = \mathcal{E}/\Pi$ to Π establishes a well-known bijective correspondence [4, p. 65, Theorem 5.4] between equivalence classes of Bieberbach groups and isometry types of compact flat Riemannian manifolds. Bieberbach groups Π and $\hat{\Pi}$ in Euclidean affine spaces $\mathcal{E}$ and $\hat{\mathcal{E}}$ are called *equivalent* here if some affine isometry $\mathcal{E} \to \hat{\mathcal{E}}$ conjugates Π onto $\hat{\Pi}$. Furthermore, Π and H in (8) serve as the fundamental and holonomy groups of $\mathcal{M}$, while Π also acts via deck tranformations on the Riemannian universal covering space of $\mathcal{M}$, isometrically identified with $\mathcal{E}$.

Since the lattice subgroup L of the fundamental group Π of $\mathcal{M}$ gives rise to a covering projection $\mathcal{E}/L \to \mathcal{M}$, Lemmas 8 and 9 combined with Remark 4 have the following obvious consequence.

Corollary 2 *In any compact flat Riemannian manifold, the class of parallel distributions with compact leaves is closed under spans and intersections.*

7 Lattice-Reducibility

A Bieberbach group Π in a Euclidean affine space $\mathcal{E}$ or, the compact flat Riemannian manifold $\mathcal{M} = \mathcal{E}/\Pi$ corresponding to Π, in the sense of Remark 4, will be called *lattice-reducible* if, for $\mathcal{V}$, H and L associated with $\mathcal{E}$, and Π as in Sect. 6, there exists $\mathcal{V}'$ such that

$$\mathcal{V}' \text{ is a nonzero proper } H\text{-invariant } L\text{-subspace of } \mathcal{V}. \tag{10}$$

(See Definition 1.) To emphasize the role of $\mathcal{V}'$ in (10), we also say that

$$\text{the lattice-reducibility condition (10) holds for } (\mathcal{V}, H, L, \mathcal{V}'). \tag{11}$$

As shown by Hiss and Szczepański [10], *every compact flat Riemannian manifold of dimension greater than one is lattice-reducible*. For details, see the appendix.

Given a Bieberbach group Π in a Euclidean affine space $\mathcal{E}$ and an affine subspace $\mathcal{E}'$ of $\mathcal{E}$ parallel, as in (3), to a vector subspace $\mathcal{V}'$ of its translation space $\mathcal{V}$, satisfying (10)—(11), we denote by Σ' the *stabilizer group of* $\mathcal{E}'$ *in* Π, so that

$$\Sigma' \text{ consists of all the elements of } \Pi \text{ mapping } \mathcal{E}' \text{ onto itself.} \tag{12}$$

Let $\gamma \in \Pi$. As the foliation of $\mathcal{E}$ formed by the cosets of $\mathcal{V}'$ is Π-invariant, cf. (10), it follows that

$$\gamma \in \Sigma' \text{ if and only if } \gamma(\mathcal{E}') \text{ intersects } \mathcal{E}'. \tag{13}$$

Theorem 1 *For a lattice-reducible Bieberbach group* Π *in a Euclidean affine space* $\mathcal{E}$ *and a vector subspace* $\mathcal{V}'$ *of* $\mathcal{V}$ *with (10), the following three conclusions necessarily hold:*

i. *The affine subspaces of dimension* $\dim \mathcal{V}'$ *in* $\mathcal{E}$*, parallel to* $\mathcal{V}'$ *in the sense of (3), are the leaves of a foliation* $F_{\mathcal{E}}$ *on* $\mathcal{E}$*, projectable under the covering projections* $\mathrm{pr} : \mathcal{E} \to \mathcal{M} = \mathcal{E}/\Pi$ *and* $\mathcal{E} \to \mathcal{T} = \mathcal{E}/L$ *onto foliations* $F_{\mathcal{M}}$ *of* $\mathcal{M}$ *and* $F_{\mathcal{T}}$ *of the torus* $\mathcal{T} = \mathcal{E}/L$*, both of which have compact totally geodesic leaves, tangent to a parallel distribution.*
ii. *The leaves* $\mathcal{M}'$ *of* $F_{\mathcal{M}}$ *coincide with the* pr*-images of the leaves* $\mathcal{E}'$ *of* $F_{\mathcal{E}}$*, and the restrictions* $\mathrm{pr} : \mathcal{E}' \to \mathcal{M}'$ *are covering projections. The same remains true if one replaces* $\mathcal{M}$ *and* pr *with* $\mathcal{T}$ *and the projection* $\mathcal{E} \to \mathcal{T}$*. Any such* $\mathcal{M}'$*, being a compact flat Riemannian manifold, corresponds via Remark 4 to a Bieberbach group* Π' *in the Euclidean affine space* $\mathcal{E}'$*. For* L'*,* H' *appearing in the* $\mathcal{M}'$*-analog* $L' \to \Pi' \to H'$ *of (8), with* Σ' *defined by (12):*
   a. Π' *consists of the restrictions to* $\mathcal{E}'$ *of all the elements of* Σ'*.*
   b. H' *is formed by the restrictions to* $\mathcal{V}'$ *of the linear parts of elements of* Σ'*.*
   c. $L' = \Pi' \cap \mathcal{V}'$*, as in (7), and* $L \cap \mathcal{V}' \subseteq L'$*.*
iii. *The restriction homomorphism* $\Sigma' \to \Pi'$ *of (ii-a) is an isomorphism.*

The last inclusion of (ii-c) may be proper; see the end of Sect. 15.

We chose to format the proof Theorem 1 as the whole next section, since some parts of it are of independent interest, and can in this way be more comfortably cited later in the text (Sects. 10–11, 13–14, 16).

8 Proof of Theorem 1

The projectability of the foliation $F_{\mathcal{E}}$ under both covering projections $\mathrm{pr} : \mathcal{E} \to \mathcal{M}$ and $\mathcal{E} \to \mathcal{T}$ follows as a trivial consequence from the fact that, due to the H-invariance of $\mathcal{V}'$:

$$F_{\mathcal{E}} \text{ is } \Pi\text{-invariant and, obviously, } L\text{-invariant,} \tag{14}$$

while Lemma 1(ii) implies the integrability of the image distribution. Next,

pr is the composition $\mathcal{E} \to \mathcal{T} \to \mathcal{M}$ of two mappings: the universal-covering projection of the flat torus $\mathcal{T} = \mathcal{E}/L$, and the quotient projection for the action of Π on $\mathcal{T}$, (15)

the latter action clearly becoming free if one replaces Π with $\Pi/L \cong H$. Both factor mappings, $\mathcal{E} \to \mathcal{T}$ and $\mathcal{T} \to \mathcal{M}$, are covering projections—the first since L is a lattice in $\mathcal{V}$ and the second due to Remark 1(a). Parts (iii)–(iv) of Lemma 1, along with Lemma 8, may now be applied to the foliations $F_{\mathcal{T}}$ and $F_{\mathcal{M}}$ of the torus $\mathcal{T}$ and of $\mathcal{M}$ obtained as projections of $F_{\mathcal{E}}$, proving the last (compact-leaves) claim of (i), as well as the first two sentences of (ii).

We now fix a leaf $\mathcal{E}'$ of $F_{\mathcal{E}}$ and choose a leaf $\mathcal{M}'$ of $F_{\mathcal{M}}$ containing $\mathrm{pr}(\mathcal{E}')$, cf. Lemma 1(i). It follows that

$$\mathrm{pr} : \mathcal{E}' \to \mathcal{M}' \text{ is a (surjective) covering projection,} \tag{16}$$

since (15) decomposes $\mathrm{pr} : \mathcal{E}' \to \mathcal{M}'$ into the composition $\mathcal{E}' \to \mathcal{T}' \to \mathcal{M}'$, in which the first mapping is the universal-covering projection of the torus $\mathcal{T}' = \mathcal{E}'/L'$, and the second one must be a covering due to Remark 1(b).

Two points of $\mathcal{E}'$ have the same pr-image if and only if one is transformed into the other by an element of the group Π' described in assertion (ii-a); namely, the "only if" part follows since, given $x, y \in \mathcal{E}'$ with $\mathrm{pr}(x) = \mathrm{pr}(y)$ in $\mathcal{M} = \mathcal{E}/\Pi$, the element of Π sending x to y must lie in Σ' by (13). Furthermore, Π' acts on $\mathcal{E}'$ freely since Π does so on $\mathcal{E}$ (Remark 3). Thus, Π' coincides with the deck transformation group for the universal covering projection (16) and satisfies (ii-a). Next, (ii-b) and (ii-c) are consequences of the definitions of H' and L'.

Finally, (iii) follows since nontrivial elements of Σ', being fixed-point free (Remark 3), have nontrivial restrictions to $\mathcal{E}'$.

9 Geometric Consequences of Lemma 8 and Proposition 1

Hiss and Szczepański's result mentioned in the Introduction, combined with Remark 4, Proposition 1, and Theorem 1, has the following immediate consequence.

Theorem 2 *Every compact flat Riemannian manifold* $\mathcal{M}$ *of dimension* $n \geq 2$ *admits two proper parallel distributions* D *and* $\hat{D}$ *with compact leaves, which are complementary to each other in the sense that* $T\mathcal{M} = D \oplus \hat{D}$.

Theorem 1(i), Lemma 8, and Corollary 1 also easily imply that the tangent bundle $T\mathcal{M}$ of any compact flat Riemannian manifold $\mathcal{M}$ admits a maximal direct-sum decomposition into parallel subbundles (distributions) with compact leaves, maximality meaning that none of the summand subbundles can be further decomposed in the same manner. Hiss and Szczepański's result [10] guarantees that, unless $\dim \mathcal{M} < 2$, at least two such summand distributions are present.

Decompositions just mentioned may be quite far from unique: when $\mathcal{M}$ is an n-torus, they stand in a bijective correspondence with decompositions of a n-dimensional rational vector space into a direct sum of lines. This is why one probably should not expect them to have interesting general properties.

10 Geometries of Individual Leaves

Throughout this section, we adopt the assumptions and notation of Theorem 1. The Π-invariance of the foliation $F_{\mathcal{E}}$, cf. (14), trivially gives rise to the obvious:

$$\text{isometric action of } \Pi \text{ on the quotient Euclidean affine space } \mathcal{E}/\mathcal{V}' \tag{17}$$

(i.e., on the leaf space of $F_{\mathcal{E}}$, whose points coincide with the affine subspaces $\mathcal{E}'$ of $\mathcal{E}$ parallel to $\mathcal{V}'$). Whenever $\mathcal{E}' \in \mathcal{E}/\mathcal{V}'$ is fixed, its stabilizer group Σ' in (12) obviously coincides with the isotropy group of $\mathcal{E}'$ for (17). The action (17) is not effective, as the kernel of the corresponding homomorphism $\Pi \to \mathrm{Iso}\,[\mathcal{E}/\mathcal{V}']$ clearly contains the group $L' = L \cap \mathcal{V}'$ forming a lattice in $\mathcal{V}'$, cf. Definition 1 and Lemma 10(a). Now, the H-invariance of L—see (9)—combined with the H-invariance of $\mathcal{V}'$ shows that $L' = L \cap \mathcal{V}'$ is a normal subgroup of Π, which leads to a further homomorphism $\Pi/L' \to \mathrm{Iso}\,[\mathcal{E}/\mathcal{V}']$ (still in general noninjective, cf. (24) below). Let pr again stand for the covering projection $\mathcal{E} \to \mathcal{M} = \mathcal{E}/\Pi$.

Given $\mathcal{E}' \in \mathcal{E}/\mathcal{V}'$ and a vector $v \in \mathcal{V}$ orthogonal to $\mathcal{V}'$, we set $\mathcal{M}'_v = \mathrm{pr}(\mathcal{E}'+v)$, so that, according to (16), $\mathcal{M}'_0 = \mathcal{M}'$. Again by (16)

$$\mathrm{pr} : \mathcal{E}' + v \to \mathcal{M}'_v \text{ is a locally-isometric universal-covering projection.} \tag{18}$$

and $\mathcal{M}'_v$ must be a (compact) leaf of $F_{\mathcal{M}}$. We also

$$\begin{array}{l}\text{choose } \delta \text{ as in Lemma 3 and Lemma 12 for the submanifold}\\ \mathcal{M}' = \mathrm{pr}(\mathcal{E}') \text{ with (16) of the compact flat manifold } \mathcal{M} = \mathcal{E}/\Pi\\ \text{and denote by } \Sigma'_v \subseteq \Pi \text{ the stabilizer group of } \mathcal{E}' + v\text{, cf. (12).}\end{array} \tag{19}$$

Lemma 13 *Under the above hypotheses, for any $\mathcal{E}' \in \mathcal{E}/\mathcal{V}'$, there exists $\delta \in (0, \infty)$ such that, whenever $u \in \mathcal{V}$ is a unit vector orthogonal to $\mathcal{V}'$ and $r, s \in (0, \delta)$, the isometries $\mathcal{E}'+ru \to \mathcal{E}'+su$ and $\mathcal{E}'+ru \to \mathcal{E}'$ acting via translations by the vectors $(s-r)u$ and, respectively, $-ru$, descend under the universal-covering projections (18), with v equal to ru, su, or 0, to an isometry $\mathcal{M}'_{ru} \to \mathcal{M}'_{su}$ or, respectively, a k-fold covering projection $\mathcal{M}'_{ru} \to \mathcal{M}'$, where the integer $k = k(u) \geq 1$ may depend on u—see the end of Sect. 15—but not on r.*

Proof For δ selected in (19) and any $c \in [0, 1]$, let $\psi_c : \mathcal{M}_\delta \to \mathcal{M}_\delta$ correspond, via the $\mathrm{Exp}^\perp$-diffeomorphic identification of Lemma 3(a), to the mapping $\mathcal{N}_\delta \to \mathcal{N}_\delta$ which multiplies vectors normal to $\mathcal{M}'$ by the scalar c. With ϕ denoting our isometry $\mathcal{E}'+ru \to \mathcal{E}'+su$ (or, $\mathcal{E}'+ru \to \mathcal{E}'$), we now have $\mathrm{pr}\circ\phi = \psi_c \circ \mathrm{pr}$ on $\mathcal{E}'+ru$, where $c = s/r$ (or, respectively, $c = 0$) since, given $x \in \mathcal{E}'$, the pr-image of the line segment $\{x + tu : t \in [0, \delta)\}$ in $\mathcal{E}$ is the length δ minimizing geodesic in $\mathcal{M}'_\delta$ emanating from the point $y = \mathrm{pr}(x) \in \mathcal{M}'$ in a direction normal to $\mathcal{M}'$, and $\mathrm{pr}\circ\phi$ sends $x+tu$, in both cases, to $\mathrm{pr}(x+ctu) = \psi_c(\mathrm{pr}(x+tu))$. The pr-image of $\phi(z)$, for any $z \in \mathcal{E}'+ru$, thus depends only on $\mathrm{pr}(z)$ (by being its ψ_c-image), and so both original isometries ϕ descend to (necessarily locally isometric) mappings $\mathcal{M}'_{ru} \to \mathcal{M}'_{su}$ and $\mathcal{M}'_{ru} \to \mathcal{M}'$, which constitute finite coverings (Remark 1(b)). The former is also bijective, its inverse arising when one switches r and s. As the composition $\mathcal{M}'_{su} \to \mathcal{M}'_{ru} \to \mathcal{M}'$ clearly equals the analogous covering projection $\mathcal{M}'_{su} \to \mathcal{M}'$ (with s rather than r), the coverings $\mathcal{M}'_{ru} \to \mathcal{M}'$ and $\mathcal{M}'_{su} \to \mathcal{M}'$ have the same multiplicity, which completes the proof. □

Remark 5 Replacing δ of (19) with 1/4 times its original value, we can also require it to have the following property: *if $\gamma \in \Pi$ and $x \in \mathcal{E}$ are such that both x and $\gamma(x)$ lie in the δ-neighborhood of $\mathcal{E}'$, defined as in Sect. 5, then $\gamma \in \Sigma'$ for the stabilizer group Σ' of $\mathcal{E}'$ given by* (12). In fact, letting $\mathcal{E}''$ be the leaf of $F_\mathcal{E}$ through x, we see from (14) that its γ-image $\gamma(\mathcal{E}'')$ is also a leaf of $F_\mathcal{E}$, while both leaves are within the distance δ from $\mathcal{E}'$, which yields $\mathrm{dist}(\mathcal{E}'', \gamma(\mathcal{E}'')) < 2\delta$ and so, due to the triangle inequality, $\mathrm{dist}(\mathcal{E}', \gamma(\mathcal{E}')) \leq \mathrm{dist}(\mathcal{E}', \mathcal{E}'') + \mathrm{dist}(\mathcal{E}'', \gamma(\mathcal{E}'')) + \mathrm{dist}(\gamma(\mathcal{E}''), \gamma(\mathcal{E}')) < \delta+2\delta+\delta = 4\delta$. Thus, $x+ru \in \gamma(\mathcal{E}')$ for some $x \in \mathcal{E}'$, some unit vector $u \in \mathcal{V}$ orthogonal $\mathcal{V}'$, and $r = \mathrm{dist}(\mathcal{E}', \gamma(\mathcal{E}')) \in [0, 4\delta)$. Assuming now (19) with δ replaced by 4δ, one gets $r = 0$, that is, $\gamma(\mathcal{E}') = \mathcal{E}'$ and $\gamma \in \Sigma'$. Namely, the pr-image of the curve $[0, 4\delta) \ni t \mapsto x + tu$ is a geodesic in the image of the diffeomorphism $\mathrm{Exp}^\perp$ of Lemma 3(a), which intersects $\mathcal{M}'$ only at $t = 0$, while $\mathcal{M}' = \mathrm{pr}(\mathcal{E}') = \mathrm{pr}(\gamma(\mathcal{E}'))$, since $\mathcal{M} = \mathcal{E}/\Pi$.

Lemma 14 *Let there be given $\mathcal{V}', \mathcal{E}'$ as in Lemma 13, δ having the additional property of Remark 5, any $r \in (0, \delta)$, and any unit vector $u \in \mathcal{V}$ orthogonal to the subspace $\mathcal{V}'$:*

a. *The stabilizer group Σ'_{ru} in (19) does not depend on $r \in (0, \delta)$.*
b. *The linear part of each element of Σ'_{ru} keeps u fixed.*
c. *Σ'_{ru} is a subgroup of Σ'_0 with the finite index $k = k(u) \geq 1$ of Lemma 13.*
d. *$\mathrm{pr} : \mathcal{E} \to \mathcal{M}$ maps the δ-neighborhood $\mathcal{E}_\delta$ of $\mathcal{E}'$ in $\mathcal{E}$ onto $\mathcal{M}'_\delta$ of Lemma 3(a).*

e. *$\mathcal{E}_\delta$ and $\mathcal{M}'_\delta$ are unions of leaves of, respectively, $F_{\mathcal{E}}$ and $F_{\mathcal{M}}$.*
f. *The preimage under* $\mathrm{pr} : \mathcal{E}_\delta \to \mathcal{M}_\delta$ *of the leaf* $\mathcal{M}'_{ru} = \mathrm{pr}(\mathcal{E}' + ru)$ *of* $F_{\mathcal{M}}$ *equals the union of the images* $\gamma(\mathcal{E}' + ru)$ *over all* $\gamma \in \Sigma'_0$.

Proof By (16) and (19), $\mathcal{M}'_v = (\mathcal{E}' + v)/\Pi'_v$, if one lets Π'_v denote the image of Σ'_v under the injective homomorphism of restriction to $\mathcal{E}'$, cf. Theorem 1(iii). Fixing $s \in [0, \delta)$ and $r \in (0, \delta)$, we therefore conclude from Lemma 13 and (13) that, whenever $x \in \mathcal{E}' + ru$ and $\gamma \in \Sigma'_{ru}$, there exists $\hat\gamma \in \Sigma'_{su}$ satisfying the condition:

$$\gamma(x) + v = \hat\gamma(x + v), \quad \text{where } v = (s - r)u, \quad \text{and } \hat\gamma = \gamma \text{ when } s = r, \tag{20}$$

the last clause being obvious since $\gamma, \hat\gamma \in \Pi$ and the action of Π is free. With u and γ fixed as well, for each given $\hat\gamma \in \Sigma'_{su}$, the set of all $x \in \mathcal{E}' + ru$ having Property (20) is closed in $\mathcal{E}' + ru$, while, as we just saw, the union of these sets over all $\hat\gamma \in \Sigma'_{su}$ equals $\mathcal{E}' + ru$. Thus, by Baire's theorem (Lemma 4), some $\hat\gamma \in \Sigma'_{su}$ satisfies (20) with all x from some non-empty open subset of $\mathcal{E}' + ru$, and hence—by real-analyticity—for all $x \in \mathcal{E}' + ru$. In terms of the translation τ_v by the vector v, we consequently have $\hat\gamma = \tau_v \circ \gamma \circ \tau_v^{-1}$ on $\mathcal{E}' + su$, so that, due to Theorem 1(iii), γ uniquely determines $\hat\gamma$, while the assignment $\gamma \mapsto \hat\gamma$ is a homomorphism $\Sigma'_{ru} \to \Sigma'_{su} \subseteq \Pi$, and $\zeta = \hat\gamma \circ \tau_v \circ \gamma^{-1} \circ \tau_v^{-1}$ equals the identity on $\mathcal{E}' + su$. If we now allow s to vary from r to 0, the resulting curve $s \mapsto \zeta$ consists of affine extensions, defined as in (5), of linear self-isometries of the orthogonal complement of $\mathcal{V}'$, and $\hat\gamma = \zeta \circ \tau_v \circ \gamma \circ \tau_v^{-1}$ on $\mathcal{E}$. As Π is discrete, the curve $s \mapsto \hat\gamma \in \Pi$, with $v = (s - r)u$, must be constant and can be evaluated by setting $s = r$ (or, $v = 0$). Thus, $\hat\gamma = \gamma$ on $\mathcal{E}$ from the last clause of (20), and so $\Sigma'_{ru} \subseteq \Sigma'_{su}$. For $s > 0$, switching r with s, we get the opposite inclusion, and (a) follows. Also, taking the linear part of the resulting relation $\gamma = \zeta \circ \tau_v \circ \gamma \circ \tau_v^{-1}$, we see that ζ equals the identity, for all s. Hence, $\gamma = \tau_v \circ \gamma \circ \tau_v^{-1}$, that is, γ commutes with τ_v which, by (9), amounts to (b). Setting $s = 0$, we obtain the first part of (c): $\Sigma'_{ru} \subseteq \Sigma'_0$. Assertion (d) is clear as pr, being locally isometric, maps line segments onto geodesic segments. Lemma 12(a) for $D = F_{\mathcal{M}}$ gives (e). With $\mathrm{pr} : \mathcal{E} \to \mathcal{M} = \mathcal{E}/\Pi$ in Theorem 1(i), the additional property of δ (Remark 5) yields (f). Finally, for $k = k(u)$, the geodesic $[0, r] \ni t \mapsto \mathrm{pr}(x + tu)$, normal to $\mathcal{M}'$ at $y = \mathrm{pr}(x)$, is one of k such geodesics $[0, r] \ni t \mapsto \mathrm{pr}(x + tw)$, joining y to points of its preimage under the projection $\mathcal{M}'_{ru} \to \mathcal{M}'$ of Lemma 13, where w ranges over a k-element set $\mathcal{R}$ of unit vectors in $\mathcal{V}$, orthogonal to $\mathcal{V}'$. The union of the corresponding subset $C = \{\mathcal{E}' + rw : w \in \mathcal{R}\}$ of the leaf space of $F_{\mathcal{E}}$ equals the preimage in (f)—and hence an orbit for the action of Σ'_0—as every leaf in the preimage contains a point nearest x. Due to the already-established inclusion $\Sigma'_{ru} \subseteq \Sigma'_0$ and (19), Σ'_{ru} is the isotropy group of $\mathcal{E}' + ru$ relative to the transitive action of Σ'_0 on C, and so k, the cardinality of C, equals the index of Σ'_{ru} in Σ'_0, which proves the second part of (c). □

11 The Generic Stabilizer Group

Given a Bieberbach group Π in a Euclidean affine space $\mathcal{E}$ with the translation vector space $\mathcal{V}$, let us fix a vector subspace $\mathcal{V}'$ of $\mathcal{V}$ satisfying (10) for L, H introduced in (8). As long as $\dim \mathcal{E} \geq 2$, such $\mathcal{V}'$ always exists (Sect. 7). An element $\mathcal{E}'$ of $\mathcal{E}/\mathcal{V}'$, that is, a coset of $\mathcal{V}'$ in $\mathcal{E}$, will be called *generic* if its stabilizer group $\Sigma' \subseteq \Pi$, defined by (12), equals

$$\text{the kernel of the homomorphism } \Pi \to \mathrm{Iso}\,[\mathcal{E}/\mathcal{V}'] \text{ corresponding to (17).} \tag{21}$$

The pr-images of generic cosets of $\mathcal{V}'$ will be called generic leaves of $F_{\mathcal{M}}$.

Still, using the symbols L, H, and pr for the groups appearing in (7)–(8) and the universal-covering projection $\mathcal{E} \to \mathcal{M} = \mathcal{E}/\Pi$, let us also

denote by $K' \subseteq H$ the normal subgroup consisting of all elements of H that act on the orthogonal complement of $\mathcal{V}'$ as the identity and by $\mathcal{U}'$ the subset of $\mathcal{E}/\mathcal{V}'$ formed by all generic cosets of $\mathcal{V}'$ in $\mathcal{E}$. (22)

Theorem 3 *For Σ' equal to (21), under the assumptions preceding (21), with the notation of (22), one has the following conclusions:*

i. *$\mathcal{U}'$ in (22) constitutes an open dense subset of $\mathcal{E}/\mathcal{V}'$.*
ii. *The normal subgroup Σ' of Π is contained as a finite-index subgroup in the stabilizer group of every $\mathcal{E}' \in \mathcal{E}/\mathcal{V}'$ for the action (17) and equal to this stabilizer group if $\mathcal{E}' \in \mathcal{U}'$.*
iii. *The* pr-*images $\mathcal{M}', \mathcal{M}''$ of any $\mathcal{E}', \mathcal{E}'' \in \mathcal{U}'$ are isometric to each other.*
iv. *If one identifies $\mathcal{E}$ with its translation vector space $\mathcal{V}$ via a choice of an origin, Σ' becomes the set of all the elements of Π having, for K' given by (22), the form*

$$\mathcal{V} \ni x \mapsto Ax + b \in \mathcal{V}, \text{ in which } b \in \mathcal{V}' \text{ and the linear part } A \text{ lies in } K'. \tag{23}$$

v. *Whenever $\mathcal{E}' \in \mathcal{U}'$, the homomorphism which restricts elements of the generic stabilizer group Σ' to $\mathcal{E}'$ is injective, and the resulting isomorphic image Π' of Σ' constitutes a Bieberbach group in the Euclidean affine space $\mathcal{E}'$. The lattice subgroup of Π' and its holonomy group H' are the intersection $L' = L \cap \mathcal{V}'$ and the image H' of the group K' defined in (22) under the injective homomorphism of restriction to $\mathcal{V}'$.*

Proof Lemma 14(a) states that the assumptions of Lemma 2 are satisfied by the Euclidean affine space $\mathcal{W} = \mathcal{E}/\mathcal{V}'$ and the mapping F that sends $\mathcal{E}' \in \mathcal{E}/\mathcal{V}'$ to its stabilizer group Σ' with (12). The assignment $\mathcal{E}' \mapsto \Sigma'$ is thus *locally constant* on some open dense set $\mathcal{U}' \subseteq \mathcal{E}/\mathcal{V}'$. Letting Σ' be the constant value of this assignment assumed on a non-empty connected open subset $\mathcal{W}'$ of $\mathcal{U}'$, and fixing

$\gamma \in \Sigma'$, we obtain $\gamma(\mathcal{E}') = \mathcal{E}'$ for all $\mathcal{E}' \in \mathcal{W}'$, and hence, from real-analyticity, for all $\mathcal{E}' \in \mathcal{E}/\mathcal{V}'$. Thus, our Σ' is contained in the stabilizer group of every $\mathcal{E}' \in \mathcal{E}/\mathcal{V}'$. Since the same applies also to another constant value Σ'' assumed on a non-empty connected open set, $\Sigma'' = \Sigma'$ and the phrase "locally constant" may be replaced with *constant*. By Lemma 14(c), any such Σ' must be a finite-index subgroup of each stabilizer group. As Σ' consists of the elements of Π preserving every $\mathcal{E}' \in \mathcal{U}'$, real-analyticity implies that they preserve all $\mathcal{E}' \in \mathcal{E}/\mathcal{V}'$, and so Σ' coincides with (21), which also shows that Σ' is a normal subgroup of Π, and (i)–(ii) follow.

Next, (iv) holds since the set of all $\gamma \in \operatorname{Aut}\mathcal{E}$ satisfying (6) clearly constitutes a subgroup of $\operatorname{Aut}\mathcal{E}$ containing, as a normal subgroup, the set of all γ with (6) such that translational-part coset of γ (see Definition 2) is contained in $\mathcal{V}'$. To verify this claim, note that the latter set is a normal subgroup, being the kernel of the obvious homomorphism from the subgroup of all γ having the property (6) into the translation subgroup $\mathcal{V}/\mathcal{V}'$ of $\operatorname{Aut}[\mathcal{E}/\mathcal{V}']$. More precisely, γ represented by the pair (A, b) (as in Definition 2) preserves each element of $\mathcal{E}/\mathcal{V}'$ if and only if $Av + b$ differs from v, for every $v \in \mathcal{V}$, by an element of $\mathcal{V}'$ or, equivalently (as one sees setting $v = 0$), $\mathcal{V}'$ contains both b and $(A - 1)(\mathcal{V})$.

Finally, Theorem 1(ii)–(iii) yields (v), while (v) implies (iii) via Remark 4. □

The example provided by a compact flat manifold $\mathcal{M} = \mathcal{E}/\Pi$ which is a Riemannian product $\mathcal{M} = \mathcal{M}' \times \mathcal{M}''$ with $\mathcal{E} = \mathcal{E}' \times \mathcal{E}''$ and $\Pi = \Pi' \times \Pi''$ for two Bieberbach groups Π', Π'' in Euclidean affine spaces $\mathcal{E}', \mathcal{E}''$ having the translation vector spaces $\mathcal{V}', \mathcal{V}''$, while $\mathcal{M}'$ is not a torus, shows that in general

$$\text{an element of } \Pi \text{ acting trivially on } \mathcal{E}/\mathcal{V}' \text{ need not lie in } L'. \tag{24}$$

Namely, the H-invariant subspace $\mathcal{V}' \times \{0\}$ then gives rise to the $\mathcal{M}'$ factor foliation $F_{\mathcal{E}}$ of the product manifold $\mathcal{M}$, and the action of the group $\Pi' \times \{1\}$ on its leaf space is obviously trivial, even though not all elements of $\Pi' \times \{1\}$ are translations.

Lemma 15 *Using any given $\mathcal{E}' \in \mathcal{U}'$ in Theorem 3, where $\mathcal{V}'$ satisfying (10) is fixed, let us identify Σ' with Π' and H' with K' via the isomorphisms $\Sigma' \to \Pi'$ and $K' \to H'$ resulting from Theorem 3(v), which turns Π' and H' into subgroups of Π and H. These subgroups $\Pi' \subseteq \Pi$ and $H' \subseteq H$ do not depend on the choice of $\mathcal{E}' \in \mathcal{U}'$, and neither does the mapping degree $d = |H'|$ of the d-fold covering projection $\mathcal{T}' \to \mathcal{M}' = \mathcal{T}'/H'$ analogous to those mentioned in (15).*

Proof Our claims about Π' and H' are immediate from (21) and (22). □

Any lattice L in the translation vector space $\mathcal{V}$ of a Euclidean affine space $\mathcal{E}$ is, obviously, a Bieberbach group in $\mathcal{E}$. In the case of a fixed vector subspace $\mathcal{V}'$ of $\mathcal{V}$ with (10), all the general facts established about any given Bieberbach group Π in $\mathcal{E}$, the compact flat manifold $\mathcal{M} = \mathcal{E}/\Pi$, and the leaves $\mathcal{M}'$ of $F_{\mathcal{M}}$ (see Theorem 1) thus remain valid for the torus $\mathcal{T} = \mathcal{E}/L$ and the leaves $\mathcal{T}'$ of $F_{\mathcal{T}}$. Every coset of $\mathcal{V}'$ is generic if we declare the lattice L of Π to be the new Bieberbach group.

12 The Leaf Space

We again adopt the assumptions and notation of Theorem 1. Not surprisingly, the leaf space $\mathcal{M}/F_{\mathcal{M}}$ has the following property (discussed below): $\mathcal{M}/F_{\mathcal{M}}$ forms

$$\text{a flat compact orbifold, canonically identified with the quotient of the torus } [\mathcal{E}/\mathcal{V}']/[L \cap \mathcal{V}'] \text{ under the isometric action of the finite group } H. \tag{25}$$

By a *crystallographic group* [13] in a Euclidean affine space, one means a discrete group of isometries having a compact fundamental domain, cf. Remark 2.

Proposition 2 *Under the assumptions listed in the first line of Sect. 11, with* Σ' *denoting the normal subgroup (21) of* Π*, the quotient group* Π/Σ' *acts effectively by isometries on the quotient Euclidean affine space* $\mathcal{E}/\mathcal{V}'$ *and, when identified with a subgroup of* $\operatorname{Iso}[\mathcal{E}/\mathcal{V}']$*, it constitutes a crystallographic group.*

Proof A compact fundamental domain exists since Π/Σ' contains the lattice subgroup L/L' of $\mathcal{V}/\mathcal{V}'$ (see the end of Sect. 4). To verify the discreteness of Π/Σ', suppose that, on the contrary, some sequence $\gamma_k \in \Pi$, $k = 1, 2 \ldots$, has terms representing mutually distinct elements of Π/Σ' which converge in $\operatorname{Iso}[\mathcal{E}/\mathcal{V}']$. As L' is a lattice in $\mathcal{V}'$, fixing $x \in \mathcal{E}$ and suitably choosing $v_k \in L'$, we achieve boundedness of the sequence $\hat{\gamma}_k(x) = \gamma_k(x) + v_k$, while $\hat{\gamma}_k$ represent the same (distinct) elements of Π/Σ' as γ_k. The ensuing convergence of a subsequence of $\hat{\gamma}_k$ contradicts the discreteness of Π. □

The resulting quotient of $\mathcal{E}/\mathcal{V}'$ under the action of Π/Σ' is thus a flat compact orbifold [5], which may clearly be identified both with the leaf space $\mathcal{M}/F_{\mathcal{M}}$, and with the quotient of the torus $[\mathcal{E}/\mathcal{V}']/[L \cap \mathcal{V}']$ mentioned in (25). The latter identification clearly implies the Hausdorff property of the leaf space $\mathcal{M}/F_{\mathcal{M}}$.

On the other hand, for an H-invariant subspace $\mathcal{V}''$ of $\mathcal{V}$ *not* assumed to be an L-subspace, there exists an L-*closure* of $\mathcal{V}''$, meaning the smallest L-subspace $\mathcal{V}'$ of $\mathcal{V}$ containing $\mathcal{V}''$, which is obviously obtained by intersecting all such L-subspaces (Lemma 9). The leaf space $\mathcal{M}/F_{\mathcal{M}}$ corresponding to $\mathcal{V}'$ then forms a natural "Hausdorffization" of the leaf space of $\mathcal{V}''$ and may also be described in terms of Gromov-Hausdorff limits. See the recent paper [2].

13 Intersections of Generic Complementary Leaves

Throughout this section Π is a given Bieberbach group in a Euclidean affine space $\mathcal{E}$ of dimension $n \geq 2$, while $\mathcal{V}', \mathcal{V}''$ are two mutually complementary H-invariant L-subspaces of the translation vector space $\mathcal{V}$ of $\mathcal{E}$, in the sense of (1) and Definition 1, for L and H associated with Π via (7). We also fix generic cosets $\mathcal{E}'$ of $\mathcal{V}'$ and $\mathcal{E}''$ of $\mathcal{V}''$ (see the beginning of Sect. 11), which leads to the analogs $L' \to \Pi' \to H'$ and $L'' \to \Pi'' \to H''$ of (8), described by Theorem 1(ii) and, $\mathcal{E}', \mathcal{E}''$ being generic, Theorem 3(v) yields $L' = L \cap \mathcal{V}'$ and $L'' = L \cap \mathcal{V}''$. Furthermore, for these $\Pi, \Pi', \Pi'', L, L', L'', H, H', H''$

$$\text{the conclusions of Lemma 7 hold if we replace } G \text{ with } \Pi, L \text{ or } H, \tag{26}$$

since so do the assumptions of Lemma 7, provided that one uses Lemma 15 to treat Π' and Π'' (or, H' and H'') as subgroups of Π (or, respectively, H). In fact, (23) and the description of K' in (22) show that all $A \in K'$ (and, among them, the linear parts of all elements of $\Sigma' = \Pi'$) leave invariant both $\mathcal{V}'$ and $\mathcal{V}''$ and act via the identity on the latter. (We have the obvious isomorphic identifications of $\mathcal{V}/\mathcal{V}'$ with $\mathcal{V}''$ on the one hand, and with the orthogonal complement of $\mathcal{V}'$ in $\mathcal{V}$ on the other, while such A descend to the identity automorphism of $\mathcal{V}/\mathcal{V}'$.) The same is clearly the case if one switches the primed symbols with the double-primed ones, while elements of Σ' now commute with those of Σ'' in view of (23). This yields (26) and, consequently, allows us to form

$$\text{the quotient groups } \hat{\Pi} = \Pi/(\Pi'\Pi''), \quad \hat{L} = L/(L'L''), \quad \hat{H} = H/(H'H''). \tag{27}$$

Finally, let $\mathrm{pr} : \mathcal{E} \to \mathcal{M} = \mathcal{E}/\Pi$ and $\mathcal{M}', \mathcal{M}'', \mathcal{T}', \mathcal{T}''$ denote, respectively, the covering projection of Theorem 1(i), the pr-image $\mathcal{M}'$ of $\mathcal{E}'$ (or, $\mathcal{M}''$ of $\mathcal{E}''$), and the tori $\mathcal{E}'/L'$ and $\mathcal{E}''/L''$, contained in the torus $\mathcal{T} = \mathcal{E}/L$ of (15). Note that $\mathcal{M}'$ and $\mathcal{M}''$ are (compact) leaves of the parallel distributions arising, due to Theorem 1(i), on $\mathcal{M} = \mathcal{E}/\Pi$, which itself is a compact flat Riemannian manifold (Remark 3).

For a homology interpretation of parts (a) and (c) below, see Theorem 5.

Theorem 4 *Under the above hypotheses, the following conclusions hold:*

a. *$\mathcal{M}' \cap \mathcal{M}''$, or $\mathcal{T}' \cap \mathcal{T}''$, is a finite subset of $\mathcal{M}$, or $\mathcal{T}$, and stands in a bijective correspondence with the quotient group $\hat{\Pi}$ or, respectively, $\hat{L}$, of (27).*
b. *The projection $\mathcal{T} \to \mathcal{M}$ in (15) maps $\mathcal{T}' \cap \mathcal{T}''$ injectively into $\mathcal{M}' \cap \mathcal{M}''$.*
c. *The cardinality $|\mathcal{M}' \cap \mathcal{M}''|$ of $\mathcal{M}' \cap \mathcal{M}''$ equals $|\mathcal{T}' \cap \mathcal{T}''|$ times $|\hat{H}|$.*
d. *The claim about $\mathcal{T}' \cap \mathcal{T}''$ in (a) remains true whether or not $\mathcal{E}', \mathcal{E}''$ are generic.*
e. *The two bijective correspondences in (a) may be chosen so that, under the resulting identifications, the injective mapping $\mathrm{pr} : \mathcal{T}' \cap \mathcal{T}'' \to \mathcal{M}' \cap \mathcal{M}''$ of (b) coincides with the group homomorphism $\hat{L} \to \hat{\Pi}$ induced by the inclusion $L \to \Pi$.*

Proof We first prove (a) for $\mathcal{M}' \cap \mathcal{M}''$. The finiteness of $\mathcal{M}' \cap \mathcal{M}''$ follows as $\mathcal{M}', \mathcal{M}''$, and hence also $\mathcal{M}' \cap \mathcal{M}''$, are compact totally geodesic submanifolds of $\mathcal{M}$, while $\mathcal{M}' \cap \mathcal{M}''$, non-empty by (4), has $\dim(\mathcal{M}' \cap \mathcal{M}'') = 0$ due to (1). The mapping $\Psi : \Pi \to \mathcal{M}' \cap \mathcal{M}''$ with $\mathrm{pr}(\mathcal{E}' \cap \gamma(\mathcal{E}'')) = \{\Psi(\gamma)\}$ is well defined in view of (4) applied to $\gamma(\mathcal{E}'')$ rather than $\mathcal{E}''$ and clearly takes values in both $\mathcal{M}' = \mathrm{pr}(\mathcal{E}')$ and $\mathcal{M}'' = \mathrm{pr}(\mathcal{E}'') = \mathrm{pr}(\gamma(\mathcal{E}''))$. Surjectivity of Ψ follows: if $\mathrm{pr}(x'') \in \mathcal{M}' \cap \mathcal{M}''$, where $x'' \in \mathcal{E}''$, then, obviously, $\mathrm{pr}(x'') = \mathrm{pr}(x')$ and $x' = \gamma(x'')$ for some $x' \in \mathcal{E}'$

and $\gamma \in \Pi$, so that $x' \in \mathcal{E}' \cap \gamma(\mathcal{E}'')$ and $\mathrm{pr}(x'') = \mathrm{pr}(x')$ equals $\Psi(\gamma)$, the unique element of $\mathrm{pr}(\mathcal{E}' \cap \gamma(\mathcal{E}''))$. Furthermore, Ψ-preimages of elements of $\mathcal{M}' \cap \mathcal{M}''$ are precisely the cosets of the normal subgroup $\Pi'\Pi''$ of Π (which clearly implies (a) for $\mathcal{M}' \cap \mathcal{M}''$). Namely, the left and right cosets coincide, and so elements γ_1, γ_2 of Π lie in the same coset of $\Pi'\Pi''$ if and only if

$$\gamma' \circ \gamma_1 = \gamma_2 \circ \gamma'' \quad \text{for some } \gamma' \in \Pi' \text{ and } \gamma'' \in \Pi''. \tag{28}$$

Now, let γ_1, γ_2 lie in the same coset of $\Pi'\Pi''$. For γ', γ'' with (28), $\gamma'(\mathcal{E}') = \mathcal{E}'$ and $\gamma''(\mathcal{E}'') = \mathcal{E}''$ by the definition (12) of Σ', Σ'' and their identification with Π', Π'' (see above). Thus, $\{\Psi(\gamma_1)\} = \mathrm{pr}(\mathcal{E}' \cap \gamma_1(\mathcal{E}'')) = \mathrm{pr}(\gamma'(\mathcal{E}' \cap \gamma_1(\mathcal{E}''))) = \mathrm{pr}(\gamma'(\mathcal{E}') \cap \gamma'(\gamma_1(\mathcal{E}''))) = \mathrm{pr}(\mathcal{E}' \cap \gamma'(\gamma_1(\mathcal{E}''))) = \mathrm{pr}(\mathcal{E}' \cap \gamma_2(\gamma''(\mathcal{E}'')))$, and so $\{\Psi(\gamma_1)\} = \mathrm{pr}(\mathcal{E}' \cap \gamma_2(\mathcal{E}'')) = \{\Psi(\gamma_2)\}$. Conversely, if $\gamma_1, \gamma_2 \in \Pi$ and $\Psi(\gamma_1) = \Psi(\gamma_2)$, the unique points x_1 of $\mathcal{E}' \cap \gamma_1(\mathcal{E}'')$ and x_2 of $\mathcal{E}' \cap \gamma_2(\mathcal{E}'')$ both lie in the same Π-orbit, and hence, $x_2 = \gamma(x_1)$ with some $\gamma \in \Pi$. For $\gamma' = \gamma$ and $\gamma'' = \gamma_2^{-1} \circ \gamma \circ \gamma_1$, the image $\gamma'(\mathcal{E}')$ (or, $\gamma''(\mathcal{E}'')$) intersects $\mathcal{E}'$ (or, $\mathcal{E}''$), the common point being $x_2 = \gamma(x_1)$ or, respectively, $\gamma_2^{-1}(x_2) = \gamma_2^{-1}(\gamma(x_1))$. From (13), we thus obtain $\gamma' \in \Sigma' = \Pi'$ and $\gamma'' \in \Sigma'' = \Pi''$, which yields (28).

Now, (a) for $\mathcal{T}' \cap \mathcal{T}''$, and (d), follow as special cases; see the end of Sect. 11.

Except for the word "injective" the claim made in (e) is immediate if one uses the mapping $\Psi : \Pi \to \mathcal{M}' \cap \mathcal{M}''$ defined above and its analog $L \to \mathcal{T}' \cap \mathcal{T}''$ obtained by replacing $\Pi, \mathcal{M}', \mathcal{M}''$ and pr with $L, \mathcal{T}', \mathcal{T}''$ and the projection $\mathcal{E} \to \mathcal{T} = \mathcal{E}/L$. This yields (b), injectivity of the homomorphism $\hat{L} \to \hat{\Pi}$ being immediate: if an element of L lies in $\Pi'\Pi''$ (and hence has the form $\gamma' \circ \gamma''$, where $(\gamma', \gamma'') \in \Pi' \times \Pi''$), (23) implies that γ', γ'' are translations with $\gamma' \in L' = L \cap \mathcal{V}'$ and $\gamma'' \in L'' = L \cap \mathcal{V}''$ (see the lines preceding (26)); in other words, $\gamma' \circ \gamma''$ represents zero in $\hat{L}$.

Finally, $\hat{L}$ identified as above with a subgroup of $\hat{\Pi}$ is the kernel of the clearly surjective homomorphism $\hat{\Pi} \to \hat{H}$, induced by $\Pi \to H$ in (8) (which, due to (e), proves (c)). Namely, $\hat{L}$ contains the kernel (the other inclusion being obvious): if the linear part of $\gamma \in \Pi$ lies in $H'H''$, and so equals the linear part of $\gamma' \circ \gamma''$ for some $(\gamma', \gamma'') \in \Pi' \times \Pi''$, then $\gamma = \lambda \circ \gamma' \circ \gamma''$, where $\lambda \in L$. □

14 Leaves and Integral Homology

This section once again employs the assumptions and notation of Theorem 1, with $\dim \mathcal{V} = n$ and $\dim \mathcal{V}' = k$, where $0 < k < n$. As the holonomy group $H \subseteq \mathrm{Iso}\,\mathcal{V} \cong \mathrm{O}(n)$ is finite (Remark 3), $\det(H) \subseteq \{1, -1\}$. In other words, the elements of H have the determinants ± 1. Using the covering projection $\mathcal{T} \to \mathcal{M} = \mathcal{T}/H$, cf. (15) and the line following it, we see that

$$\text{the image of } H \text{ under det equals } \{1\} \text{ if and only if } \mathcal{M} \text{ is orientable.} \tag{29}$$

By Theorem 3(iii), the generic leaves of $F_{\mathcal{M}}$, defined as in the line following (21), are either all orientable or all nonorientable.

Theorem 5 *Let $\mathcal{M}$ be orientable. Then, all the generic leaves $\mathcal{M}'$ of $F_{\mathcal{M}}$ may be oriented so as to represent the same nonzero k-dimensional real homology class $[\mathcal{M}'] \in H_k(\mathcal{M}, \mathbb{R})$, while the first two cardinalities in Theorem 4(c) equal the intersection numbers of the real homology classes $[\mathcal{M}'], [\mathcal{M}'']$, or $[\mathcal{T}'], [\mathcal{T}'']$.*

Proof A fixed orientation of $\mathcal{V}'$, being preserved, due to (22)–(23) and (29), by the generic stabilizer group Σ', gives rise to orientations of all the leaves $\mathcal{T}'$ of $F_{\mathcal{T}}$ and all the generic leaves $\mathcal{M}'$ of $F_{\mathcal{M}}$, so as to make the covering projections $\mathcal{T}' \to \mathcal{M}'$ in the line following (16) orientation preserving. Since the torus group $\mathcal{V}/L$ acts transitively on the oriented leaves $\mathcal{T}'$, they all represent a single real homology class $[\mathcal{T}'] \in H_k(\mathcal{T}, \mathbb{R})$, equal to the image of the fundamental class of $\mathcal{T}'$ under the inclusion $\mathcal{T}' \to \mathcal{T}$. At the same time, for generic leaves $\mathcal{M}'$, the d-fold covering projection $\mathcal{T}' \to \mathcal{M}'$ (where $d = |H'|$ does not depend on the choice of $\mathcal{M}'$, cf. Lemma 15) sends the fundamental class of $\mathcal{T}'$ to d times the fundamental class of $\mathcal{M}'$. Thus, by functoriality, $d[\mathcal{M}'] \in H_k(\mathcal{M}, \mathbb{R})$ is the image of $[\mathcal{T}'] \in H_k(\mathcal{T}, \mathbb{R})$ under the covering projection $\mathcal{T} \to \mathcal{M}$, which makes it the same for all the generic leaves $\mathcal{M}'$. Finally, $[\mathcal{M}'] \neq 0$, since a fixed constant positive differential k-form on the oriented space $\mathcal{V}'$ descends, in view of the first line of this proof, to a parallel positive volume form on each oriented generic leaf $\mathcal{M}'$ which yields a positive value when integrated over $[\mathcal{M}']$. □

Note that the final clause in Theorem 5 is, not surprisingly, consistent with the fact that—by Theorem 4(a) and Lemma 15—the intersection numbers depend just on the two mutually complementary H-invariant L-subspaces $\mathcal{V}', \mathcal{V}''$ of $\mathcal{V}$, and not on the individual generic leaves $\mathcal{M}', \mathcal{M}'', \mathcal{T}'$ or $\mathcal{T}''$.

15 Generalized Klein Bottles

This section presents some known examples [4, p. 163] to illustrate our discussion.

The symbols $\mathbb{R}^H, \mathbb{Z}^H$ used below follow the set-theoretical notational convention: Y^X is *the set of all mappings* $X \to Y$, not the fixed-point set of some group action.

Let S^1 and $r_\theta : S^1 \to S^1$ denote the unit circle in $\mathbb{C}$ and the rotation by angle θ (multiplication by $e^{i\theta}$). For a fixed integer $n \geq 2$ and the group $H = \mathbb{Z}_n \subseteq S^1$ of nth roots of unity, $\mathbb{Z}^H \cong \mathbb{Z}^n$ is a lattice in the Euclidean space $\mathcal{V} = \mathbb{R}^H \cong \mathbb{R}^n$ with the ℓ^2 inner product, and $\mathbb{Z}_0^H = \{\psi \in \mathbb{Z}^H : \psi_{\text{avg}} = 0\}$ is a subgroup of $\mathbb{Z}^H$ isomorphic to $\mathbb{Z}^{n-1}$, where $(\,)_{\text{avg}}$ denotes the averaging functional $\mathcal{V} \to \mathbb{R}$. Setting $\Pi = [(1/n)\mathbb{Z}] \times \mathbb{Z}_0^H$, one easily sees that the assignment:

$$((t, \psi), f) \mapsto f \circ r_{2\pi t} + t + \psi, \quad \text{where } (t, \psi) \in \Pi \text{ and } f \in \mathcal{V} = \mathbb{R}^H, \tag{30}$$

defines an affine isometric action on $\mathcal{V}$ by Π treated as a group with the group operation $(t, \psi)(t', \psi') = (t + t', \psi' \circ r_{2\pi t} + \psi)$. The term t in (30) is the constant function $t : H \to \mathbb{R}$. Note that, in the right-hand side of (30), as $(t, \psi) \in \Pi$

$$t_{\text{avg}} = t, \qquad \psi_{\text{avg}} = 0, \qquad (f \circ r_{2\pi t})_{\text{avg}} = f_{\text{avg}}. \tag{31}$$

Proposition 3 *These $H, \mathcal{V}$, and Π have the following properties:*

i. *The action of Π on $\mathcal{V}$ is effective and free.*
ii. *Π is a Bieberbach group in the underlying Euclidean affine n-space of $\mathcal{V}$.*
iii. *The holonomy group and lattice subgroup of Π are our $H \cong \mathbb{Z}_n$, acting on $\mathcal{V}$ linearly by $H \times \mathcal{V} \ni (e^{i\theta}, f) \mapsto f \circ r_\theta \in \mathcal{V}$, and $L = \mathbb{Z} \times \mathbb{Z}_0^H$.*
iv. *As a transformation of $\mathcal{V}$, each $(t, \psi) \in L$ equals the translation by $t + \psi$.*
v. *L consists of all translations by vectors $\psi' \in \mathbb{Z}^H$ such that $\psi'_{\text{avg}} \in \mathbb{Z}$.*

An example of two mutually complementary H-invariant L-subspaces of $\mathcal{V}$, in the sense of (1) and Definition 1, is provided by the line $\mathcal{V}'$ of constant functions $H \to \mathbb{R}$ and the hyperplane $\mathcal{V}''$ consisting of all $f : H \to \mathbb{R}$ with $f_{\text{avg}} = 0$. The generic stabilizer groups $\Sigma', \Sigma'' \subseteq \Pi$ associated via (21) with $\mathcal{V}'$ and $\mathcal{V}''$ are the translation groups $\mathbb{Z} \times \{0\}$ and $\{0\} \times \mathbb{Z}_0^H$, both contained in L. Furthermore,

a. *Under the obvious identifications of $\mathcal{V}/\mathcal{V}'$ with $\mathcal{V}''$ and $\mathcal{V}/\mathcal{V}''$ with $\mathcal{V}'$, the quotient actions of Π become $\Pi \times \mathcal{V}'' \ni ((t, \psi), f) \mapsto f \circ r_{2\pi t} + \psi \in \mathcal{V}''$ and, respectively, $\Pi \times \mathcal{V}' \ni ((t, \psi), f) \mapsto f + t \in \mathcal{V}'$.*
b. *Every coset of the L-subspace $\mathcal{V}''$ is generic, as defined in Sect. 11.*
c. *Nongeneric cosets of $\mathcal{V}'$ are precisely those cosets containing $f : H \to \mathbb{R}$ such that $f \circ r_{2\pi t} - f$ is integer-valued for some $t \in [(1/n)\mathbb{Z}] \setminus \mathbb{Z}$.*
d. *The obvious homomorphism $\Pi = [(1/n)\mathbb{Z}] \times \mathbb{Z}_0^H \to (1/n)\mathbb{Z}$ maps the stabilizer group of each coset of $\mathcal{V}'$ isomorphically onto a subgroup of $(1/n)\mathbb{Z}$.*
e. *The subgroups of $(1/n)\mathbb{Z}$ resulting from (c) have the form $(d/n)\mathbb{Z}$, where d is a positive divisor of n, or, equivalently, are the preimages, under the homomorphism $\mathbb{R} \ni t \mapsto e^{2\pi i t} \in S^1$, of subgroups of the group $H = \mathbb{Z}_n \subseteq S^1$ formed by the nth roots of unity.*

Proof First, Π acts on $\mathcal{V}$ freely: if $f \circ r_{2\pi t} + t + \psi = f$, cf. (30), with $f : H \to \mathbb{R}$, applying $(\,)_{\text{avg}}$ to both sides, we get, by (31), $t = 0$, and hence, $f \circ r_{2\pi t} = f$, so that the equality $f \circ r_{2\pi t} + t + \psi = f$ reads $\psi = 0$. Secondly, H and L *defined* by (iii) arise from Π as required in (7): the claim about H is obvious, and so are (iv)–(v), showing that $L \subseteq \Pi \cap \mathcal{V}$. Conversely, $\Pi \cap \mathcal{V} \subseteq L$. To verify this, suppose that $f \circ r_{2\pi t} + t + \psi = f + \psi'$ for all $f \in \mathcal{V} = \mathbb{R}^H$, some $(t, \psi) \in \Pi$, and some $\psi' \in \mathcal{V}$. Taking the linear parts of both sides, we see that $t \in \mathbb{Z}$ and $(t, \psi) \in L$, as required.

Our Π has a compact fundamental domain in $\mathcal{V}$, since so does the lattice $L \subseteq \Pi$. Also, Π is torsion-free: $\Pi \ni (t, \psi) \mapsto t \in \mathbb{R}$ being a group homomorphism, any finite-order element (t, ψ) of Π has $t = 0$, and so, by (30), it acts via translation by ψ, which gives $\psi = 0$. Next, to establish the discreteness of the subset Π of

$\mathrm{Iso}\mathcal{V}$ (and, consequently, (ii)), suppose that a sequence $(t_k, \psi_k) \in \Pi$ with pairwise distinct terms yields, via (30), a sequence convergent in $\mathrm{Iso}\mathcal{V}$. Evaluating (30) on $f = 0$, we get $(t_k, \psi_k) \to (t, \psi)$ in $\mathbb{R} \times \mathbb{R}^H$ as $k \to \infty$, for some (t, ψ) and, since $(t_k, \psi_k) \in [(1/n)\mathbb{Z}] \times \mathbb{Z}_0^H$, the sequence (t_k, ψ_k) becomes eventually constant, contrary to the fact that its terms are pairwise distinct.

As for $\mathcal{V}'$ and $\mathcal{V}''$, note that, by (iv), a $\mathbb{Z}$-basis of $L \cap \mathcal{V}'$ (or, $L \cap \mathcal{V}''$) may be defined to consist just of the constant function 1 (or, respectively, of the $n-1$ functions $\psi_q : H \to \mathbb{Z}$, labeled by $q \in H \smallsetminus \{1\}$, where $\psi_q(q) = 1 = -\psi_q(1)$ and $\psi_q = 0$ on $H \smallsetminus \{1, q\}$). Specifically, $\psi' = \sum_q \psi'(q)\psi_q$ whenever $\psi' \in \mathbb{Z}^H$ and $\psi'_{\mathrm{avg}} = 0$. Our descriptions of Σ' and Σ'' are in turn immediate from (a), which itself is a trivial consequence of (30)–(31), and easily implies (b)–(c). Now, (d) follows as the relation $f \circ r_{2\pi t} + \psi = f$, with fixed $f \in \mathcal{V}''$, uniquely determines ψ, once t is given. Finally, we have (e) since the first assignment in (a) only depends on t through $e^{2\pi i t}$, which completes the proof. □

The compact flat Riemannian manifold $\mathcal{V}/\Pi$ arising here from our Bieberbach group Π as in Sect. 6 is called the n-dimensional *generalized Klein bottle* [4, p. 163]. The linear functional $\mathcal{V} \ni f \mapsto f_{\mathrm{avg}} \in \mathbb{R}$ is equivariant, due to (31), with respect to the actions of Π and $\mathbb{Z}$ (the latter, on $\mathbb{R}$, via translations by multiples of $1/n$), relative to the homomorphism $\Pi \ni (t, \psi) \mapsto t \in (1/n)\mathbb{Z}$. Thus, it descends, in view of Remark 1(c), to a bundle projection $\mathcal{V}/\Pi \to \mathbb{R}/[(1/n)\mathbb{Z}]$, making $\mathcal{V}/\Pi$ a bundle of tori over the circle. The fibers of this bundle are, obviously, the images, under the projection $\mathrm{pr} : \mathcal{V} \to \mathcal{V}/\Pi$, of cosets of the L-subspace $\mathcal{V}'' \subseteq \mathcal{V}$ mentioned in Proposition 3(b), all of them generic. On the other hand, $\mathcal{V}'$ has some nongeneric cosets—by Proposition 3(c), an example is $\mathcal{V}'$ itself, with the stabilizer group easily seen to be $[(1/n)\mathbb{Z}] \times \{0\}$. The pr-images of the cosets of $\mathcal{V}'$ are embedded circles, forming the leaves of the foliation $F_{\mathcal{M}}$ of $\mathcal{M} = \mathcal{V}/\Pi$ arising as in Theorem 1.

For the foliation $F_{\mathcal{M}}$ of $\mathcal{M} = \mathcal{V}/\Pi$ obtained in the general case of Theorem 1, the stabilizer group of a leaf $\mathcal{M}'$ of $F_{\mathcal{M}}$ is only defined as a conjugacy class of subgroups of Π (the subgroups being the stabilizer groups Σ' of leaves $\mathcal{E}'$ of $F_{\mathcal{E}}$ with $\mathrm{pr}(\mathcal{E}') = \mathcal{M}'$). Genericity of $\mathcal{M}'$, mentioned in the line following (21), amounts to genericity of all such $\mathcal{E}'$, and Σ' is then uniquely associated with $\mathcal{M}'$ (due to its being the generic stabilizer group, normal in Π). In the subsequent discussion, Σ' is also treated as uniquely defined, for a entirely different reason: each Σ' is replaced by its image under a specific homomorphism from Π into the Abelian group $(1/n)\mathbb{Z}$.

We denote by $\mid$ the divisibility relation in the set Δ_n of all positive divisors of n and by $\mathcal{M}$ the n-dimensional generalized Klein bottle $\mathcal{V}/\Pi$. Let us also use Proposition 3(d)–(e) and the final sentence of the last paragraph to treat

$$\text{the stabilizer groups of leaves of } F_{\mathcal{M}} \text{ as subgroups of } (1/n)\mathbb{Z} \text{ having the form } (d/n)\mathbb{Z}, \text{ where } d \in \Delta_n. \tag{32}$$

Proposition 4 *There exists a family* $\{\mathcal{M}[d] : d \in \Delta_n\}$ *of compact connected immersed submanifolds of* $\mathcal{M}$ *with the following properties, for all* $d, d' \in \Delta_n$*:*

i. *Each* $\mathcal{M}[d]$ *has the dimension* d *and is foliated by circle leaves of* $F_{\mathcal{M}}$.
ii. $\mathcal{M}[d'] \subseteq \mathcal{M}[d]$ *whenever* $d'|d$.
iii. $\mathcal{M}[d] \setminus \bigcup_k \mathcal{M}[k]$*, with* k *ranging over* $\Delta_d \setminus \{d\}$*, equals the union of all leaves of* $F_{\mathcal{M}}$ *having the stabilizer group* $(d/n)\mathbb{Z}$.
iv. $\mathcal{M}[n] = \mathcal{M}$ *and* $\mathcal{M}[1] = \mathrm{pr}(\mathcal{V}')$.
v. *If* n *is prime,* $\mathcal{M}[1] = \mathrm{pr}(\mathcal{V}')$ *is the only nongeneric leaf of* $F_{\mathcal{M}}$.

Proof The zero-average functions $h : H \to \mathbb{R}/\mathbb{Z}$, from the group $H = \mathbb{Z}_n \subseteq S^1$ of nth roots of unity into the circle $\mathbb{R}/\mathbb{Z}$, form a manifold $[\mathbb{R}/\mathbb{Z}]^H$ diffeomorphic to the torus T^{n-1}, and the covering projection $\mathcal{V}'' \to [\mathbb{R}/\mathbb{Z}]^H$, sending a zero-average function $f : H \to \mathbb{R}$ to its composition h with the projection $\mathbb{R} \to \mathbb{R}/\mathbb{Z}$, is obviously equivariant relative to the first action in Proposition 3(a):

$$\text{the action of } H \text{ on } [\mathbb{R}/\mathbb{Z}]^H \text{ given by } (q, h) \mapsto h \circ r_{2\pi t}, \tag{33}$$

for $t \in (1/n)\mathbb{Z}$ with $q = e^{2\pi i t}$, and the homomorphism $\Pi \ni (t, \psi) \mapsto e^{2\pi i t} \in H$. Thus, (33) is the translational action of $H = \mathbb{Z}_n$ on mappings $H \to \mathbb{R}/\mathbb{Z}$, and the preimages of its isotropy groups under the homomorphism $(1/n)\mathbb{Z} \ni t \mapsto e^{2\pi i t} \in H$ are precisely the stabilizer groups in (32). We now define $\mathcal{M}[d]$, for any $d \in \Delta_n$, to be the union of circle leaves of $F_{\mathcal{M}}$ having stabilizer groups contained in $(d/n)\mathbb{Z}$. Thus, $\mathcal{M}[d]$ is the image under the quotient projection $\mathrm{pr} : \mathcal{V} \to \mathcal{V}/\Pi$ of the union of all cosets $v + \mathcal{V}'$ for vectors $v \in \mathcal{V}''$ which the covering projection $\mathcal{V}'' \to [\mathbb{R}/\mathbb{Z}]^H$ sends to (zero-average) functions $h : H \to \mathbb{R}/\mathbb{Z}$ having $h \circ r_{2\pi d/n} = h$. Since such h form a manifold diffeomorphic to the torus T^{d-1}, via the assignment to h of the zero-sum d-tuple consisting of $h(e^{2\pi i k/n}) \in \mathbb{R}/\mathbb{Z}$, with $k = 0, 1, \ldots, d-1$, all our claims about $\mathcal{M}[d]$ easily follow from the fact that both projections just mentioned are locally diffeomorphic. □

The n-dimensional generalized Klein bottle, for any $n \geq 2$, is an example illustrating the fact that the last inclusion of Theorem 1(ii-c) may be proper. In fact, the stabilizer group $[(1/n)\mathbb{Z}] \times \{0\}$ of $\mathcal{V}'$, mentioned six lines before Proposition 4, although not contained in the lattice L, acts on $\mathcal{V}'$ by translations. Also, unless n is prime, Proposition 4(iii) shows that the dependence on u in the last line of Lemma 13 is actually possible, as it occurs for $\mathcal{M}' = \mathcal{M}[k]$, with any $k \in \Delta_n \setminus \{n\}$.

16 Remarks on Holonomy

The correspondence between Bieberbach groups and compact flat manifolds mentioned in Remark 4 has an extension to almost-Bieberbach groups and infra-nilmanifolds [6] obtained by using—instead of the translation vector space of a

Euclidean affine space—a connected, simply connected nilpotent Lie group $\mathcal{G}$ acting simply transitively on a manifold $\mathcal{E}$ and replacing the Bieberbach group with a torsion-free uniform discrete subgroup Π of $\mathrm{Diff}\,\mathcal{E}$ contained in a semidirect product (canonically transplanted so as to act on $\mathcal{E}$) of $\mathcal{G}$ and a maximal compact subgroup of $\mathrm{Aut}\,\mathcal{G}$. Here, "uniform" means admitting a compact fundamental domain, cf. Remark 2. The analogs of (8) and (15) remain valid, reflecting the fact that any infra-nilmanifold is the quotient of a nilmanifold under the action of a finite group H.

A somewhat similar picture may arise in some cases where $\mathcal{G}$ is not assumed nilpotent. As an example, let $\mathcal{G} \cong \mathrm{Spin}(m, 1)$ be the universal covering group of the identity component $\mathcal{G}/\mathbb{Z}_2 \cong \mathrm{SO}^+(m, 1)$ of the pseudo-orthogonal group of an $(m + 1)$-dimensional Lorentzian vector space $\mathcal{L}$, $m \geq 3$. Here, $\mathcal{E}$ is the (twofold) universal covering manifold of the orthonormal-frame bundle of the future unit pseudosphere $\mathcal{S} \subseteq \mathcal{L}$, isometric to the hyperbolic m-space, and $\mathcal{G}/\mathbb{Z}_2$ acts on $\mathcal{S}$ via hyperbolic isometries, leading to an action of $\mathcal{G}$ on $\mathcal{E}$. The orthonormal-frame bundles of compact hyperbolic manifolds obtained as quotients of $\mathcal{S}$ give rise to the required torsion-free uniform discrete subgroups Π.

The resulting compact quotient manifolds $\mathcal{M} = \mathcal{E}/\Pi$ can be endowed with various interesting Riemannian metrics coming from Π-invariant metrics on $\mathcal{E}$. For Π and $\mathcal{E}$ of the preceding paragraph, a particularly natural choice of an invariant *indefinite* metric is provided by the Killing form of $\mathcal{G}$, turning $\mathcal{M}$ into a compact locally symmetric pseudo-Riemannian Einstein manifold.

Outside of the Bieberbach-group case, however, these metrics are not flat, and finite groups H such as mentioned above cannot serve as their holonomy groups. The holonomy interpretation of H still makes sense, though, if—instead of metrics—one uses either of the two Π-invariant flat connections, with (parallel) torsion, naturally distinguished on $\mathcal{E}$. Here, $\mathcal{E}$ is, again, a manifold on which a connected, simply connected Lie group $\mathcal{G}$ acts simply transitively. Two natural bi-invariant connections with the stated properties exist on $\mathcal{G}$, rather than $\mathcal{E}$, and are characterized by the requirement that they make all the left-invariant (or, right-invariant) vector fields parallel. Due to their naturality, these two connections on $\mathcal{G}$ are also invariant under all the Lie-group automorphisms of $\mathcal{G}$. It is the two connections on $\mathcal{G}$ that induce, in an obvious way, the ones on $\mathcal{E}$, mentioned six lines earlier.

Acknowledgments Both authors' research was supported in part by a FAPESP-OSU 2015 Regular Research Award (FAPESP grant: 2015/50265-6). The authors wish to thank Andrzej Szczepański for helpful comments. We also greatly appreciate suggestions made by anonymous referees of earlier versions of the manuscript, which allowed us to make the exposition much easier to follow.

Appendix: Hiss and Szczepański's Reducibility Theorem

Consider an *abstract Bieberbach group*, that is, any torsion-free group Π containing a finitely generated free Abelian normal subgroup L of finite index, which is at the same time a maximal Abelian subgroup of Π. As shown by Zassenhaus [14], up to isomorphisms these groups coincide with the Bieberbach groups of Sect. 6, we again summarize their structure using the short exact sequence:

$$L \to \Pi \to H, \quad \text{where } H = \Pi/L. \tag{34}$$

For Abelian groups G_1, G_2, and G', one has canonical isomorphisms:

$$\mathbb{Z} \otimes G \cong G, \ (G_1 \oplus G_2) \otimes G' \cong (G_1 \otimes G') \oplus (G_2 \otimes G'), \ L \otimes \mathbb{Q} \cong \mathrm{Hom}(L^*, \mathbb{Q}), \tag{35}$$

where $L^* = \mathrm{Hom}(L, \mathbb{Z})$ and, for simplicity, L is assumed to be finitely generated and free. In the last case, with a suitable integer $n \geq 0$, there are further noncanonical isomorphisms:

$$\text{a)}\ L \cong \mathbb{Z}^n, \qquad \text{b)}\ L \otimes \mathbb{Q} \cong \mathbb{Q}^n, \tag{36}$$

while, using the injective homomorphism $L \ni \lambda \mapsto \lambda \otimes 1 \in L \otimes \mathbb{Q}$ to treat L as a subgroup of $L \otimes \mathbb{Q}$, we see that under suitably chosen identifications (36)

$$\text{the inclusion } L \subseteq L \otimes \mathbb{Q} \text{ corresponds to the standard inclusion } \mathbb{Z}^n \subseteq \mathbb{Q}^n. \tag{37}$$

Finally, if L as above is a (full) lattice in an finite-dimensional real vector space $\mathcal{V}$, a further canonical isomorphic identification arises:

$$L \otimes \mathbb{Q} \cong \mathrm{Span}_{\mathbb{Q}} L, \tag{38}$$

that is, we may view $L \otimes \mathbb{Q}$ as the rational vector subspace of $\mathcal{V}$ spanned by L.

Let Π now be an abstract Bieberbach group. Hiss and Szczepański [10, the corollary in Sect. 1] proved that, if L in (34) satisfies (36.a) with $n \geq 2$, then *the* (obviously $\mathbb{Q}$-linear) *action of H on $L \otimes \mathbb{Q}$ must be reducible*, in the sense of admitting a nonzero proper invariant rational vector subspace $\mathcal{W}$.

Next, using (37), we may write $L = \mathbb{Z}^n \subseteq \mathbb{Q}^n = L \otimes \mathbb{Q}$, so that $\mathcal{W} \subseteq \mathbb{Q}^n \subseteq \mathbb{R}^n$, and the closure $\mathcal{V}'$ of $\mathcal{W}$ in $\mathbb{R}^n$ has the real dimension $\dim_{\mathbb{Q}} \mathcal{W}$ (any $\mathbb{Q}$-basis of $\mathcal{W}$ being, obviously, an $\mathbb{R}$-basis of $\mathcal{V}'$). By clearing denominators, one can replace such a $\mathbb{Q}$-basis with one consisting of vectors in $L = \mathbb{Z}^n$, and so, by Lemma 10(a), the intersection $L' = L \cap \mathcal{W} = L \cap \mathcal{V}'$ is a lattice in $\mathcal{V}'$. We thus obtain (11).

A stronger version of Hiss and Szczepański's reducibility theorem, established more recently by Lutowski [12], states that the rational holonomy representations of any compact flat manifold other than a torus have at least two nonequivalent irreducible subrepresentations.

References

1. D. M. ARNOLD, *Finite Rank Torsion-Free Abelian Groups and Rings*, Lecture Notes in Mathematics **931**, Springer-Verlag, New York, 1982.
2. R. G. BETTIOL, A. DERDZINSKI, R. MOSSA, P. PICCIONE, *On the collapse of compact flat manifolds*, Math. Nachrichten **295** (2022), 2338–2356.
3. N. Bourbaki, *Topologie générale*, 3ème éd., Chap. V–VIII, Actualités Sci. Ind., No. 1235, Hermann, Paris, 1963.
4. L. S. CHARLAP, *Bieberbach Groups and Flat Manifolds*, Universitext, Springer-Verlag, New York, 1986.
5. M. W. DAVIS, *Lectures on orbifolds and reflection groups*, in: Advanced Lectures in Mathematics **16**: Transformation Groups and Moduli Spaces of Curves (L. Ji and S.-T. Yau, eds.), International Press, Somerville, MA, USA/Higher Education Press, Beijing, China, 2011, 63–93.
6. K. DEKIMPE, *Almost-Bieberbach Groups : Affine and Polynomial Structures*, Lecture Notes in Mathematics **1639**. Springer-Verlag, Berlin, 1996.
7. A. DERDZINSKI, P. PICCIONE, *Flat manifolds and reducibility*, arXiv:1903.10479v2.
8. B. I. DUNDAS, *A Short Course in Differential Topology*, Cambridge Mathematical Textbooks. Cambridge University Press, Cambridge, 2018.
9. R. E. EDWARDS, *Fourier Series : A Modern Introduction, Volume 1*, 2nd ed., Springer-Verlag, New York, 1979.
10. G. HISS, A. SZCZEPAŃSKI, *On torsion free crystallographic groups*. J. Pure Appl. Algebra **74** (1991), 39–56.
11. S. KOBAYASHI, K. NOMIZU, *Foundations of Differential Geometry, Vol. I*. Interscience, New York, 1963.
12. R. LUTOWSKI, *Flat manifolds with homogeneous holonomy representation*, Publ. Math. Debrecen, **99** (2021), no. 1–2, art. 6.
13. A. SZCZEPAŃSKI, *Geometry of Crystallographic Groups*. Algebra and Discrete Mathematics, **4**. World Scientific, Hackensack, NJ, 2012.
14. H. ZASSENHAUS, *Beweis eines Satzes über diskrete Gruppen*. Abh. Math. Sem. Univ. Hamburg **12** (1938), 289–312.

Elementary Proof of a Lemma for Parallel Mean Curvature Surfaces in Complex Space Forms

Katsuei Kenmotsu

To my dear friend Marcos Dajczer on his 75th birthday.

Abstract We provide an elementary proof of a lemma that plays an important role in the classification of parallel mean curvature surfaces in two-dimensional complex space forms.

1 Introduction

In the theory of parallel mean curvature surfaces in two-dimensional complex space forms, Lemma 1 of Kenmotsu [4] plays an important role in the classification of these surfaces (see Kenmotsu [5]). The lemma claims that a part of complexified second fundamental tensors of the surface is a function of the Kaehler angle of the surface if the ambient space is non-Euclidean and the Kaehler angle is not constant. The proof of the lemma in [4] uses the exterior derivatives of forms and a skillful substitution of the second fundamental tensor. In this paper, we provide an elementary proof of Lemma 1 of [4] by using the Kaehler angle as a coordinate function of the surface and partial differentiations only.

2 The Lemma

The notations and structure equations in this paper are consistent with those in Kenmotsu [4]. In this context, M^2 represents a two-dimensional Riemannian manifold, and x denotes an isometric immersion of M^2 into a complex space form

K. Kenmotsu (✉)
Tohoku University, Sendai, Japan
e-mail: kenmotsu-math@tohoku.ac.jp

© The Author(s), under exclusive license to Springer Nature Switzerland AG 2025

R. Tojeiro (ed.), *Advances in Submanifold Theory and Related Topics*,
Latin American Mathematics Series – UFSCar subseries,
https://doi.org/10.1007/978-3-032-00918-0_9

$\overline{M}[4\rho]$ of complex dimension two, where 4ρ represents the holomorphic sectional curvature of the complex space form. In this paper, we study the immersion x such that the mean curvature vector of the immersion, denoted as H, is nonzero and parallel when considering the normal connection on the normal bundle of the immersion.

Since x is not minimal, x is neither holomorphic nor anti-holomorphic on M^2. Hence, the Kaehler angle of x, denoted as α, satisfies $\sin\alpha \neq 0$ in a neighborhood of any point on M^2. Since our study here only pertains to local properties of the immersion, we may consider this neighborhood to be the entirety of M^2.

Given that we utilize formulas (2.1) through (2.6) (excluding (2.3)) from Kenmotsu [4], for the sake of completeness, we provide them below. There exists a local field of complex one form, denoted as ϕ, on M^2 such that the Riemannian metric ds^2 on M^2 can be expressed as $ds^2 = \phi\overline{\phi}$ (cf. Chern and Wolfson [1]). Additionally, let a and c be complex-valued functions on M^2 that determine the second fundamental tensors of the immersion. Then, the Kaehler angle α and the complex 1-form ϕ satisfy

$$d\alpha = (a+b)\phi + (\overline{a}+b)\overline{\phi}, \tag{1}$$

$$d\phi = (\overline{a}-b)\cot\alpha \cdot \phi \wedge \overline{\phi}, \tag{2}$$

where $2b = |H| > 0$. The two functions a and c satisfy

$$da \wedge \phi = -\left(2a(\overline{a}-b)\cot\alpha + \frac{3}{2}\rho\sin\alpha\cos\alpha\right)\phi \wedge \overline{\phi}, \tag{3}$$

$$dc \wedge \overline{\phi} = 2c(a-b)\cot\alpha \cdot \phi \wedge \overline{\phi}, \tag{4}$$

$$|c|^2 = |a|^2 + \frac{\rho}{2}(-2 + 3\sin^2\alpha), \tag{5}$$

where (3) and (4) are the Codazzi-Mainardi equations, and (5) is the Ricci equation of x. The Gauss equation of the immersion is not used in this paper.

Now, we prove that the Kaehler angle α can be taken as a coordinate function on M^2, if α is not constant.

Lemma 1 *Let x be an isometric immersion of M^2 into $\overline{M}[4\rho]$ with nonzero parallel mean curvature vector. If the Kaehler angle α is not constant, then there exists a real-valued function β on M^2 such that (α, β) is a local coordinate system on M^2.*

Proof Let (u, v) be the isothermal local coordinates on M^2 such that $\phi = \lambda(du + idv)$, $(i^2 = -1)$. Suppose that the Kaehler angle α on M^2 is not constant on M^2, so we may assume $d\alpha \neq 0$ on M^2, because we study here only local properties of the immersion. By (1), we note that $a + b \neq 0$ on M^2. Let a real-valued function $k(u, v)$ $(\neq 0)$ defined on a neighborhood of any point of M^2 be a solution of a linear first-order partial differential equation $\alpha_u k_u + \alpha_v k_v + (\alpha_{uu} + \alpha_{vv})k = 0$. (There exists always such a solution of the equation by the famous existence theorem.)

We consider a system of partial differential equations defined by $\beta_u = k\alpha_v$, $\beta_v = -k\alpha_u$. Since this is completely integrable by the property of $k(u, v)$, there exists a real-valued solution $\beta = \beta(u, v)$ of the system in a neighborhood of any point of M^2. Since $\alpha_u = \lambda(a+b) + \overline{\lambda}(\overline{a}+b)$, $\alpha_v = i(\lambda(a+b) - \overline{\lambda}(\overline{a}+b))$ by (1), it holds that $d\beta = ik((a+b)\phi - (\overline{a}+b)\overline{\phi})$, which implies

$$\phi = \frac{1}{2k(a+b)}(kd\alpha - id\beta), \quad \phi \wedge \overline{\phi} = \frac{i}{2k|a+b|^2} d\alpha \wedge d\beta, \tag{6}$$

on a neighborhood of any point in M^2. Hence, (α, β) is a local coordinate system on the neighborhood, proving Lemma 1. □

Hereafter, the quantities k, a, and c are considered as functions of α and β.

The local structure of the immersion satisfying $a = \overline{a}$ on M^2 has been determined in Kenmotsu and Zhou [3] and Kenmotsu [4]. Hirakawa [2] gave another proof for this case. Also, the immersions with $d\alpha = 0$ on M^2 are classified by Hirakawa [2]. From this point on, we will study an immersion satisfying $d\alpha \neq 0$ and $a \neq \overline{a}$ at a point of M^2. An immersion x is called of general type if it satisfies these two conditions on M^2. The Lemma 1 in [4] is stated as

Lemma 2 *Let x be an isometric immersion of M^2 into $\overline{M}[4\rho]$ with nonzero parallel mean curvature vector. Suppose that $\rho \neq 0$ and x is of general type. Then, $d\alpha \wedge da = 0$ on M^2.*

Lemma 2 implies that if x is of general type, then a is a function of α only, and this played a crucial role to determine the local structures of these surfaces (see Kenmotsu [5]).

3 Elementary Proof

The original proof of Lemma 2 (referred to as Lemma 1 in [4]) utilizes the theory of exterior differential forms and employs a skillful substitution for the partial derivatives of a. In this paper, we present an elementary proof of Lemma 2, relying solely on the use of the Kaehler angle as a coordinate function of the surface and on partial differentiations, except for the proof of equation (7). Our main idea involves utilizing the complex conjugate to calculate the difference in the higher-order derivatives for α and β of functions a and c (see, e.g., the proof of Lemma 3). It shall be remarked that all rigorous computations of the derivatives for functions p_i ($i = 1, 2, \ldots$), as defined in this paper, were carried out using the symbolic manipulation program Mathematica by Wolfram. We do not provide the Mathematica program codes as they are all simple, consisting only of rational functions and their differentiations.

Let us clear the first equation in (6) of fractions and calculate its exterior derivatives. It follows from (1), (2), and (3) that

$$k_\alpha = -p_1 k, \tag{7}$$

where p_i, $(i = 1, 2, \ldots)$, are functions of α, a, and $\overline{a}$ listed in the Appendix.

Equations (3) and (4) are rewritten as functions of α and β:

$$a_\alpha - ika_\beta = p_2, \tag{8}$$

$$c_\alpha + ikc_\beta = 2cp_3. \tag{9}$$

Lemma 3 *Let x be an isometric immersion of M^2 into $\overline{M}[4\rho]$ with nonzero parallel mean curvature vector. If the Kaehler angle α is not constant, then it holds that*

$$\overline{c}c_\alpha - a\overline{a}_\alpha = p_4. \tag{10}$$

Proof Differentiating partially (5) for β and multiplying the formula by ik, we see that $\overline{c}(ikc_\beta) - c\overline{(ikc_\beta)} = \overline{a}(ika_\beta) - a\overline{(ika_\beta)}$. By (8) and (9), this is written as

$$-\overline{c}c_\alpha + c\overline{c}_\alpha = \overline{a}a_\alpha - a\overline{a}_\alpha - 2|c|^2(p_3 - \overline{p}_3) - \overline{a}p_2 + a\overline{p}_2.$$

Differeitating partially (5) for α yields $\overline{c}c_\alpha + c\overline{c}_\alpha = \overline{a}a_\alpha + a\overline{a}_\alpha + 3\rho\sin\alpha\cos\alpha$. Lemma 3 follows from these two formulas. □

It follows from (8), (9), and (10) that

$$\overline{c}c_\beta - a\overline{a}_\beta = \frac{i}{k}p_5. \tag{11}$$

Differentiating partially (10) for α proves that

$$|c_\alpha|^2 - |a_\alpha|^2 + \overline{c}c_{\alpha\alpha} - a\overline{a_{\alpha\alpha}} = \frac{\partial p_4}{\partial \alpha} + \frac{\partial p_4}{\partial a}a_\alpha + \frac{\partial p_4}{\partial \overline{a}}\overline{a_\alpha}. \tag{12}$$

We want to find a formula of $\overline{c}c_{\alpha\alpha} - a\overline{a_{\alpha\alpha}}$ in terms of a_α. Calculating partial derivatives of (9) and (8) for α, we have

$$c_{\alpha\alpha} + ikc_{\beta\alpha} = p_1 ikc_\beta + 2p_3 c_\alpha + 2c\left(\frac{\partial p_3}{\partial \alpha} + \frac{\partial p_3}{\partial a}a_\alpha + \frac{\partial p_3}{\partial \overline{a}}\overline{a_\alpha}\right), \tag{13}$$

$$\overline{a_{\alpha\alpha}} + ik\overline{a_{\beta\alpha}} = \overline{p_1}ik\overline{a_\beta} + \overline{\frac{\partial p_2}{\partial \alpha}} + \overline{\frac{\partial p_2}{\partial \overline{a}}}a_\alpha + \overline{\frac{\partial p_2}{\partial a}}\overline{a_\alpha}. \tag{14}$$

It is remarked here that

$$p_1 - \overline{p_1} = 0, \quad \frac{\partial p_3}{\partial \overline{a}} = 0, \quad \overline{\frac{\partial p_2}{\partial a}} = 2p_3. \tag{15}$$

It follows from (10), (11), (13), (14), and (15) that

$$\begin{aligned}\overline{c}c_{\alpha\alpha} - a\overline{a_{\alpha\alpha}} &= ik(a\overline{a_{\beta\alpha}} - \overline{c}c_{\beta\alpha}) + \left(-a\overline{\frac{\partial p_2}{\partial \overline{a}}} + 2|c|^2\frac{\partial p_3}{\partial a}\right)a_\alpha \\ &\quad -p_1p_5 + 2p_3p_4 - a\overline{\frac{\partial p_2}{\partial \alpha}} + 2|c|^2\frac{\partial p_3}{\partial \alpha}.\end{aligned}$$

Let the both sides of (10) differentiate partially for β. Then, multiplying the formula by ik, (8) and (9) imply

$$\begin{aligned}ik(a\overline{a_{\alpha\beta}} - \overline{c}c_{\alpha\beta}) &= |c_\alpha|^2 - |a_\alpha|^2 - \frac{\partial p_4}{\partial a}a_\alpha + \left(p_2 - 2a\overline{p}_3 + \frac{\partial p_4}{\partial \overline{a}}\right)\overline{a_\alpha} \\ &\quad -2\overline{p}_3p_4 + p_2\frac{\partial p_4}{\partial a} - \overline{p}_2\frac{\partial p_4}{\partial \overline{a}}.\end{aligned}$$

It follows from above two formulas that

$$\begin{aligned}&\overline{c}c_{\alpha\alpha} - a\overline{a_{\alpha\alpha}} \\ &\quad = |c_\alpha|^2 - |a_\alpha|^2 + \left(-a\overline{\frac{\partial p_2}{\partial \overline{a}}} + 2|c|^2\frac{\partial p_3}{\partial a} - \frac{\partial p_4}{\partial a}\right)a_\alpha \\ &\quad + \left(p_2 - 2a\overline{p_3} + \frac{\partial p_4}{\partial \overline{a}}\right)\overline{a_\alpha} + 2(p_3 - \overline{p_3})p_4 - p_1p_5 \\ &\quad -a\overline{\frac{\partial p_2}{\partial \alpha}} + 2|c|^2\frac{\partial p_3}{\partial \alpha} + p_2\frac{\partial p_4}{\partial a} - \overline{p_2}\frac{\partial p_4}{\partial \overline{a}}.\end{aligned}$$

Substituting this for (12) shows that

$$|c_\alpha|^2 - |a_\alpha|^2 = \left(\frac{1}{2}a\overline{\frac{\partial p_2}{\partial \overline{a}}} - |c|^2\frac{\partial p_3}{\partial a} + \frac{\partial p_4}{\partial a}\right)a_\alpha + \left(-\frac{1}{2}p_2 + a\overline{p}_3\right)\overline{a_\alpha} + p_6.$$

The coefficients of a_α and $\overline{a_\alpha}$ are simplified as

$$\frac{1}{2}a\overline{\frac{\partial p_2}{\partial \overline{a}}} - |c|^2\frac{\partial p_3}{\partial a} + \frac{\partial p_4}{\partial a} = -\frac{1}{2}\overline{p_2} + \overline{a}p_3 = -\frac{3\rho\sin\alpha\cos\alpha}{4(a+b)},$$

so that

$$|c_\alpha|^2 - |a_\alpha|^2 = -\frac{3\rho \sin\alpha \cos\alpha}{4(a+b)} a_\alpha - \frac{3\rho \sin\alpha \cos\alpha}{4(\overline{a}+b)} \overline{a_\alpha} + p_6.$$

Since c_α is expressed by $\overline{a_\alpha}$, a, $\overline{a}$ and $\overline{c}$ by Lemma 3, the above formula proves that

Lemma 4 *Let x be an isometric immersion of M^2 into $\overline{M}[4\rho]$, $(\rho \neq 0)$, with nonzero parallel mean curvature vector. If x is of general type, then we have*

$$|a_\alpha|^2 = p_7 a_\alpha + \overline{p_7 a_\alpha} + p_8. \tag{16}$$

Note that p_7 and p_8 are well defined under the assumptions of $\rho \neq 0$ and $d\alpha \neq 0$, and for later use

$$\overline{\frac{\partial p_2}{\partial \overline{a}}} = 2\frac{\partial p_7}{\partial a}, \quad p_8 = \overline{p}_8. \tag{17}$$

The next task is to find representation formulas of $a_{\alpha\alpha}$ and $a_{\alpha\beta}$ in terms of a_α and $\overline{a_\alpha}$. Differentiating partially (16) for α proves that

$$a_{\alpha\alpha}(\overline{a_\alpha} - p_7) + \overline{a_{\alpha\alpha}}(a_\alpha - \overline{p_7}) = \frac{\partial p_7}{\partial a} a_\alpha^2 + \overline{\frac{\partial p_7}{\partial a} a_\alpha^2} + p_{9a} a_\alpha + \overline{p_{9a} a_\alpha} + p_{10a}. \tag{18}$$

Differentiating partially (16) for β and multiplying the formula by ik prove that

$$\begin{aligned} &ika_{\alpha\beta}(\overline{a_\alpha} - p_7) - \overline{(ika_{\alpha\beta})}(a_\alpha - \overline{p_7}) \\ &= \left(\frac{\partial p_7}{\partial a} ika_\beta - \frac{\partial p_7}{\partial \overline{a}} \overline{(ika_\beta)}\right) a_\alpha - \left(\overline{\frac{\partial p_7}{\partial a} ika_\beta} - \overline{\frac{\partial p_7}{\partial \overline{a}}} ika_\beta\right) \overline{a_\alpha} \\ &\quad + \frac{\partial p_8}{\partial a} ika_\beta - \frac{\partial p_8}{\partial \overline{a}} \overline{(ika_\beta)}. \end{aligned} \tag{19}$$

Differentiating partially (8) for α and using (7) and (8) prove that

$$ika_{\beta\alpha} = a_{\alpha\alpha} + \left(p_1 - \frac{\partial p_2}{\partial a}\right) a_\alpha - \frac{\partial p_2}{\partial \overline{a}} \overline{a_\alpha} - p_1 p_2 - \frac{\partial p_2}{\partial \alpha}. \tag{20}$$

Substituting (8) and (20) for (19) and then using (16) show that

$$\begin{aligned} &a_{\alpha\alpha}(\overline{a_\alpha} - p_7) - \overline{a_{\alpha\alpha}}(a_\alpha - \overline{p_7}) \\ &= -\left(\overline{\frac{\partial p_2}{\partial \overline{a}}} - \frac{\partial p_7}{\partial a}\right) a_\alpha^2 + \left(\frac{\partial p_2}{\partial \overline{a}} - \overline{\frac{\partial p_7}{\partial a}}\right) \overline{a_\alpha}^2 + p_{9b} a_\alpha - \overline{p_{9b} a_\alpha} + p_{10b}. \end{aligned}$$

Adding this to (18) and using (17) show that

$$2a_{\alpha\alpha}(\overline{a_\alpha}-p_7)=\frac{\partial p_2}{\partial \overline{a}}\overline{a_\alpha}^2+(p_{9a}+p_{9b})a_\alpha+(\overline{p_{9a}}-\overline{p_{9b}})\overline{a_\alpha}+p_{10a}+p_{10b}. \tag{21}$$

By the identity $(\overline{a_\alpha}-p_7)(a_\alpha-\overline{p_7})=|p_7|^2+p_8$, when $\overline{a_\alpha}-p_7\neq 0$, the above formula is written as

$$2a_{\alpha\alpha}=\frac{(a_\alpha-\overline{p_7})}{|p_7|^2+p_8}\left(\frac{\partial p_2}{\partial \overline{a}}\overline{a_\alpha}^2+(p_{9a}+p_{9b})a_\alpha+(\overline{p_{9a}}-\overline{p_{9b}})\overline{a_\alpha}+p_{10a}+p_{10b}\right).$$

Expanding the right-hand side formula above, the identity

$$a_\alpha\overline{a_\alpha}^2=|a_\alpha|^2\overline{a_\alpha}=p_7^2a_\alpha+\overline{p}_7\overline{a_\alpha}^2+(|p_7|^2+p_8)\overline{a_\alpha}+p_7p_8 \tag{22}$$

proves the following Lemma 5.

Lemma 5 *Let x be an isometric immersion of M^2 into $\overline{M}[4\rho]$, $(\rho\neq 0)$, with nonzero parallel mean curvature vector. Suppose that x is of general type and $\overline{a_\alpha}-p_7\neq 0$. Then*

$$a_{\alpha\alpha}=p_{11}a_\alpha^2+p_{12}a_\alpha+\frac{1}{2}\frac{\partial p_2}{\partial \overline{a}}\overline{a_\alpha}+p_{13}. \tag{23}$$

When $\overline{a_\alpha}-p_7=0$ on an open set of M^2, it follows from (8) that

$$\begin{aligned}ik\frac{\partial a_\alpha}{\partial \beta}&=ik\left(\frac{\partial \overline{p_7}}{\partial a}a_\beta+\frac{\partial \overline{p_7}}{\partial \overline{a}}\overline{a_\beta}\right)=\frac{\partial \overline{p_7}}{\partial a}(ika_\beta)-\frac{\partial \overline{p_7}}{\partial \overline{a}}\overline{(ika_\beta)}\\&=\frac{\partial \overline{p_7}}{\partial a}(\overline{p_7}-p_2)-\frac{\partial \overline{p_7}}{\partial \overline{a}}(p_7-\overline{p_2}).\end{aligned} \tag{24}$$

On the other hand, by (20), we have

$$\begin{aligned}ik\frac{\partial a_\beta}{\partial \alpha}&=\frac{\partial \overline{p_7}}{\partial \alpha}+\frac{\partial \overline{p_7}}{\partial a}\overline{p_7}+\frac{\partial \overline{p_7}}{\partial \overline{a}}p_7+\left(p_1-\frac{\partial p_2}{\partial a}\right)\overline{p_7}\\&\quad-\frac{\partial p_2}{\partial \overline{a}}p_7-p_1p_2-\frac{\partial p_2}{\partial \alpha}.\end{aligned} \tag{25}$$

Since $a_{\alpha\beta}=a_{\beta\alpha}$, it follows from (24), (25), (15), and (17) that

$$p_1p_2-p_1\overline{p_7}+2\overline{p_3p_7}+\frac{\partial p_2}{\partial \alpha}+\frac{1}{2}\overline{p_2}\frac{\partial p_2}{\partial \overline{a}}-\frac{\partial \overline{p_7}}{\partial \alpha}-p_2\frac{\partial \overline{p_7}}{\partial a}=0, \tag{26}$$

which is an equation for α, a, and $\overline{a}$.

Lemma 6 *Let x be an isometric immersion of M^2 into $\overline{M}[4\rho]$, $(\rho \neq 0)$, with nonzero parallel mean curvature vector. Suppose that x is of general type and $\overline{a_\alpha} - p_7 = 0$ on an open set of M^2. Then, a is a function of α.*

Proof Let $F(\alpha, a, \overline{a})$ be the left-hand side formula of (26). We observe that $F(\alpha, a, \overline{a})$ is nontrivial because of $F(\pi/4, 0, 0) = 15\rho/8b \neq 0$. If $F_a = F_{\overline{a}} = 0$, then $F(\alpha, a, \overline{a})$ is a function of α only. Equation (26) implies that α is constant, leading to a contradiction. So, F_a or $F_{\overline{a}}$ cannot be zero. When $F_{\overline{a}}$ is not zero, $\overline{a}$ is a function of α and a, denoted as $\overline{a} = f(\alpha, a)$, by the implicit function theorem. The function $f(\alpha, a)$ is not real valued; otherwise, we would have $\overline{a} = a$, which results in a contradiction. Taking the complex conjugate of the equation $\overline{a} = f(\alpha, a)$ yields $a = \overline{f(\alpha, a)} = g(\alpha, \overline{a})$ for some function g. Define $K(\alpha, a) = a - g(\alpha, f(\alpha, a))$. By replacing $F(\alpha, a, \overline{a})$ with $K(\alpha, a)$ and using similar arguments as above, $K(\alpha, a) = 0$ is solved with respect to a, proving Lemma 6. When $F_a \neq 0$, Lemma 6 is proved in the same mannar as before. □

From now on, we study the case of $\overline{a_\alpha} - p_7 \neq 0$. It follows from Lemma 5 and (20) that

$$ika_{\beta\alpha} = p_{11}a_\alpha^2 + p_{14}a_\alpha - \frac{1}{2}\frac{\partial p_2}{\partial \overline{a}}\overline{a_\alpha} + p_{15}. \tag{27}$$

In order to get the difference in the second derivatives of a_α, we calculate $a_{\alpha\alpha\beta}$ by (23) as follows:

$$\begin{aligned} ik\frac{\partial}{\partial\beta}a_{\alpha\alpha} &= (2p_{11}a_\alpha + p_{12})ika_{\alpha\beta} - \frac{1}{2}\frac{\partial p_2}{\partial\overline{a}}\overline{(ika_{\alpha\beta})} \\ &+ \left(\frac{\partial p_{11}}{\partial a}ika_\beta - \frac{\partial p_{11}}{\partial \overline{a}}\overline{(ika_\beta)}\right)a_\alpha^2 + \frac{\partial p_{13}}{\partial a}ika_\beta - \frac{\partial p_{13}}{\partial\overline{a}}\overline{(ika_\beta)} \\ &+ \left(\frac{\partial p_{12}}{\partial a}ika_\beta - \frac{\partial p_{12}}{\partial\overline{a}}\overline{(ika_\beta)}\right)a_\alpha + \frac{1}{2}\left(\frac{\partial^2 p_2}{\partial\overline{a}\partial a}ika_\beta - \frac{\partial^2 p_2}{\partial\overline{a}^2}\overline{(ika_\beta)}\right)\overline{a_\alpha}. \end{aligned}$$

It follows from (8), (27), and the complex conjugate of (22) that

$$\begin{aligned} ik\frac{\partial}{\partial\beta}a_{\alpha\alpha} &= \left(2p_{11}^2 + \frac{\partial p_{11}}{\partial a}\right)a_\alpha^3 + p_{16a}a_\alpha^2 - \frac{1}{2}\left(\overline{p_{11}}\frac{\partial p_2}{\partial\overline{a}} + \frac{\partial^2 p_2}{\partial\overline{a}^2}\right)\overline{a_\alpha}^2 \\ &+ p_{17a}a_\alpha + p_{18a}\overline{a_\alpha} + p_{19a}. \end{aligned}$$

By the similar way, we show by (7), (23), and (27) that

$$\begin{aligned} ik\frac{\partial}{\partial\alpha}a_{\beta\alpha} &= \frac{\partial}{\partial\alpha}(ika_{\beta\alpha}) - ik_\alpha a_{\beta\alpha} \\ &= \left(2p_{11}^2 + \frac{\partial p_{11}}{\partial a}\right)a_\alpha^3 + p_{16b}a_\alpha^2 - \frac{1}{2}\left(\overline{p_{11}}\frac{\partial p_2}{\partial\overline{a}} + \frac{\partial^2 p_2}{\partial\overline{a}^2}\right)\overline{a_\alpha}^2 \\ &+ p_{17b}a_\alpha + p_{18b}\overline{a_\alpha} + p_{19b}. \end{aligned}$$

Since $(a_{\alpha\alpha})_\beta = (a_{\beta\alpha})_\alpha$, the above two formulas imply

$$p_{16}a_\alpha^2 + p_{17}a_\alpha + p_{18}\overline{a_\alpha} + p_{19} = 0. \tag{28}$$

This is a nontrivial equation for a_α, because our system consiting of (3), (4), and (5) is overdetermined. The rigorus proof that (28) is nontrivial is shown by $p_{16}(\pi/4, 0, 0) \neq 0$ or $p_{16}(\pi/3, 0, 0) \neq 0$. By (28), a_α is determined by α, a, and $\overline{a}$, denoted as $a_\alpha = p_{23}(\alpha, a, \overline{a})$, because all coefficients of (28) are functions of α, a, and $\overline{a}$. Furthermore, p_{23} is unique up to the complex conjugate that is later shown in the proof of Lemma 8.

The following is the main result of this paper.

Lemma 7 *Let x be an isometric immersion of M^2 into $\overline{M}[4\rho]$, $(\rho \neq 0)$, with nonzero parallel mean curvature vector. If x is of general type, then a is a function of α.*

Proof By Lemma 6, we may assume $a_\alpha - \overline{p_7} \neq 0$. Replacing $\overline{p_7}$ in (24), and (25) with p_{23}, we have

$$p_1 p_2 + \frac{\partial p_2}{\partial \alpha} - \left(p_1 - \frac{\partial p_2}{\partial a}\right) p_{23} + \frac{\partial p_2}{\partial \overline{a}}\overline{p_{23}} - \frac{\partial p_{23}}{\partial \alpha} - p_2 \frac{\partial p_{23}}{\partial a} + (\overline{p_2} - 2\overline{p_{23}}) \frac{\partial p_{23}}{\partial \overline{a}} = 0.$$

Let $G(\alpha, a, \overline{a})$ be the left-hand side formula above. This is nontrivial, because the system of (3), (4), and (5) is overdetermined. For the rigorus proof, see Lemma 8 in the appendix. We prove Lemma 7 in the same manner as the proof of Lemma 6 replacing $F(\alpha, a, \overline{a})$ with $G(\alpha, a, \overline{a})$. □

By Lemma 7, if x is of general type, then $a_\beta = 0$ and so a satisfies the ordinary differential equation $a'(\alpha) = p_2(\alpha, a, \overline{a})$ by (8). Based on this, the classification of such immersions is carried out in [5].

Appendix

$$p_1(\alpha, a, \overline{a}) = \frac{1}{|a + b|^2}\left(|a - b|^2 + \frac{3}{2}\rho \sin^2 \alpha\right) \cot \alpha,$$

$$p_2(\alpha, a, \overline{a}) = \frac{1}{\overline{a} + b}\left(2a(\overline{a} - b) + \frac{3}{2}\rho \sin^2 \alpha\right) \cot \alpha,$$

$$p_3(\alpha, a, \overline{a}) = \frac{a - b}{a + b} \cot \alpha,$$

$$p_4(\alpha, a, \overline{a}) = \left(a\overline{a} + \frac{\rho}{2}(-2 + 3\sin^2 \alpha)\right)(p_3 - \overline{p_3}) + \frac{1}{2}(\overline{a}p_2 - a\overline{p_2})$$

$$+\frac{3}{2}\rho\sin\alpha\cos\alpha,$$

$$p_5(\alpha,a,\overline{a}) = a\overline{p_2} - 2\left(a\overline{a} + \frac{\rho}{2}(-2+3\sin^2\alpha)\right)p_3 + p_4,$$

$$p_6(\alpha,a,\overline{a}) = (\overline{p_3} - p_3)p_4 - \left(a\overline{a} + \frac{\rho}{2}(-2+3\sin^2\alpha)\right)\frac{\partial p_3}{\partial\alpha}$$
$$+\frac{1}{2}\left(p_1p_5 + a\overline{\frac{\partial p_2}{\partial\alpha}} + \frac{\partial p_4}{\partial\alpha} - p_2\frac{\partial p_4}{\partial a} + \overline{p_2}\frac{\partial p_4}{\partial\overline{a}}\right),$$

$$p_7(\alpha,a,\overline{a}) = \frac{\overline{a}}{\frac{\rho}{2}(-2+3\sin^2\alpha)}\left(p_4 + \frac{3\rho\sin\alpha\cos\alpha}{4(a+b)}a\right) + \frac{3\rho\sin\alpha\cos\alpha}{4(a+b)},$$

$$p_8(\alpha,a,\overline{a}) = \frac{1}{\frac{\rho}{2}(-2+3\sin^2\alpha)}\left(|p_4|^2 - \left(a\overline{a} + \frac{\rho}{2}(-2+3\sin^2\alpha)\right)p_6\right),$$

$$p_{9a}(\alpha,a,\overline{a}) = \frac{\partial p_7}{\partial\alpha} + \left(\frac{\partial p_7}{\partial\overline{a}} + \overline{\frac{\partial p_7}{\partial\overline{a}}}\right)p_7 + \frac{\partial p_8}{\partial a},$$

$$p_{9b}(\alpha,a,\overline{a}) = p_1(-\overline{p_2} + p_7) - \overline{\frac{\partial p_2}{\partial\alpha}} - p_2\frac{\partial p_7}{\partial a} + \overline{p_2}\frac{\partial p_7}{\partial\overline{a}} - p_7\overline{\frac{\partial p_2}{\partial a}} + \overline{p_7}\overline{\frac{\partial p_2}{\partial\overline{a}}}$$
$$-p_7\left(\frac{\partial p_7}{\partial\overline{a}} - \overline{\frac{\partial p_7}{\partial\overline{a}}}\right) + \frac{\partial p_8}{\partial a},$$

$$p_{10a}(\alpha,a,\overline{a}) = p_8\left(\frac{\partial p_7}{\partial\overline{a}} + \overline{\frac{\partial p_7}{\partial\overline{a}}}\right) + \frac{\partial p_8}{\partial\alpha},$$

$$p_{10b}(\alpha,a,\overline{a}) = p_1(-p_2p_7 + \overline{p_2p_7}) - p_2\frac{\partial p_8}{\partial a} + \overline{p_2}\frac{\partial p_8}{\partial\overline{a}} - p_7\frac{\partial p_2}{\partial\alpha} + p_7\overline{\frac{\partial p_2}{\partial\alpha}}$$
$$+p_8\left(\frac{\partial p_2}{\partial a} - \overline{\frac{\partial p_2}{\partial a}} - \frac{\partial p_7}{\partial\overline{a}} + \overline{\frac{\partial p_7}{\partial\overline{a}}}\right),$$

$$p_{11}(\alpha,a,\overline{a}) = \frac{p_{9a} + p_{9b}}{2(|p_7|^2 + p_8)},$$

$$p_{12}(\alpha,a,\overline{a}) = \frac{p_7^2\frac{\partial p_2}{\partial\overline{a}} + p_7(\overline{p_{9a}} - \overline{p_{9b}}) - \overline{p_7}(p_{9a} + p_{9b}) + p_{10a} + p_{10b}}{2(|p_7|^2 + p_8)}$$

$$p_{13}(\alpha,a,\overline{a}) = \frac{p_7p_8\frac{\partial p_2}{\partial\overline{a}} - \overline{p_7}(p_{10a} + p_{10b}) + p_8(\overline{p_{9a}} - \overline{p_{9b}})}{2(|p_7|^2 + p_8)}$$

$$p_{14}(\alpha,a,\overline{a}) = p_1 - \frac{\partial p_2}{\partial a} + p_{12},$$

$$p_{15}(\alpha,a,\overline{a}) = -p_1p_2 - \frac{\partial p_2}{\partial\alpha} + p_{13},$$

$$p_{16a}(\alpha, a, \overline{a}) = p_{11}(p_{12} + 2p_{14}) - p_2\frac{\partial p_{11}}{\partial a} + \overline{p_2}\frac{\partial p_{11}}{\partial \overline{a}} - p_7\frac{\partial p_{11}}{\partial \overline{a}} + \frac{\partial p_{12}}{\partial a},$$

$$p_{16b}(\alpha, a, \overline{a}) = p_{11}(p_1 + 2p_{12} + p_{14}) + \frac{\partial p_{11}}{\partial \alpha} + p_7\frac{\partial p_{11}}{\partial \overline{a}} + \frac{\partial p_{14}}{\partial a},$$

$$p_{16}(\alpha, a, \overline{a}) = p_{16a} - p_{16b},$$

$$p_{17a}(\alpha, a, \overline{a}) = 2p_{11}p_{15} + p_{12}p_{14} + \frac{1}{4}\left|\frac{\partial p_2}{\partial \overline{a}}\right|^2 + \frac{1}{2}p_7\frac{\partial^2 p_2}{\partial \overline{a}\partial a} - p_7p_{11}\frac{\partial p_2}{\partial \overline{a}}$$
$$-(|p_7|^2 + p_8)\frac{\partial p_{11}}{\partial \overline{a}} - p_2\frac{\partial p_{12}}{\partial a} + \overline{p_2}\frac{\partial p_{12}}{\partial \overline{a}} - p_7\frac{\partial p_{12}}{\partial \overline{a}} + \frac{\partial p_{13}}{\partial a},$$

$$p_{17b}(\alpha, a, \overline{a}) = p_1p_{14} + 2p_{11}p_{13} + p_{12}p_{14} - \frac{1}{4}\left|\frac{\partial p_2}{\partial \overline{a}}\right|^2 - \frac{1}{2}p_7\frac{\partial^2 p_2}{\partial \overline{a}\partial a}$$
$$+p_7p_{11}\frac{\partial p_2}{\partial \overline{a}} + (|p_7|^2 + p_8)\frac{\partial p_{11}}{\partial \overline{a}} + p_7\frac{\partial p_{14}}{\partial \overline{a}} + \frac{\partial p_{14}}{\partial \alpha} + \frac{\partial p_{15}}{\partial a}$$

$$p_{17}(\alpha, a, \overline{a}) = p_{17a} - p_{17b},$$

$$p_{18a}(\alpha, a, \overline{a}) = -\frac{1}{2}(2\overline{p_7} + 2\overline{p_7}p_{11} + p_{12} + \overline{p_{14}})\frac{\partial p_2}{\partial \overline{a}} + \frac{1}{2}(-p_2 + \overline{p_7})\frac{\partial^2 p_2}{\partial \overline{a}\partial a}$$
$$+\frac{1}{2}\overline{p_2}\frac{\partial^2 p_2}{\partial \overline{a}^2} - \overline{p_7}^2\frac{\partial p_{11}}{\partial \overline{a}} - \frac{\partial p_{13}}{\partial \overline{a}}$$

$$p_{18b}(\alpha, a, \overline{a}) = -\frac{1}{2}(p_1 - 2\overline{p_7}p_{11} + \overline{p_{12}} - p_{14})\frac{\partial p_2}{\partial \overline{a}} - \frac{1}{2}\frac{\partial^2 p_2}{\partial \alpha\partial a} - \frac{1}{2}\overline{p_7}\frac{\partial^2 p_2}{\partial \overline{a}\partial a}$$
$$+\overline{p_7}^2\frac{\partial p_{11}}{\partial \overline{a}} + \overline{p_7}\frac{\partial p_{14}}{\partial \overline{a}} + \frac{\partial p_{15}}{\partial \overline{a}},$$

$$p_{18}(\alpha, a, \overline{a}) = p_{18a} - p_{18b},$$

$$p_{19a}(\alpha, a, \overline{a}) = p_{12}p_{15} - \frac{1}{2}(2p_8p_{11} + \overline{p_{15}})\frac{\partial p_2}{\partial \overline{a}} + \frac{1}{2}p_8\frac{\partial^2 p_2}{\partial \overline{a}\partial a} - \overline{p_7p_8}\frac{\partial p_{11}}{\partial \overline{a}}$$
$$-p_8\frac{\partial p_{12}}{\partial \overline{a}} - p_2\frac{\partial p_{13}}{\partial a} + \overline{p_2}\frac{\partial p_{13}}{\partial \overline{a}},$$

$$p_{19b}(\alpha, a, \overline{a}) = p_1p_{15} + p_{13}p_{14} + \frac{1}{2}(2p_8p_{11} - \overline{p_{13}})\frac{\partial p_2}{\partial \overline{a}} - \frac{1}{2}p_8\frac{\partial^2 p_2}{\partial \overline{a}\partial a}$$
$$+\overline{p_7p_8}\frac{\partial p_{11}}{\partial \overline{a}} + p_8\frac{\partial p_{14}}{\partial \overline{a}} + \frac{\partial p_{15}}{\partial \alpha},$$

$$p_{19}(\alpha, a, \overline{a}) = p_{19a} - p_{19b},$$

$$p_{20}(\alpha, a, \overline{a}) = -\overline{p_7}p_{16} + p_{17},$$

$$p_{21}(\alpha, a, \overline{a}) = -\overline{p_7}p_{17} + p_7p_{18} + p_{19},$$

$$p_{22}(\alpha, a, \overline{a}) = p_8p_{18} - \overline{p_7}p_{19},$$

$$p_{23}(\alpha, a, \overline{a}) = \text{a solution of (28)},$$

$$p_{24}(\alpha, a, \overline{a}) = \frac{\frac{\partial p_{16}}{\partial \alpha} p_{23}^3 + \frac{\partial p_{20}}{\partial \alpha} p_{23}^2 + \frac{\partial p_{21}}{\partial \alpha} p_{23} + \frac{\partial p_{22}}{\partial \alpha}}{3p_{16}p_{23}^2 + 2p_{20}p_{23} + p_{21}},$$

$$p_{25}(\alpha, a, \overline{a}) = \frac{\frac{\partial p_{16}}{\partial a} p_{23}^3 + \frac{\partial p_{20}}{\partial a} p_{23}^2 + \frac{\partial p_{21}}{\partial a} p_{23} + \frac{\partial p_{22}}{\partial a}}{3p_{16}p_{23}^2 + 2p_{20}p_{23} + p_{21}},$$

$$p_{26}(\alpha, a, \overline{a}) = \frac{\frac{\partial p_{16}}{\partial \overline{a}} p_{23}^3 + \frac{\partial p_{20}}{\partial \overline{a}} p_{23}^2 + \frac{\partial p_{21}}{\partial \overline{a}} p_{23} + \frac{\partial p_{22}}{\partial \overline{a}}}{3p_{16}p_{23}^2 + 2p_{20}p_{23} + p_{21}},$$

$$\overline{p_i}(\alpha, a, \overline{a}) = p_i(\alpha, \overline{a}, a), \ (i = 1, 2, 3, \ldots).$$

Lemma 8 *$G(\alpha, a, \overline{a})$ is not identically zero.*

Proof In order to determine whether $G(\pi/4, 0, 0) \neq 0$, we provide a procedure for writing the Mathematica (by Wolfram) program codes. First, we find the value $p_{23}(\pi/4, 0, 0)$ by solving (28) as follows. Since all coefficients of (28) are real at $(\pi/4, 0, 0)$ and $a_\alpha \neq \overline{a_\alpha}$, $p_{23}(\pi/4, 0, 0)$ is not zero and uniquely determined up to the complex conjugate. Next, by (16) and (28), we have $p_{16}a_\alpha^3 + p_{20}a_\alpha^2 + p_{21}a_\alpha + p_{22} = 0$. Partially differentiating this for α, we obtain $\partial p_{23}/\partial \alpha = -p_{24}$. We note that $3p_{16}p_{23}^2 + 2p_{20}p_{23} + p_{21}$ at $(\pi/4, 0, 0)$ is not zero. By the similar way, we get $\partial p_{23}/\partial a = -p_{25}$, and $\partial p_{23}/\partial \overline{a} = -p_{26}$. Consequently

$$G(\alpha, a, \overline{a}) = p_1p_2 + \frac{\partial p_2}{\partial \alpha} - \left(p_1 - \frac{\partial p_2}{\partial a}\right) p_{23} + \frac{\partial p_2}{\partial \overline{a}} \overline{p_{23}} + p_{24}$$
$$+ p_2p_{25} - (\overline{p_2} - 2\overline{p_{23}})p_{26}.$$

By this formula, we observe $G(\pi/4, 0, 0) \neq 0$, thereby proving Lemma 8. □

References

1. S. S. Chern and J. G. Wolfson, Minimal surfaces by moving frames, Amer. J. Math. 105, 59-83 (1983)
2. S. Hirakawa, Constant Gaussian Curvature Surfaces with Parallel Mean curvature Vector in Two-Dimensional Complex Space Forms, Geom. Dedicata 118, 229-244 (2006)
3. K. Kenmotsu and D. Zhou, The classification of the surfaces with parallel mean curvature vector in two dimensional complex space forms, Amer. J. Math. 122, 295-317 (2000)
4. K. Kenmotsu, Correction to "The classification of the surfaces with parallel mean curvature vector in two-dimensional complex space forms", Amer. J. Math. 138, 395-402 (2016)
5. K. Kenmotsu, Complete parallel mean curvature surfaces in two-dimensional complex space-forms, Bull Braz Math Soc, New Series 49, 775-788 (2018)

MIX
Papier aus verantwortungsvollen Quellen
Paper from responsible sources
FSC® C105338

If you have any concerns about our products,
you can contact us on
ProductSafety@springernature.com

In case Publisher is established outside the EU,
the EU authorized representative is:
Springer Nature Customer Service Center GmbH
Europaplatz 3, 69115 Heidelberg, Germany

Printed by Libri Plureos GmbH
in Hamburg, Germany